Nuclear Magnetic Resonance in Chemistry

NUCLEAR MAGNETIC RESONANCE IN CHEMISTRY

Proceedings of a Symposium held at Cagliari, Sardinia, Italy in September, 1964 under the auspices of the Società Italiana per il Progresso delle Scienze

Edited by BIAGIO PESCE

ISTITUTO DI CHIMICA-FISICA ED ELETTROCHIMICA
UNIVERSITÀ DEGLI STUDI DI ROMA
ROMA, ITALY

1965

ACADEMIC PRESS NEW YORK and LONDON

ACADEMIC PRESS INC.
111 Fifth Avenue, New York, New York 10003

United Kingdom Edition published by
ACADEMIC PRESS INC. (LONDON) LTD.
Berkeley Square House, London W.1

LIBRARY OF CONGRESS CATALOG CARD NUMBER: 64-24812

PRINTED IN THE UNITED STATES OF AMERICA

List of Contributors

Numbers in parentheses indicate the pages on which the authors' contributions begin.

R. J. Abraham (111), The Robert Robinson Laboratories, The University of Liverpool, Liverpool, England

W. A. Anderson (11), Varian Associates, Analytical Instrument Division, Palo Alto, California

M. Anteunis (119), State University of Ghent, Ghent, Belgium

G. Aranda (299), Institut d'Electronique et Laboratoire de Chimie Structurale Organique, Faculté des Sciences, Orsay, France

K. D. Bartle (69), School of Chemistry, The University of Leeds, Leeds, England

H. Benoit (53), Institut d'Electronique, Faculté des Sciences d'Orsay, Orsay, France

Gerhard Binsch (269),[1] Gates and Crellin Laboratories of Chemistry, California Institute of Technology, Pasadena, California

F. Bloch (1), Stanford University, Stanford, California

N. Boden (149), The Donnan Laboratories, The University of Liverpool, Liverpool, England

W. Bonthrone (263), Milstead Laboratory of Chemical Enzymology, Sittingbourne, Kent, England

A. J. R. Bourn (277),[2] Queen Mary College, London, England

Pietro Bucci (133), Istituto di Chimica-Fisica dell'Università di Pisa, Pisa, Italy

A. D. Buckingham (35), Inorganic Chemistry Laboratory, Oxford University, Oxford, England

Giovanni Casini (255), Istituto di Chimica Farmaceutica e Tossicologica dell'Università di Roma, Roma, Italy

[1] *Present address*: Institut für Organische Chemie der Universität München, München, Germany.

[2] *Present address*: Department of Chemistry, University of California, Los Angeles, California.

L. Cavalli (111),[3] The Robert Robinson Laboratories, The University of Liverpool, Liverpool, England

R. J. Chuck (189), Queen Mary College, London, England

Mario Cola (249), Istituto di Chimica Generale dell'Università di Pavia, Pavia, Italy

T. M. Connor (319), Basic Physics Division, The National Physical Laboratory, Teddington, England

Filippo Conti (349, 363), Istituto di Chimica-Fisica, Università di Roma, Roma, Italy

Camillo Dejak (363), Istituto Chimico Policattedra, Università di Cagliari, Cagliari, Italy

S. S. Dharmatti (305), Tata Institute of Fundamental Research, Bombay, India

P. Diehl (77), Physics Department, University of Basel, Basel, Switzerland

H. Dreeskamp (143), Laboratorium für Physikalische Chemie, Technische Hochschule, Stuttgart, Germany

J. W. Emsley (149),[4] The Donnan Laboratories, The University of Liverpool, Liverpool, England

J. Feeney (149),[5] The Donnan Laboratories, The University of Liverpool, Liverpool, England

Cafiero Franconi (349, 363), Istituto di Chimica-Fisica, Università di Roma, Roma, Italy

R. Freeman (11), Varian Associates, Analytical Instrument Division, Palo Alto, California

René Freymann (13), Faculté des Sciences de Paris, Paris, France

H. Friebolin (125), Institut für Elektrowerkstoffe Freiburg, Chemisches Institut der Universität Freiburg, Freiburg, Germany

D. Gagnaire (165), Laboratoire de Chimie Organique Physique et Centre d'Etudes Nucléaires, Grenoble, France

[3] *Present address*: Laboratori Ricerche di Bollate, Societa Edison, Milan, Italy.

[4] *Present address*: Chemistry Department, The University of Durham, Durham City, England.

[5] *Present address*: Varian Research Laboratory, Walton-on-Thames, Surrey, England.

G. Giacometti (173), Istituto di Chimica Fisica, Università di Padova, Padova, Italy

D. G. Gillies (277),[6] Queen Mary College, London, England

L. C. D. Groenweghe (283), Monsanto Company, Central Research Department, St. Louis, Missouri

H. G. Hertz (199), Institut für physikalische Chemie der Universität Münster, Münster, Germany

M. J. R. Hoch (343),[7] Department of Physics, University of St. Andrews, St. Andrews, Fife, Scotland

D. W. Jones (331), Department of Chemical Technology, Institute of Technology, Bradford, England

J. Jullien (299), Institut d'Electronique et Laboratoire de Chimie Structurale Organique, Faculté des Sciences, Orsay, France

J. L. Jungnickel (83), Shell Development Company, Emeryville, California

S. Kabuss (125), Institut für Elektrowerkstoffe Freiburg, Chemisches Institut der Universität Freiburg, Freiburg, Germany

G. Keller (199), Institut für physikalische Chemie der Universität Münster, Münster, Germany

G. Klose (237), Physical and Inorganic Chemistry Institute, University of Leipzig, Leipzig, Germany

Dora G. DeKowalewski (159), Department of Physics, Faculty of Exact Sciences, University of Buenos Aires, Buenos Aires, Argentina

Valdemar J. Kowalewski (159), Department of Physics, Faculty of Exact Sciences, University of Buenos Aires, Buenos Aires, Argentina

Joseph B. Lambert (269),[8] Gates and Crellin Laboratories of Chemistry, California Institute of Technology, Pasadena, California

G. Lanzi (325), Istituto di Fisica dell'Università di Pavia, Pavia, Italy

[6] *Present address*: Division of Pure Chemistry, National Research Council, Ottawa, Canada.

[7] *Present address*: Department of Physics, University of Natal, Pieter Maritzburg, South Africa.

[8] *Present address*: Department of Chemistry, Northwestern University, Evanston, Illinois.

A. Loewenstein (229), Department of Chemistry, Technion—Israel Institute of Technology, Haifa, Israel

E. L. Mackor (59), Koninklijke/Shell-Laboratorium Amsterdam (Shell Research N. V.), Amsterdam, The Netherlands

K. A. McLauchlan (313, 319), Basic Physics Division, National Physical Laboratory, Teddington, England

C. MacLean (59), Koninklijke/Shell-Laboratorium Amsterdam (Shell Research N. V.), Amsterdam, The Netherlands

J. A. Martin (299), Institut d'Electronique et Laboratoire de Chimie Structurale Organique, Faculté des Sciences, Orsay, France

A. G. Massey (189), Queen Mary College, London, England

Lucio Merlini (243), Istituto di Chimica, Politecnico di Milano, Milano, Italy

Rosanna Mondelli (243), Istituto di Chimica, Politecnico di Milano, Milano, Italy

G. Navon (229), Department of Physical Chemistry, Hebrew University, Jerusalem, Israel

H. Ottavi (53), Institut d'Electronique, Faculté des Sciences d'Orsay, Orsay, France

K. G. R. Pachler (111), The Chemical Physics Group, South African Council for Scientific and Industrial Research, Pretoria, Republic of South Africa

E. Payo-Subiza (165), Laboratoire de Chimie Organique Physique et Centre d'Etudes Nucléaires, Grenoble, France

J. E. Pearson (331), Department of Chemical Technology, Institute of Technology, Bradford, England

Angelo Perotti (249), Istituto di Chimica Generale dell'Università di Pavia, Pavia, Italy

N. J. Poulis (339), Kamerlingh Onnes Laboratorium, Rijks-Universiteit, Leiden, The Netherlands

W. G. Proctor (7), Varian AG, Zurich, Switzerland

R. Ramanadin (299),[9] Institut d'Electronique et Laboratoire de Chimie Structurale Organique, Faculté des Sciences, Orsay, France

E. W. Randall (189, 277), Queen Mary College, London, England

D. H. Reid (263), Department of Chemistry, St. Salvator's College, University of St. Andrews, St. Andrews, Fife, Scotland

C. A. Reilly (83), Shell Development Company, Emeryville, California

G. Rigatti (173), Istituto di Chimica Fisica, Università di Padova, Padova, Italy

Bryan W. Roberts (269),[10] Gates and Crellin Laboratories of Chemistry, California Institute of Technology, Pasadena, California

John D. Roberts (269), Gates and Crellin Laboratories of Chemistry, California Institute of Technology, Pasadena, California

Renzo Rossi (133), Istituto di Chimica Organica Industriale dell'Università di Pisa, Pisa, Italy

A. Rousseau (165), Laboratoire de Chimie Organique Physique et Centre d'Etudes Nucléaires, Grenoble, France

F. A. Rushworth (343), Department of Physics, University of St. Andrews, St. Andrews, Fife, Scotland

E. Sackmann (143), Laboratorium für Physikalische Chemie, Technische Hochschule, Stuttgart, Germany

Maria Letizia Salvi (255), Istituto di Chimica dell'Università Cattolica del Sacro Cuore, Roma, Italy

V. Saraswati (305), Tata Institute of Fundamental Research, Bombay, India

H. Schmidbaur (185), Institut für Anorganische Chemie der Universität Marburg, Marburg, Germany

W. G. Schneider (63), Division of Pure Chemistry, National Research Council, Ottawa, Canada

D. Shaw (189), Queen Mary College, London, England

[9] *Present address*: Laboratoire des Recherches, Péchiney Saint-Gobain, Salindres (Gard), France.

[10] *Present address*: Department of Chemistry, University of Southern California, Los Angeles, California.

M. Shporer (229), Department of Chemistry, Technion—Israel Institute of Technology, Haifa, Israel

F. S. Skelton (263), Milstead Laboratory of Chemical Enzymology, Sittingbourne, Kent, England

J. A. S. Smith (69),[11] School of Chemistry, The University of Leeds, Leeds, England

P. J. Stephens (35), Inorganic Chemistry Laboratory, Oxford University, Oxford, England

L. H. Sutcliffe (149), The Donnan Laboratories, The University of Liverpool, Liverpool, England

F. Taddei (179), Istituto di Chimica Fisica ed Elettrochimica, Università di Bologna, Bologna, Italy

D. Tavernier (119), State University of Ghent, Ghent, Belgium

E. Uhlemann (237), Physical and Inorganic Chemistry Institute, University of Leipzig, Leipzig, Germany

R. Vijayaraghavan (305), Tata Institute of Fundamental Research, Bombay, India

J. R. Van Wazer (283), Monsanto Company, Central Research Department, St. Louis, Missouri

H. E. Weaver (7), Varian AG, Zurich, Switzerland

S. Wittekoek (339), Kamerlingh Onnes Laboratorium, Rijks-Universiteit, Leiden, The Netherlands

C. Zauli (179), Istituto di Chimica Fisica ed Elettrochimica, Università di Bologna, Bologna, Italy

R. F. Zürcher (45), Physical Laboratories, Ciba Limited, Basel, Switzerland

[11]*Present address*: School of Molecular Sciences, The University of Warwick, Coventry, England.

Preface

Within a few years of the discovery of the phenomenon of nuclear magnetic resonance in 1945, it became apparent that its usefulness was not confined to a study of the magnetic properties of the nucleus, but rather that it was of supreme interest and had enormous potential in the field of chemistry as well.

This followed from the discovery that the magnetic behavior of a nucleus with a permanent magnetic moment, under specific conditions involving both radiofrequency oscillations and static magnetic fields, is not dependent solely on the nuclear properties, but also on the electronic distribution in the molecule containing the nucleus—in other words, on the chemical environment of the atom concerned. This feature, the chemical shift, was discovered almost simultaneously in 1949 by W. G. Proctor and F. C. Yu, W. C. Dickinson, G. Lindström, and H. A. Thomas. A short time later, a second characteristic dependent on the chemical bonding of the molecule was reported, namely spin-spin splitting.

The result of these two effects is a resolution of the magnetic properties of the nucleus into a magnetic resonance spectrum of fine and hyperfine structure that is highly characteristic of the molecular structure. It is precisely this feature that makes nuclear magnetic resonance an invaluable tool to chemists, who, despite the high cost and complicated nature of the apparatus, are making ever-increasing use of it to resolve problems in physical chemistry, analytical chemistry, and particularly molecular studies in both inorganic and, more extensively, organic fields. Within the fifteen years of its use by chemists, this new means of investigation has produced a vast output of fruitful results, and it is destined to be of expanding utility and fertile application in chemistry in the future.

The Società Italiana per il Progresso delle Scienze felt it opportune to take stock of the development of NMR spectroscopy as applied to chemical problems, and organized an International Symposium as part of the 48th Meeting of the Society, held in Sardinia in 1964. A considerable number of specialists in various branches of chemistry took part. Many of the participants, notably Prof. F. Bloch (one of the discoverers of the phenomenon), who presided at the Symposium, are well known for their pioneering contributions in this field.

The Symposium was quite successful in its aim of surveying the development of NMR in chemical studies, both past achievements and, more particularly, the current trends and future possibilities, and in presenting selectively much of the wide range of problems to which NMR has been applied.

The S.I.P.S. and the Sardinian Organizing Committee wish to express their appreciation to Academic Press for undertaking the publication of the papers presented at the Symposium. The Editor also wishes to express his thanks to Dr. C. W. Fryer for his valuable aid in editing this volume.

Rome, Italy
September 1965

BIAGIO PESCE

Contents

Introduction: Some Fundamental Aspects of NMR

F. BLOCH

Stanford University, Stanford, California

The investigation of nuclear magnetism in matter of normal density began almost 20 years ago. Going beyond the earlier research on atomic beams, the purely electromagnetic method of nuclear induction was introduced at that time and it was immediately evident that this method offered new possibilities for the measurement of nuclear magnetic moments. Apart from the obvious application to elemental and isotope analysis it took several years, however, before the technique had reached the precision necessary to develop into the important tool for chemical research which it represents today. It may be useful, as an introduction to this volume on nuclear magnetic resonance in chemistry, to briefly recall the basic features of this development.

The connection between nuclear magnetic resonance and chemistry primarily rests upon the fact that the magnetic resonance frequency of the nucleus is affected by the atomic electrons. In the original discussions, this relatively small effect had been neglected and the environment was taken into account only inasfar as it is responsible for the phenomenon of relaxation. The first effect of this kind was established by Knight (*1*), who discovered a shift of the resonance frequency in metals associated with the paramagnetism of the conduction electrons. In fact, their magnetic moments contribute to the field acting upon the nuclei by an amount proportional to the externally applied magnetic field and, typically, comparable to 1%. Shortly afterward a shift between nonmetallic compounds of the same element was found by Proctor and Yu (*2*) and by Dickinson (*3*). Although considerably smaller than the Knight shift, it is this effect which is of particular importance to chemistry and is justly known as the chemical shift.

The decisive step toward the explanation had already been made by Lamb (*4*) almost 10 years before the discovery of the chemical shift. He pointed out that the external magnetic field is to a certain extent shielded by the atomic electrons so that its effective magnitude at the location of the nucleus is reduced by the fractional amount

$$\sigma = \frac{e^2}{3mc^2} \int \frac{\rho}{r} d\tau, \tag{1}$$

where ρ is the mean density of electrons at the distance r from the nucleus. The value of the screening constant σ is easily found in the case of the hydrogen atom since the integral in Eq. (1) is here given by the reciprocal of the Bohr radius $a_0 = \hbar^2/me^2$ so that

$$\sigma_H = \alpha^2/3 \cong 2 \times 10^{-5} \tag{2}$$

with the well-known fine-structure constant $\alpha = e^2/\hbar c \cong 1/137$. For higher atomic numbers, the screening constant increases both because of the greater number of electrons and owing to their smaller average distance from the nucleus so that, for example in the case of lead, one finds $\sigma_{Pb} \cong 10^{-2}$. While the Knight shift originates from the action of the magnetic moment of the electrons upon the nucleus and is, in this sense, related to atomic hyperfine structure, the screening remains effective in the absence of such a moment in close analogy to diamagnetism as compared with paramagnetism.

So far, the discussion of the screening effect has referred to isolated atoms and their bonding in molecules calls for additional considerations. In the first place, it is only the *difference* in the screening constant for a given isotope which is directly observable either as chemical shift between different compounds of the same element or as "internal" chemical shift between nonequivalent sites within the molecule of a single compound. It rests upon the varying manner in which the orbits of the valency electrons and, hence, their screening are modified in the establishment of the corresponding bonds while the screening of the inner electrons remains practically unaffected. In the second place, the effect of this modification upon the magnetic field at the location of the nucleus is not simply related to the mean density ρ but involves another feature of purely molecular origin. In the realm of magnetic susceptibilities it is known as the induced paramagnetism which arises from the fact that the angular momentum of the electrons in the molecule is no longer a constant of motion. This feature is particularly pronounced in the presence of low excitation energies and it is of corresponding significance in the interpretation of chemical shifts so that their measured values contain a rather subtle message in regard to the bonding electrons. On the other hand, the considerably simpler case of the atom can still be used as a guide in the estimate of magnitudes.

In particular, this holds for hydrogen compounds with screening constants close to the value σ_H of Eq. (2) and with chemical shifts, frequently amounting to one tenth or less of this small quantity. Thus in the classical example of ethyl alcohol the internal chemical shift between hydrogen atoms amounts to about one part in a million. It was observed by Arnold *et al.* (*5*) as a fine structure in the proton magnetic resonance, consisting of three separate lines of intensity ratio 3 : 2 : 1 in accordance with the number of hydrogen atoms in the three nonequivalent groups of CH_3CH_2OH. Compared with the more sizable chemical shift for elements of higher atomic number this observation had to be preceded by a considerable improvement of the resolution through better homogeneity of the magnetic field. Apart from such instrumental requirements, however, a more deep-seated condition underlies the possibility of distinguishing such fine details.

Indeed, this would not be possible were it not for the small natural width of magnetic resonance lines in liquids. A measure for this width is obtained from the observed relaxation time T_1 of the order of magnitude of several seconds. Applying the principle of indeterminancy $\Delta E\ \Delta t \cong h/2\pi$ and letting $E = h\nu$, $\Delta t \cong T_1$, it follows in fact with $\Delta E = h\Delta\nu$ that

$$\Delta\nu \cong \frac{1}{2\pi T_1}\,. \tag{3}$$

If one identifies this indeterminancy of the frequency with the natural linewidth one is thus led to expect the latter to be of the order of magnitude of one tenth of a cycle per second which, even at a relatively low frequency of 10 Mc, amounts to no more than about one part in hundred millions. This identification, however, is not *a priori* permissible since there exists another important source of line broadening. It is due to the variation $\Delta H'$ of the magnetic field, acting upon a given nucleus, which arises from the magnetic moment μ of neighboring nuclei at a distance d. With $\Delta H' \cong \mu/d^3$, $\mu \cong 10^{-23}$ cgsu, $d \cong 10^{-8}$ cm, and with a gyromagnetic ratio $\gamma \cong 10^4\ \mathrm{sec}^{-1}G^{-1}$, this would lead to a linewidth

$$\Delta\nu' \cong \gamma\Delta H'/2\pi \cong 10^4\ \mathrm{sec}^{-1}\ , \tag{4}$$

approximately 10^5 times larger than the value estimated above. Such widths of many kilocycles per second are indeed perfectly normal in solid matter and it is due to the fortunate circumstance of motional narrowing that they are not observed in liquids. In fact, the rotational Brownian motion of the molecules, typically carried out with frequencies in the infrared region, has the effect of averaging out the field variation $\Delta H'$. Except for the case of rather viscous liquids it is here permissible, therefore, to consider the value given by Eq. (3) as a valid estimate for the natural width of nuclear magnetic resonance lines.

While such extremely narrow lines offer the possibility of resolving spectra well beyond the part in a million necessary for the observation of normal chemical shifts in hydrogen compounds, such a high resolution imposes severe instrumental conditions. In fact, it has not been possible so far to obtain sufficiently homogeneous magnetic fields to vary over the dimensions of the sample by as little as one part in 10^8 which would be required to approach the natural limit of resolution. Nevertheless, these conditions can be fulfilled by providing an artificial motional narrowing superimposed upon the Brownian process mentioned above. Similarly to the latter, the mechanical spinning of the sample at a sufficiently high frequency effectively averages out the variations of the external field perpendicular to the axis of rotation and greatly facilitates the reduction of those along the axis by means of suitably adapted small shim currents.

The incentive to increase the resolution to below one cycle per second stems primarily from another cause of fine structure in magnetic resonance spectra, first recognized by Gutowsky *et al.* (*6*) and by Hahn and Maxwell (*7*). It was stated before that the interaction of neighboring nuclear magnetic moments which leads to a strong line broadening in solids is averaged out by the rapid rotational motion of the molecules in the liquid. This statement, however, is valid only to the extent to which the effect of the electrons upon this interaction is neglected. Actually, the field produced by the magnetic moment of a nucleus undergoes a slight modification by the bonding electrons, similar to that of the external field which leads to the chemical shift. This modification is not averaged out by the rotation of the molecule since the bonding electrons themselves participate in the rotation, thus leaving an invariant part of the interaction of two nuclei A and B, which can be written in the form $\hbar J_{A,B}(\mathbf{I}_A \cdot \mathbf{I}_B)$, where $\mathbf{I}_{A,B}$ stands for their respective spin vectors. To estimate the frequency J, one has to multiply the field $\Delta H'$ of Eq. (4) with a screening constant of comparable magnitude to that which leads to the chemical shift. With A and B representing both protons it thus follows from Eqs. (2) and (4) that

$$J \cong \alpha^2 \Delta\nu'. \tag{5}$$

With $\Delta\nu' \sim 10^4$ sec^{-1} one is thus led to a value of J of the order of magnitude of one cycle per second, which is indeed comparable with the corresponding fine structure observed in proton resonance spectra.

It is significant for the intimate connection between nuclear magnetic resonance and molecular structure that the wealth of information contained in the observed spectra of liquids can be reduced to the knowledge of chemical shifts and J-coupling constants. While this fact is of great importance in order to interpret the observations, one has to remember that it is due to the remarkably favorable circumstances characteristic of the liquid state. On the one hand, it shares its high density with that of solids so that one obtains relatively strong signals. On the other hand, it shares the disordered motion of the molecules with that found in gases. Far from causing complications, the rapid rotation, interrupted by frequent irregular collisions, is not only responsible for the fact that resonance lines in liquids are usually sharper than in gases and, particularly, in solids but it has also a greatly simplifying influence upon the spectra. Indeed, the complicated effects of the incessant strong interaction between neighboring molecules are, as it were, "erased" by being effectively averaged out in their rapid relative motion and the only neighborhood relations, clearly manifested in the fine structure due to *J*-coupling, are those which are internal to the molecule. They are preserved for the very same reason for which the molecule preserves its chemical identity regardless of the interaction with others in the liquid, that is, by the persistent relative arrangement of its constituent atoms.

There is yet another fundamental aspect of nuclear resonance which must be mentioned because of its direct bearing upon chemical processes. It is based upon the fact that the frequency differences, measured in a spectrum, determine a time scale with which the rate of these processes can be compared. An early example of such a comparison was found by Arnold (*8*) in the variable fine structure of the hydroxyl proton resonance in ethyl alcohol. While, in the absence of exchange, it consists of three separate lines due to *J* coupling with the neighboring methyl group, the transition to a single line is observed when the exchange of hydroxyl protons between molecules occurs at a rate comparable with the frequency separation of the individual lines. An analogous phenomenon occurs when the rate of interchange between nonequivalent groups within the molecule is comparable with their internal chemical shift. Even when this rate is so high as to cause the appearance of a single line it still has a noticeable effect upon its width and can be measured by comparison with a time scale, provided by the spacing of repeated radio-frequency pulses (*9*).

The preceding brief account does not by any means claim completeness. A great deal more could be said, for example, about spin decoupling and other effects obtained by the simultaneous application of different radio frequencies. However, such methods will be considered in their context during the course of the symposium and it may be appropriate to conclude this introduction with a more general remark: The progressing high specialization of our time confronts science with the grave danger of an estrangement of its many branches. It is gratifying, therefore, to witness here a fine example of the opposite tendency which resulted in the blending of originally separated branches into a fruitful and mutually beneficial unit.

References

1. W. D. Knight, *Phys. Rev.* **76**, 1259 (1949).
2. W. G. Proctor and F. C. Yu, *Phys. Rev.* **77**, 717 (1950).
3. W. C. Dickinson, *Phys. Rev.* **77**, 736 (1950).

4. W. E. Lamb, *Phys. Rev.* **60**, 817 (1941).
5. J. T. Arnold, S. S. Dharmatti, and M. E. Packard, *J. Chem. Phys.* **19**, 1608 (1951).
6. H. S. Gutowsky, D. W. McCall, and C. P. Slichter, *Phys. Rev.* **84**, 589 (1951).
7. E. L. Hahn and D. E. Maxwell, *Phys. Rev.* **84**, 1246 (1951).
8. J. T. Arnold, *Phys. Rev.* **102**, 136 (1956).
9. Z. Luz and S. Meiboom, *J. Chem. Phys.* **39**, 366 (1963); A. Allerhand and H. S. Gutowsky, *J. Chem. Phys.* **41**, 2115 (1964).

The Development and Application of Superconducting Solenoids to High-Resolution NMR

W. G. PROCTOR and H. E. WEAVER
Varian AG, Zurich, Switzerland

Introduction

Since the original discovery in 1911 by K. Onnes that certain metals lost all measurable trace of resistance at temperatures near absolute zero, it has been the dream of scientists to use those materials in the form of wire-wound solenoids to generate very high magnetic fields without the continual expenditure of large amounts of electrical energy and the associated need for exhorbitant amounts of coolant. These early attempts were frustrated by the totally unexpected incompatibility between the state of zero resistance and a certain maximum value of an applied magnetic field. It was quickly established that the magnetic field above which the resistivity was restored amounted to only a few hundred gauss.

It was not until the recent discovery of the very high field tolerance of the intermetallic NbSn, NbTi, and NbZr alloys that an intensive investigation of superconducting materials for solenoids has been reestablished. As a consequence of their mechanical properties, the fabrication of these materials requires subtle metallurgical techniques: that these materials are now commercially available in wire or tape form in practical lengths of several kilometers is a tribute to the success of present-day industrial research. It is now believed that the current is carried by superconducting "filaments" while most of the wire remains normal.

Some Characteristics of Superconducting Solenoids

Hysteresis

Current loops or vortex currents are induced by changing magnetic fields and the behavior of a solenoid will then depend upon its magnetic history. This effect manifests itself in a manner reminiscent of iron coil magnets in that fairly large remnant fields—several hundred gauss—are found even though the driving current has been reduced to zero. When this phenomenon was first reported in 1962, it was thought that the completely unpredictable distribution of these currents would make use of superconducting coils for highly homogeneous fields impossible. However, as experiments have now shown, homogeneous fields can be achieved irrespective of magnetic and thermal history at the relatively high fields needed by NMR. Indeed, homogeneities of a few parts per million have been achieved over a 2-cm distance along the axis of symmetry. By employing the well-established techniques of spinning the sample tube and using gradient-canceling coils, homogeneities for high resolution have reached a few parts in 10^9, or better than 1 cps resolution at 200 Mcps for protons.

Quenching

It is reasonable to expect that wire, when wound into solenoids and buried in the interior with poor thermal conduction to the helium bath, will get warm as the field is changed, since there will be Joule heating from the normal metal parts. This can be controlled by minimizing the rate of change of applied current. If, owing to such heating, or to the collapse of vortex currents, a section of wire experiences too great a rise in temperature and goes normal, this section of wire thus provides a very lossy mechanism for the dissipation of magnetic energy in the solenoid. The normal section, becoming more heated, grows rapidly along the wire, at a rate of several meters per second, and to a lesser extent across wire boundaries. The process is very rapid, and the field will collapse within a fraction of a second. The heat, however, is dissipated comparatively slowly, ultimately vaporizing a corresponding amount of liquid helium. Therefore an explosive situation does not occur despite the final conversion of large amounts of stored energy (up to 10^5 joules). It has been found to be of considerable advantage to coat the superconducting wire with a thin copper coating in order to carry the current in case a portion of the superconductor goes normal and to increase the heat capacity.

Training

When solenoids of reasonable size were wound and tested, it was discovered that wire which had an excellent performance in short pieces quenched, i.e., went normal, at disappointingly low currents. However, in reenergizing the solenoid after an initial quenching, it was found that the critical current became significantly higher until after one or two cycles the limiting current-field performance would closely approximate that of a short sample. Such improvement in solenoid performance is known as "training."

Stability

Stability is automatically achieved in that the coil, in its superconducting state, maintains a constant total flux through it. The current is brought to the desired level—generally about 20 amp—by an exterior power supply, which is in effect disconnected with a superconducting switch. The current continues to flow in the coil and will flow perpetually, resisting total changes of flux. Imposed gradients, which, however, do not add to the total flux may be superimposed. This permits the use of shim coils, but does not guard against flux changes due to movement of magnetic objects nearby which produce linear gradients. Despite this, excellent results have been achieved with unshielded solenoids, although the use of a high-permeable shield is an easy solution to the problem.

Design of a High-Resolution Solenoid

By altering the cross section of the coil, so that there are more turns nearer the ends of the coil than near the center, one may calculate that remarkably good homogeneities—such as one part in 10^7 over a centimeter axial length—are possible. Such homogeneities, however, are not in practice attainable because construction techniques fail to reproduce the current geometry perfectly.

To design a coil for proton resonance at 200 Mc, a field intensity of 47 kilogauss with a homogeneity and field stability of a few parts in 10^9 were necessary. This was achieved with an NbZr coil of 35.7 cm length, inner diameter 4.25 cm, outer diameter 12.7 cm, weighing 13.2 kg. When mounted in a dewar with an interior access, the room temperature working diameter was 2.6 cm. This solenoid was shown in operation together with the rest of a 200-Mc high-resolution system at the Pittsburgh Spectroscopy Conference of March 1964. Performance met expectations in that satisfactory spectra with a resolution of about one cps were repeatedly demonstrated; however, sideband intensity was about 5%.

The basic homogeneity was improved by handling residual gradients by exterior gradient generating coils and by spinning the sample tube with an air turbine. In the language of the coil designers, the demonstrated coil was an outside notch, sixth-order solenoid. This means that the extra ampere-turns at the end of the coil were provided by extra turns on the exterior of the coil, rather than the interior, and that, from the geometrical center of the coil, the field increased only as the 6th power of the distance from that point.

Some New Problems in Spectrometer Design

It is clear that the sample tubes must now rest parallel to the polarizing field. Hence it is not possible to wrap the receiver coil around the sample in the usual fashion, but both it and the transmitter coil must be "split." Rather than attempt to put the sample into place in the probe, which is located about a meter from the top of the dewar, it was found to be more convenient to remove the probe each time the sample was changed.

Because of increased losses in coils and leads at higher frequencies, it appears that the increase in sensitivity is approximately linear with the field, affording useful but not dramatic increases in sensitivity.

Modulation and sweeps are also possible, provided that the coils which generate these small fields are located well on the interior of the solenoid and do not seriously couple with it. However, when using side-band methods for high resolution, in order to preserve a given modulation index, the modulation amplitude must be greater, since the increased chemical shift requires a higher modulation frequency. This has not so far been a problem with protons, but it may be serious with F^{19} or other nuclei with large chemical shifts. Coupling to the solenoid means not only mechanical forces but greater heat dissipation at the normal surfaces.

Transmission of Modulation Information Through the Nuclear Spin Coupling

W. A. ANDERSON and R. FREEMAN

Varian Associates, Analytical Instrument Division, Palo Alto, California

Carbon-13 chemical shift information can be an important tool in the determination of chemical structure. Unfortunately the combination of factors of low isotopic abundance, relatively small gyromagnetic ratio, and very long relaxation times all conspire to make difficult the detection of ^{13}C NMR signals. If, however, one can monitor ^{13}C transitions indirectly by observing resonance signals of protons that are coupled to the ^{13}C, one overcomes the latter two difficulties. This indirect detection scheme was first illustrated by Royden (*1*) when she determined the gyromagnetic ratio of ^{13}C nuclei in an enriched sample of methyl iodide. She observed the proton resonance of the methyl group while applying a second perturbing field near the ^{13}C resonance frequency. When the strong *rf* field was adjusted to the exact ^{13}C resonance condition a collapse of the proton multiplet was observed.

The same type of experiment has been performed using samples containing only naturally abundant ^{13}C nuclei; however, such observations are sometimes quite difficult because of large nearby resonance signals associated with protons in the much more abundant molecules containing ^{12}C nuclei.

In the experiments described here, a modulated *rf* field is applied near the ^{13}C resonance frequency causing these spins to nutate at the modulation frequency. A spin-spin coupling will transmit this modulation information to any proton group in the same molecule. This modulation is equivalent to a small local magnetic field modulation at the proton sites, and is present only in molecules containing a ^{13}C nucleus. This type of modulation transfer permits selective detection of proton groups that are coupled to the ^{13}C group which is being perturbed (*2*).

The use of this technique to determine ^{13}C chemical shifts is illustrated in Fig. 1 with acetic acid containing only the 1.1% naturally abundant ^{13}C. This spectrum was made by sweeping a frequency-modulated *rf* field through the ^{13}C region of the spectrum while the 60-Mc proton frequency remained close to the sideband resonance frequency of the methyl group. The scale at the bottom of Fig. 1 corresponds to the number of kilocycles by which the irradiation frequency was above 15 Mc. The multiplet at the left arises from the methyl ^{13}C resonances, and the

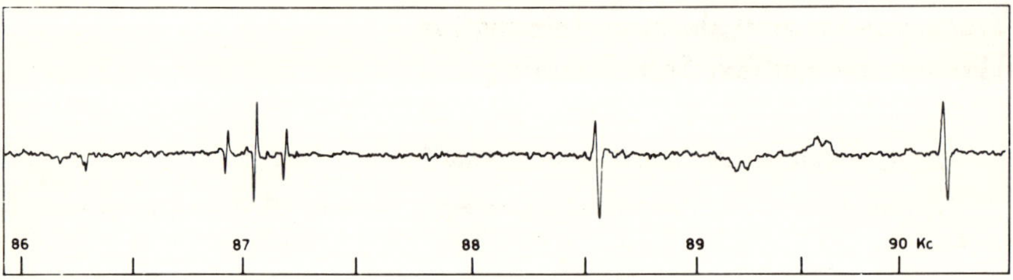

FIG. 1.

center of this multiplet corresponds to the ^{13}C chemical shift of this methyl carbon atom. The strength of the *rf* field applied to the ^{13}C nuclei is comparable with the methyl coupling constant. The pattern centered near 89.4 kc corresponds to the chemical shift of the carboxyl ^{13}C resonance. The central lines are greatly broadened because the *rf* field is very large compared with this coupling constant. The two narrow satellite lines near 88.6 and 90.2 kc arise from the modulation sidebands stimulating the carboxyl ^{13}C resonance.

References

1. V. Royden, *Phys. Rev.* **96**, 543 (1954).
2. W. A. Anderson, *J. Chem. Phys.* **37**, 1373 (1962); R. Freeman and W. A. Anderson, *J. Chem. Phys.* **42**, 1199 (1965).

Résonance magnétique nucléaire et spectroscopie infrarouge

RENÉ FREYMANN

Faculté des Sciences de Paris, Paris, France

Problèmes intramoléculaires: Études en résonance magnétique nucléaire (RMN) et en infrarouge (IR)

Un exemple de différence apparente entre les données de l'infrarouge et de la RMN sur un problème de liaison intramoléculaire: La perturbation de CH par les orbitales d'un atome O ou N voisin

Spectres d'absorption dans le proche infrarouge des éthers oxydes et composés analogues. La Fig. 1 montre des fréquences harmoniques 4ν et 3ν de $(C_2H_5)_2O$. Deux types distincts de bandes apparaissent: les unes, bandes normales, 4ν (CH) ou 3ν (CH) vers les hautes fréquences; les autres, bandes "perturbées" 4ν (CH ⟵ O) ou 3ν (CH ⟵ O) vers les basses fréquences. Nous attribuons ces dernières à la perturbation des oscillateurs CH par les paires libres de l'atome O voisin. Ceci est confirmé par l'expérience suivante:

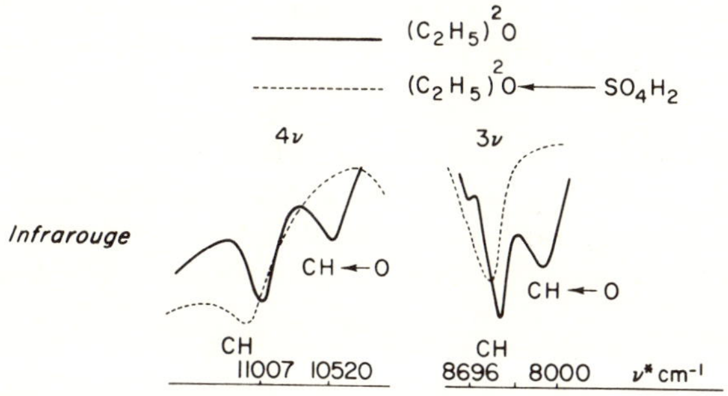

FIG. 1. Perturbation des fréquences infrarouges CH (par les orbitales de O voisin) pour $(C_2H_5)^2O$.

Lorsqu'on sature $(C_2H_5)_2O$ par ClH, SO_4H_2 (*5*), ou mieux BF_3 (*17*), la bande normale est peu affectée, mais *la bande perturbée disparaît* (courbe en pointillés, Fig. 1).

Ces résultats ont été étendus, pour 4ν et 3ν (*16*), à $(CH_3)_2O$, $[(CH_3)_2CH]_2O$, $(C_4H_9)_2O$, aux acétals $ROCH_2OR$, aux aldéhydes RCHO. Des résultats analogues ont été obtenus (*17*) pour le fondamental ν et l'harmonique 2ν de $(CH_3)_2O$ et $(CH_3)_2O$, BF_3.

Ainsi, l'infrarouge semble affirmer *l'existence de deux types de vibrateurs CH, dont l'un résulte de la perturbation intramoléculaire par les paires libres* des atomes O (ou N) voisins.

Résonance magnétique nucléaire de $(CH_3)_2O$ et $(C_2H_5)_2O$. Au contraire, en résonance nucléaire, à 20°C, on n'observe pas deux raies distinctes caractéristiques d'un CH normal et d'un CH perturbé ; en outre, il n'y a pas de différence entre vapeur, liquide pur et liquide infiniment dilué dans CCl_4. Nous nous proposons de rechercher s'il y a dédoublement à basse température (afin de modifier d'éventuels phénomènes d'échanges).

	δ (CH_2)	δ (CH_3)
$(CH_3)_2O$		
Vapeur		3.22
Infiniment dilué dans CCl_4		3.22
$(C_2H_5)_2O$		
Vapeur	1.13	3.32
Liquide pur	1.15	3.34
Infiniment dilué dans CCl_4	1.13	3.32

Rappel de relations entre les données intramoléculaires de la RMN (effet δ et effet J) et celles de diverses méthodes physiques autres que l'IR

Notre but est d'établir des relations directes entre les données de la RMN et de l'IR. Mais nous montrerons auparavant que ces liens ont été établis par: (1) la comparaison entre RMN (δ ou J) et diverses grandeurs physiques ; (2) la comparaison entre IR et ces mêmes grandeurs physiques.

Relation entre effet δ (en RMN) et diverses grandeurs physiques.

(*a*) *Effet δ et densités électroniques.* Rappelons par exemple les relations trouvées entre effet δ et charges π par A. Veillard et B. Pullmann [*Compt. Rend.* **253**, 2418 (1961)] pour divers composés d'intérêt biologique.

(*b*) *Effet δ et hybridation des liaisons.* Nous n'insisterons pas sur ce point, car l'étude de l'effet J (dont il est question au paragraphe suivant) est en général plus efficace.

(c) *Effet δ et électronégativité (ou moment dipolaire) des groupements substitués.*

(d) *Effet δ et réactivité chimique (constante de Hammett).*

(e) *Effet δ et susceptibilité magnétique.*

Relation entre effet J et diverses grandeurs physiques. L'effet J caractérise l'hybridation des liaisons des noyaux couplés, tels que $H \cdots H$; $^{13}C \cdots H$; $F \cdots H$, etc. Voici, par exemple (Fig. 2), quelques valeurs de $J_{P \cdots H}$ (en Hertz):

CH_3PCl_2 (17.5); $CH_3P(O)Cl_2$ (16.25); $CH_3P(S)Cl_2$ (14.8); $CH_3P(Se)Cl_2$ (13.5)

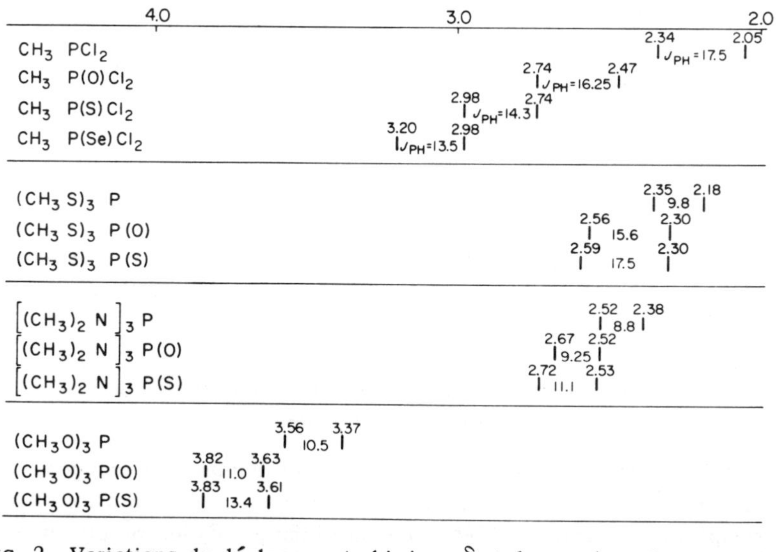

FIG. 2. Variations du déplacement chimique δ et du couplage J_{PH} avec les liaisons : CH_3P ; CH_3S ; CH_3N ; CH_3O.

D'autre part, la Fig. 3a (9), relie la valeur du couplage $J_{^{13}C \cdots H}$ à l'indice de liaison de divers composés organophosphorés. Soulignons que les couplages $J_{^{13}C \cdots H}$ sont plus faciles à interpréter que les couplages $J_{P \cdots H}$.

Relations entre effet δ et effet J. Nous venons successivement de rapprocher l'effet δ, puis l'effet J de diverses grandeurs; il était logique de comparer entre eux l'effet δ et l'effet J. Par exemple, un parallèle est établi entre $J_{P \cdots H}$ et δ en ce qui concerne les composés organophosphorés par les données de la Fig. 2 [d'après G.L.A. Martin et G. Mavel (9)].

Mlle G. L. Martin (8) a d'ailleurs pu tracer des courbes reliant $J_{^{13}CH}$ et δ pour certains des nombreux composés organophosphorés qu'elle avait étudiés.

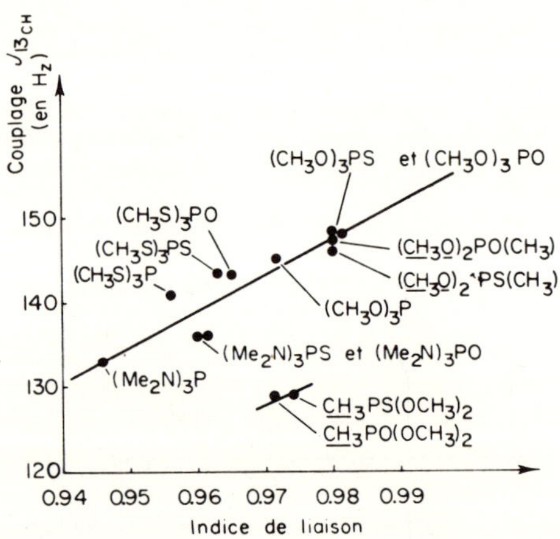

FIG. 3a. Valeur du couplage $J_{^{13}CH}$ reliée à l'indice de liaison de divers composés organophosphorés, d'après G. L. A. Martin et G. Mavel (9).

Relations entre données RMN et données IR sur des problèmes intramoléculaires

Comparaison entre δ(RMN) et position ν des bandes infrarouges. Si l'on essaie de porter les *positions* ν des fréquences de vibration, (CH) par exemple, en fonction du déplacement chimique δ, les résultats sont peu satisfaisants ; c'est ainsi que pour les composés acétyléniques, la fréquence infrarouge ν ($\equiv$CH) et ses harmoniques ne varient pratiquement pas avec la structure du reste de la molécule.

Liens entre $J_{^{31}PH}$ (RMN) et position des fréquences infrarouges ν (PH). On doit à Wolf (*19*), du laboratoire de M. le Professeur Mathis (Toulouse, France), la courbe reliant le couplage spin-spin de ^{31}P et de H à la fréquence fondamentale ν (PH), de nombreux composés organophosphorés (Fig. 3b).

Liens entre δ (RMN) et intensité infrarouge |A| pour les composés acétyléniques. Par contre, pour les mêmes composés acétyléniques, Jouve (7) a montré que *l'intensité A'* de la bande infrarouge ν ($\equiv$CH) (et

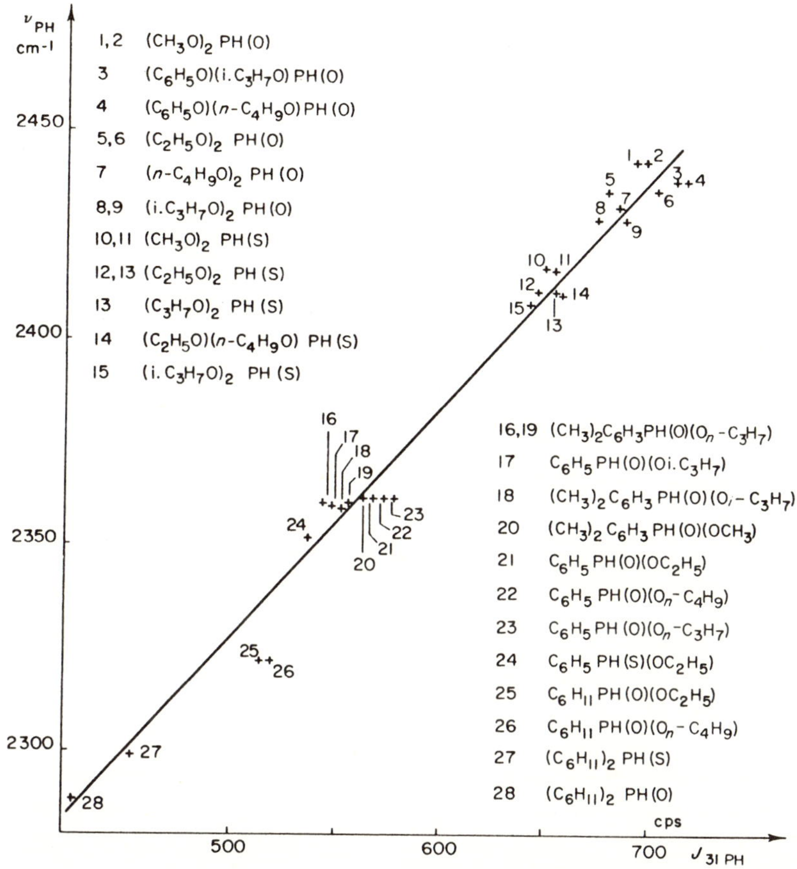

FIG. 3b. Courbe reliant le couplage spin-spin de ^{31}P et de H à la fréquence fondamentale ν_{PH} de divers composés organophosphorés d'après WOLF (*19*).

de ses harmoniques), varie beaucoup d'un composé à l'autre, et *il a établi un lien avec le déplacement chimique δ en RMN.**

On a porté (Fig. 4) en ordonnée A′ relatif à l'infrarouge ($C = 0{,}4$ mole/litre) pour la fréquence fondamentale ν ($\equiv$CH); en abscisse, la valeur en RMN de δ_∞ dans CCl_4 (dilution infinie pour éliminer les effets de susceptibilité ou d'anisotropie magnétiques). On note une liaison remarquable entre A′ et δ_∞ pour une dizaine de composés acétyléniques distincts pour lesquels l'autoassociation est faible (et du même ordre de grandeur).

*On sait que l'on distingue intensité apparente A′ et intensité absolue $|A|$. Etant donné les difficultés rencontrées, nous précisons quelques détails techniques (p. 18).

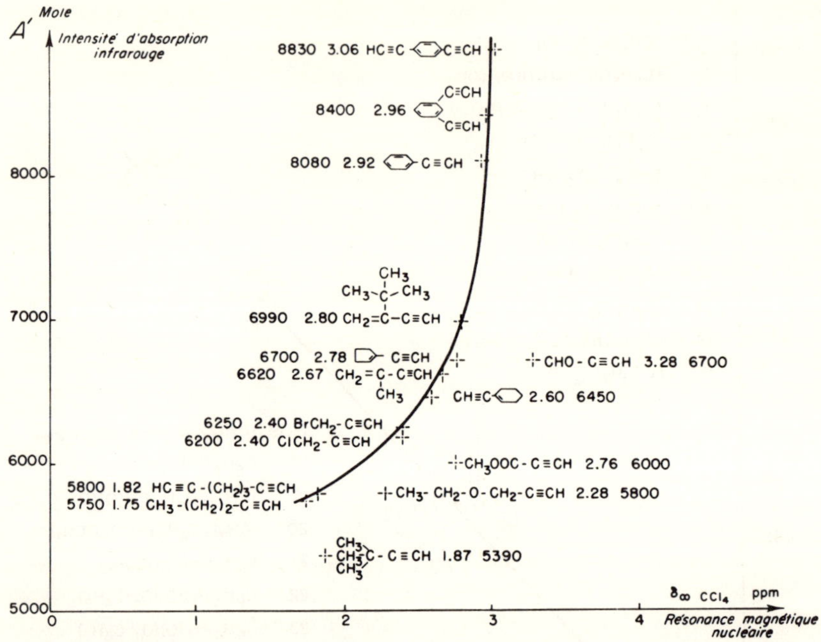

FIG. 4. Relation entre A′ (intensité de l'absorption infrarouge) et δ_∞ (RMN): Acétyléniques (*F*).

Les spectres ont été enregistrés par transmission (à l'aide du spectromètre Perkin-Elmer 125; fente occupant 0.72 cm^{-1} pour une largeur de bande à mi-hauteur de 12 à 22 cm^{-1}). L'intensité apparente intégrée A′ portée sur la Fig. 4 est donnée par:

$$A' = \frac{1}{Cl}\int_0^\infty \left(\log \frac{T_0}{T}\right) d\nu^*$$

ν^* nombre d'ondes, T_0 transmission hors de la bande, T transmission pour ν^*, *l* longueur de la cuve (voisine de 0.1 mm pour le fondamental), *C* concentration du produit (dans CCl_4, à 0.4 mole/litre).

Afin de contrôler les données de la Fig. 4, Jouve (7) a déterminé ensuite l'intensité absolue $|A|$ en extrapolant à concentration nulle les valeurs de A mesurées pour diverses valeurs de *C*. La linéarité des relations obtenues en portant $\log T_0/T$ et $\int_0^\infty (\log T_0/T)\, d\nu^*$ en fonction de *C* confirme la validité de la loi de Beer-Lambert. Pour cela, des cuves de 5 et 10 cm d'épaisseur (et des concentrations correspondantes de 10^{-4} et 10^{-2} mole/litre) ont dû être utilisées respectivement pour ν ($\equiv$CH) et 2ν ($\equiv$CH).

On notera que la valeur de l'intensité $|A|$ mesurée est en partie faussée par la présence d'une bande faible vers les basses fréquences (à 20 cm^{-1} environ de la bande principale; transition chaude), d'intensité 8 à 10% de l'intensité totale.

Par contre, si $\delta_{pur} - \delta_{\infty}$ est plus élevé—et, c'est le cas des composés oxygénés—les points s'écartent de la courbe.*

*Quant au *tertio*-butylacétylène $(CH_3)_3C{\equiv}CH$, sa position hors de la courbe ne surprend pas: la fréquence de valence ν ($C{\equiv}C$) est de 10 cm^{-1} plus faible que pour tous les autres carbures; ce composé est anormal.

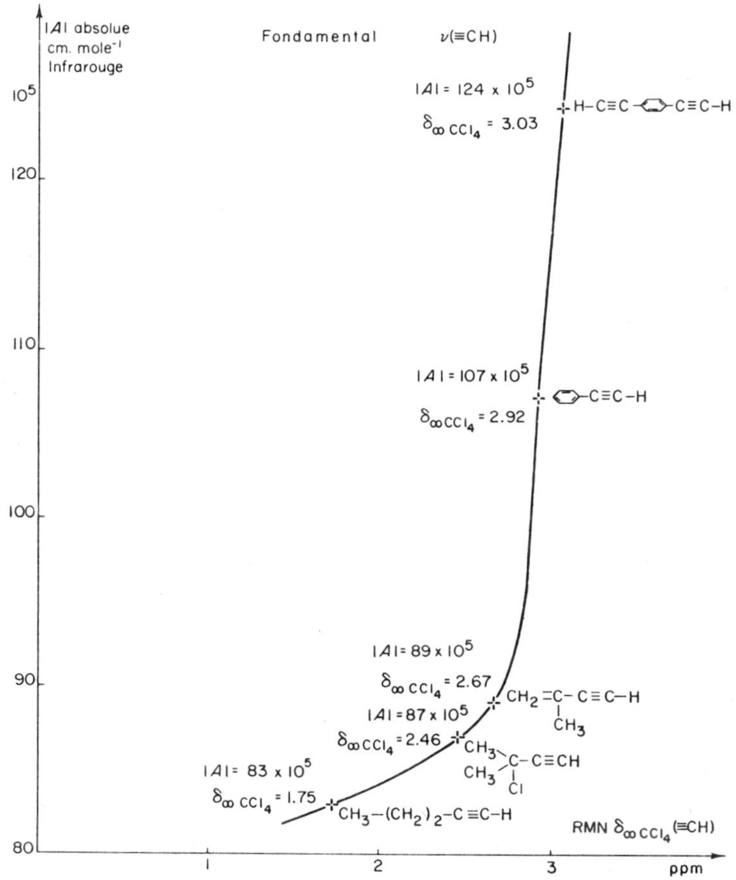

FIG. 5. Comme la Fig. 4, pour $|A|$ intensité *absolue* et pour le fondamental ν($\equiv$CH).

	$\delta_{pur} - \delta_{\infty CCl_4}$ (ppm)
Non autoassocié	0.12
Ether	0.30
Ester	0.54
Aldéhyde	0.76

Les résultats précédents ont été confirmés pour les harmoniques 2ν et 3ν des mêmes composés acétyléniques.

Il importait de savoir si l'allure de la courbe de la Fig. 4 restait la même pour l'intensité absolue $|A|$ (et non plus A′). C'est ce que con-

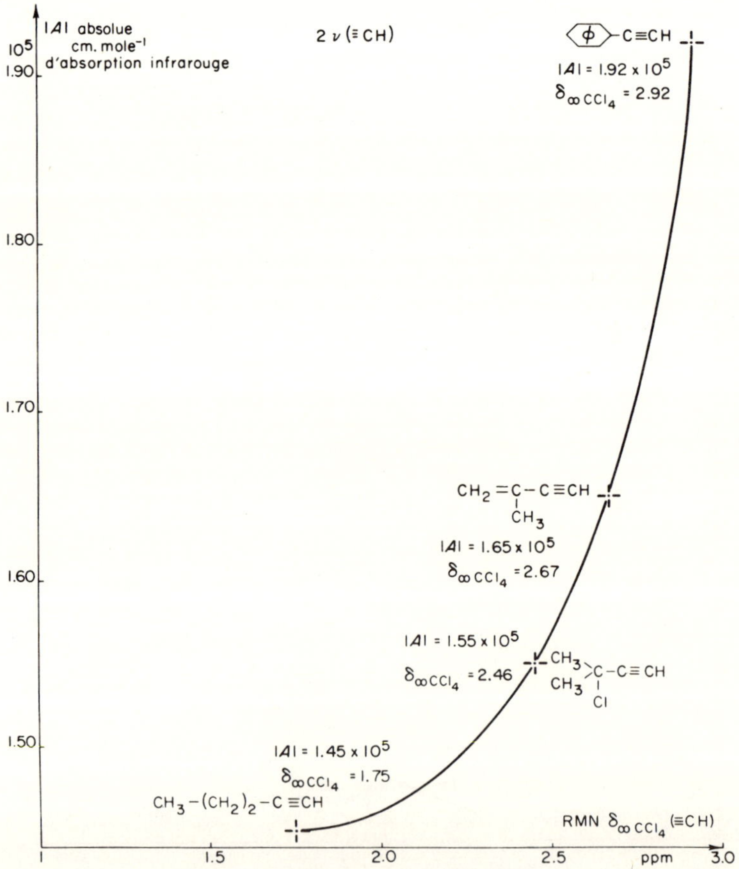

FIG. 6. Comme Fig. 4, pour $|A|$ intensité *absolue* et pour l'harmonique $2\nu(\equiv CH)$.

firment les Figs. 5 et 6 relatives à ν ($\equiv$CH) et 2ν ($\equiv$CH) (le nombre limité de mesures est dû à leur difficulté technique).

Liens entre δ (RMN) et intensité infrarouge $|A|$ pour des composés halogénés. On doit également à Jouve (7) d'avoir souligné l'existence d'une relation entre $|A|$ et (δ) pour d'autres séries organiques.

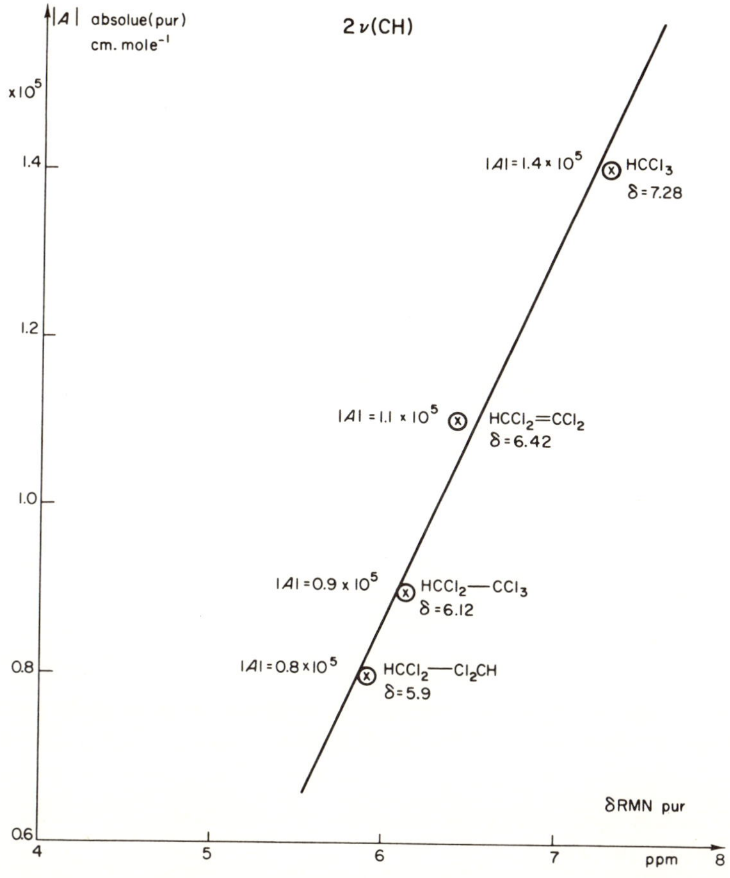

FIG. 7. Données infrarouges 2ν (CH) pour l'intensité absolue $|A|$ [d'après Amat (*1*)] reliées à celles de δ (en RMN) [d'après Mme. Martin (*10*)].

La Fig. 7 relative à divers composés halogénés relie les données infrarouges 2ν (CH) de G. Amat (*1*) pour $|A|$ et celles de δ_∞ (en RMN) de M. Martin (*10*).

De même (Fig. 8), pour la série CH_3X, Jouve (*F*) a relié les données infrarouges ν (CH) de Barrow et McKean (*2*), et celles, RMN, de Shool-

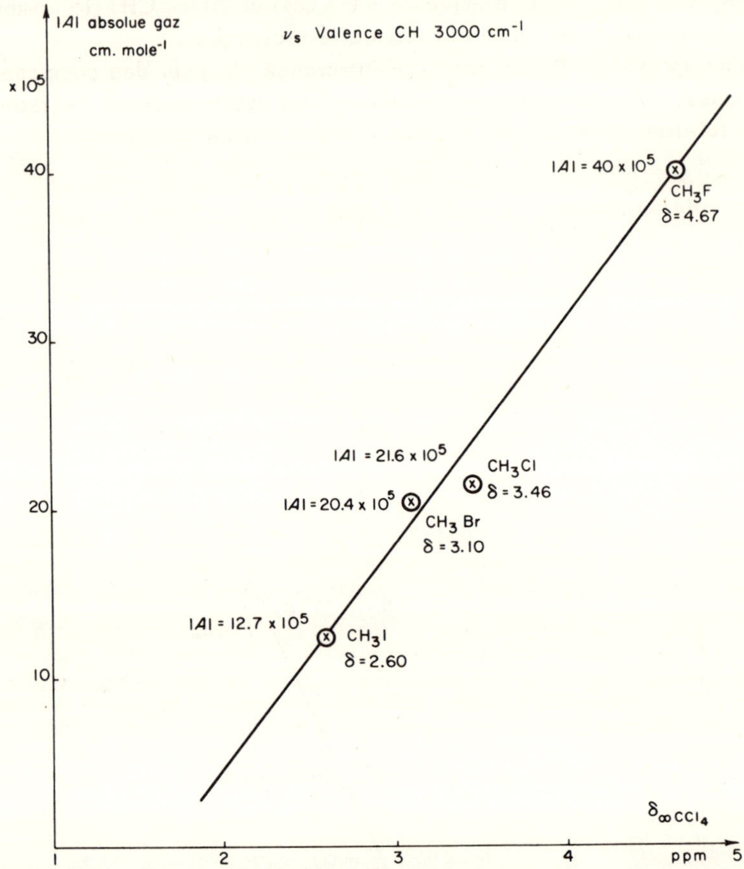

FIG. 8. Données infrarouges ν (CH) de G. M. Barrow et D. C. McKean (*2*) reliées à celles (en RMN) de J. N. Shoolery (*15*).

ery and Dailey (*15*). Ici, dans les deux cas, la relation $|A| = f(\delta)$ est sensiblement linéaire.

Enfin, la Fig. 9 apporte, pour la série HF, HCl, HBr, HI, l'exemple le plus important, du point de vue théorique, de relation entre les données infrarouges sur $|A|$ de Penner et Weber (*13*) et d'Oksengorn (communication privée) et les données de la RMN d'après Schneider.

Conclusion. Nous rappellerons les conclusions de Jouve: les courbes des Figs. 4 à 9 confirment l'existence d'une relation entre déplacement chimique RMN et intensité absolue des fréquences infrarouges de vibration (de valence) correspondantes.

Or, en RMN, Pople (*14*) fait intervenir dans sa théorie du déplacement

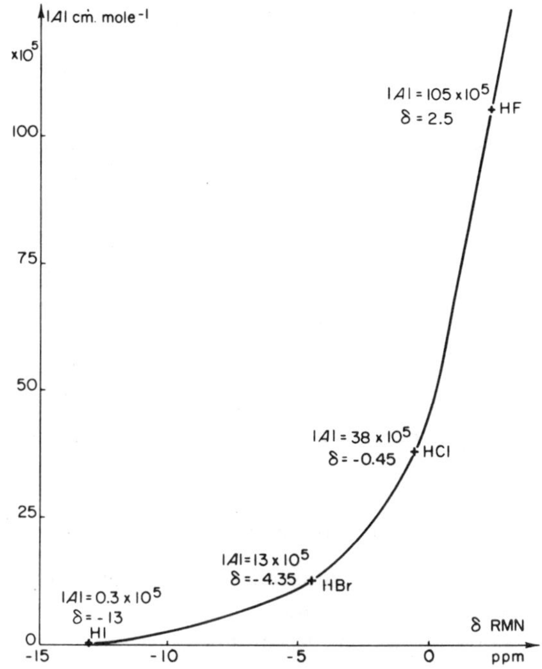

FIG. 9. Relation entre les données infrarouges sur $|A|$ de Penner et Weber (*13*) et d'Oksengorn (communication privée) et les données de la RMN (d'après W. G. Schneider).

chimique, la densité électronique $\rho(r)$ autour du proton ; d'autre part, l'intensité absolue infrarouge $|A|$ est liée par

$$|A| = \frac{N\pi}{3C}\left(\frac{\partial\mu}{\partial Q}\right)^2$$

aux variations du moment dipolaire μ de la liaison CH.

Donc, δ et $|A|$ sont tous deux liés à la densité électronique au voisinage de la liaison CH.

Toutefois, pour que les relations $|A| = f(\delta_\infty)$ puissent être reliées entre elles quand on change de série, il faut faire intervenir deux phénomènes propres à chaque technique :

(1) en infrarouge, introduire le couplage entre vibrateurs, et envisager, dans le cas de $\equiv$CH, la somme

$$A_{\equiv CH}/\nu^2_{\equiv CH} + A_{C\equiv C}/\nu^2_{C\equiv C}\,.$$

(2) en RMN, il faut tenir compte du décalage dû à l'anisotropie magnétique de la triple liaison.

Problèmes intermoléculaires étudiés en RMN et en IR

Etude de phénomènes intermoléculaires à l'état gazeux par IR et par RMN : ClH...

Nous donnerons un seul exemple de comparaison des deux techniques. Des travaux que nous avions abordés en 1937 avec Vodar (*4*) et qui furent repris par Vu Hai (*18*), ont montré que les spectres de vibration-rotation infrarouge constituent un procédé de choix pour l'étude des actions intermoléculaires : interactions des molécules HCl entre elles, interaction (sous pression) entre molécules HCl et molécules d'argon, CO_2, etc.

MM. Dayan et Widenlocher (*3*) ont repris *les mêmes études par résonance magnétique nucléaire* de gaz purs ou mélangés (mesurés en haute résolution par une nouvelle technique très simple ; étude de H_2, HD, CH_4, C_2H_6, C_2H_4, HCl, etc.). Ils ont établi un parallèle entre les données IR et RMN.

Distinction par RMN entre divers phénomènes intermoléculaires

Ce problème a fait l'objet de plusieurs communications antérieures (*5–12*). Nous rappellerons donc seulement quelques cas typiques d'étude par RMN. Nous ne parlerons pas ici du précieux apport, bien connu, de l'IR.

(1) *"Effets de solvant"* : Liaisons intermoléculaires "faibles" ; forces de Van der Waals ; comparaisons liquide pur, solution, vapeur (*3, 7*).

(2) *"Liaison hydrogène"* (intermoléculaire) (*6, 10, 12*) :

A. entre molécules de même nature
  - (i) Association (des corps purs)
  - (ii) Désassociation (par solvant non polaire) ;

B. entre molécules de natures différentes
  Complexation (par solvant polaire) = solvatation

C. Phénomènes d'échanges.

(3) *Effet* π (composés benzéniques ou hétérocycliques) (*6*).

(4) *Complexes à transfert de charge.*

(5) *Phénomènes d'ionisation.*

(6) Interaction entre RMN et propriétés électroniques [*effet de traces de composés paramagnétiques : sels de cuivre, oxygène,* etc., sur la RMN (*5*)].

Relations entre les données intermoléculaires de la RMN et celles de diverses méthodes physiques (autres que l'IR)—Notion de basicité et méthode du $CHCl_3$

Un grand nombre de recherches sur les spectres infrarouges avait déjà conduit à relier à diverses grandeurs physiques *les déplacements de fréquence infrarouge*

$$\Delta\nu = \nu_{gaz} - \nu_{solution}$$

ou, parfois,

$$\frac{\Delta\nu}{\nu} = \frac{\nu_{gaz} - \nu_{solution}}{\nu_{gaz}}$$

De même, en résonance magnétique nucléaire, de nombreux auteurs, notamment G. Mavel (*12*) et Mme. Martin (*10*), étudient

$$\Delta\delta = \delta_{gaz} - \delta_{\text{dilution } \infty \text{ dans le solvant}}$$

en fonction de la permittivité, du pK, etc., donc, en dernière analyse, en fonction de la *basicité* du solvant.

$\Delta\delta$ pour le proton de H_2O (*12*).

$\Delta\delta_{H_2O}$ *et* $(\varepsilon' - 1)^2/(2\varepsilon' + 1)^2$. On a également relié $\Delta\delta$ (et cela est d'ailleurs préférable) à $(n^2 - 1)^2/(2n^2 + 1)^2$ (Fig. 10).

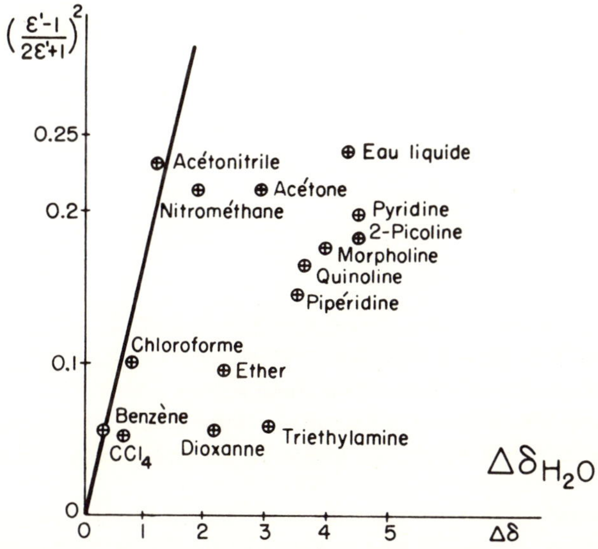

FIG. 10. Diagramme $\Delta\delta_{H_2O}$ en fonction de la permittivité ε.

$\Delta\delta_{H_2O}$ *et* pK (*12*). (L'anomalie aux faibles pK serait peut-être liée aux phénomènes d'ionisation) (Fig. 11).

$\Delta\delta$ pour le proton de $CHCl_3$: La "méthode du chloroforme" de Mme. Martin permet, mieux que le pK, de caractériser la basicité d'un solvant.

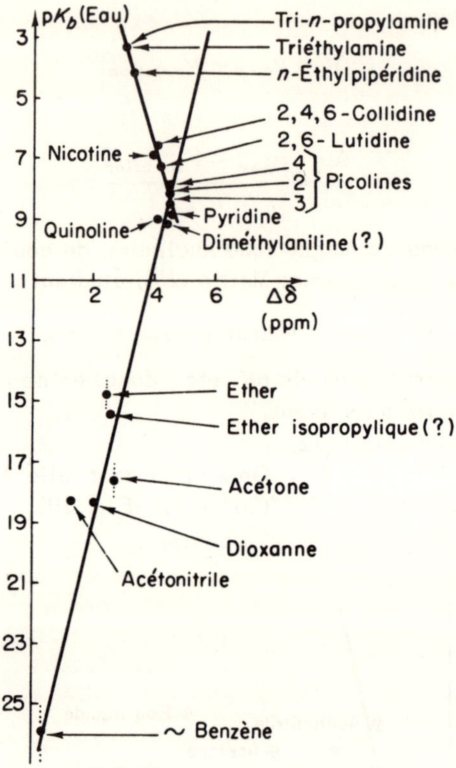

FIG. 11. Diagramme $\Delta\delta_{H_2O}$ en fonction du pK (*12*).

$\Delta\delta_{CHCl_3}$ *et* pK *dans l'eau.* Mme Martin a essayé de relier $\Delta\delta_{CHCl_3}$ au pK (dans l'eau) de divers composés. Cette comparaison (Fig. 12) est peu satisfaisante ; Mme Martin l'explique ainsi :

—les valeurs du pK sont déterminées *dans l' eau, et non dans* $CHCl_3$; par conséquent, le pK ne permet guère de caractériser les solvants organiques.

— RMN et IR mesurent l'aptitude des bases à former des *liaisons hydrogène,* alors que le pK traduit leur *capacité à accepter complètement un proton* (Figs. 12 and 13).

Hypothèse de Mme. Martin: $\Delta\delta_{CHCl_3}$ *peut servir, au lieu du* pK, *à caractériser la basicité.* En effet,

— alors que pK_{H_2O} donne des résultats décevants (Fig. 12), les données $\Delta\delta_{CHCl_3}$ et $\Delta\delta_{H_2O}$ sont compatibles entre elles (Fig. 13). $\Delta\delta_{H_2O}$ (Mavel) est lié de façon très satisfaisante à $\Delta\delta_{\infty CHCl_3}$ (Mme. Martin).

— les données de $\Delta\delta_{CHCl_3}$ et celles de $\Delta\delta_{CHCl_2CCl_3}$, $\Delta\delta_{CHBr_3}$, $\Delta\delta_{CHCl=CCl_2}$, $\Delta\delta_{CH_2Cl_2}$ sont d'ailleurs compatibles pour de nombreux solvants.

La "méthode du chloroforme" de Mme. Martin caractérise bien la basicité des oxygènes. En conclusion de ce qui précède, Mme. Martin a tracé les courbes reliant $\Delta\delta_{CHCl_3}$ et (δ_{CHO}) (elle a d'ailleurs confirmé ces données, en collaboration avec J. Cantacuzène) par l'étude de l'absorption dans le spectre visible de solutions d'iode.

Le résultat très satisfaisant de cette comparaison confirme qu'il est préférable de caractériser les basicités par $\Delta\delta_{CHCl_3}$ plutôt que par le p*K*.

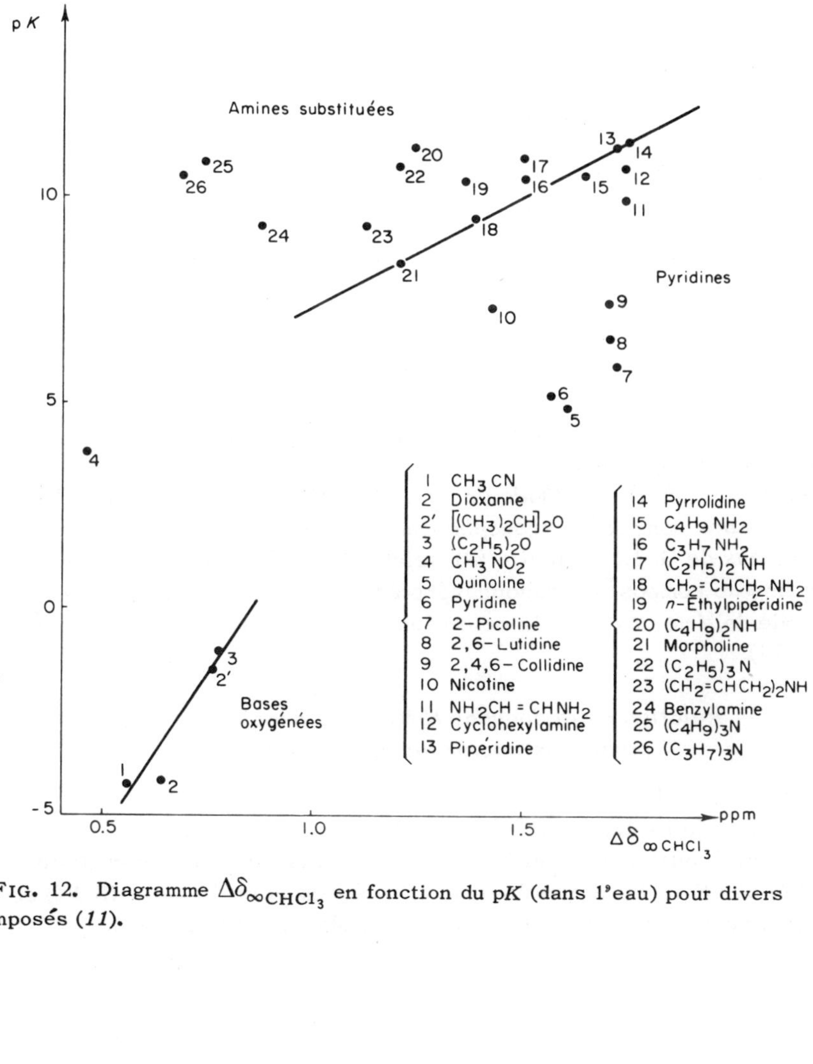

FIG. 12. Diagramme $\Delta\delta_{\infty CHCl_3}$ en fonction du p*K* (dans l'eau) pour divers composés (*11*).

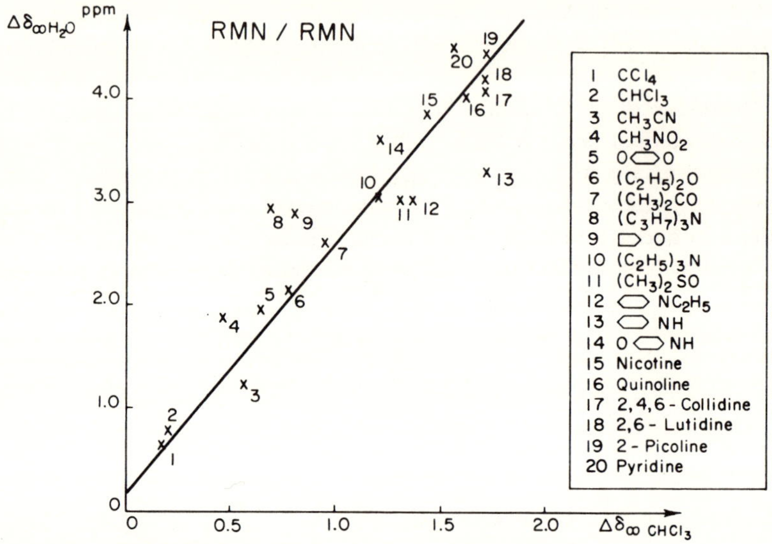

FIG. 13. Données $\Delta\delta_{\infty CHCl_3}$ reliées à celles $\Delta\delta_{\infty H_2O}$ pour divers composés. (*11, 12*).

Remarque: Une difficulté théorique résulte de ce que, en toute rigueur (*12, 10*), il faudrait utiliser

$$\Delta\delta_C = \frac{K+1}{K}\,\Delta\delta_\infty$$

et non pas $\Delta\delta_\infty$; mais il est difficile, par RMN, de mesurer K avec suffisamment de précision.

Application de la "méthode du chloroforme" à la détermination des basicités des composés organophosphorés. Comme le montre la Fig. 14, les travaux de G. L. A. Martin (*8*), relatifs à l'effet sur $CHCl_3$ d'une trentaine de composés organophosphorés, permettent d'appliquer la méthode de Mme. Martin à la définition de la basicité de ces composés (Fig. 14).

Relation entre données RMN et données IR sur des problèmes intermoléculaires (solvants)

Nous venons de voir que, pour la RMN, Mme. Martin avait préféré caractériser la basicité par $\Delta\delta_{CHCl_3}$ plutôt que par le pK.

Comme, d'autre part, pour l'infrarouge, divers auteurs avaient relié $\Delta\nu$ (IR) au pK, il lui a paru logique (*10*) de comparer $\Delta\delta_{CHCl_3}$ (RMN) et $\Delta\nu$ (IR), poursuivant ainsi le travail de G. Mavel sur l'eau (*12*).

		7 ppm — 8 ppm — 9 ppm
1	C_6H_{12}	7.08
2	$CHCl_3$	7.28
3	$CH_3P(\rightarrow S)Cl_2$	7.43
4	$(CH_3S)_3P$	7.46
5	$Cl_3P\rightarrow O$	7.53
6	$(CH_3S)_3P\rightarrow S$	7.55
7	$CH_3SP(\rightarrow O)Cl_2$	7.61
8	$(CH_3O)_3P\rightarrow S$	7.67
9	$(CH_3O)_3P$	7.68
10	$CH_3P(\rightarrow S)(SCH_3)_2$	7.70
11	$CH_3OP(\rightarrow O)Cl_2$	7.71
12	$CH_3P(\rightarrow O)F_2$	7.73
13	$[(CH_3)_2N]_3P$	7.74
14	$CH_3P(\rightarrow O)Cl_2$	7.75
15	$CH_3P(\rightarrow S)(OCH_3)_2$	7.77
16	$(CH_3O)_3P\rightarrow Se$	7.78
17	$CH_3P(\rightarrow S)(OC_3H_7\,iso)_2$	7.81
18	$(CH_3)_2NP(\rightarrow O)Cl_2$	7.84
19	$(CH_3O)_2P(\rightarrow O)Cl$	7.87
20	$(CH_3S)_3P\rightarrow O$	8.00
21	$(CH_3O)_2P(\rightarrow O)SCH_3$	8.08
22	$[(CH_3)_2N]_3P\rightarrow S$	8.18
23	$(CH_3O)_2P(\rightarrow O)H$	8.22
24	$(CH_3O)_3P\rightarrow O$	8.25
25	$(C_4H_9O)_3P\rightarrow O$	8.36
26	$(C_3H_7O)_3P\rightarrow O$	8.39
27	$(C_2H_5O)_3P\rightarrow O$	8.41
28	$CH_3P(\rightarrow O)(OC_3H_7\,iso)_2$	8.49
29	$(CH_3O)_2P(\rightarrow O)N(CH_3)_2$	8.59
30	$CH_3P(\rightarrow O)(OC_2H_5)_2$	8.61
31	$[(CH_3)_2N]_3P\rightarrow O$	9.11

FIG. 14. Données $\delta_{\infty CHCl_3}$ dans des organophosphorés (*8*).

$\Delta\delta_{CHCl_3}$ remplace le p*K*: Relations avec $\Delta\nu$ des liaisons OH, OD ou NH. Les Figs. 15 à 17 résument les résultats de Mme. Martin (*10*), reliant les variations de $\Delta\nu$ des fréquences infrarouges OH, OD ou NH aux variations $\Delta\delta_{CHCl_3}$.

Soulignons que, comme nous l'avons indiqué, les courbes obtenues sont beaucoup plus satisfaisantes que celles reliant données infrarouges et p*K*.

$\Delta\delta_{CHCl_3}$ remplace le p*K* : $\Delta\nu$ et $\Delta\delta_{CHCl_3}$ pour la liaison C—Br. De même, les variations $\Delta\nu$ de la fréquence de vibration infrarouge C—Br

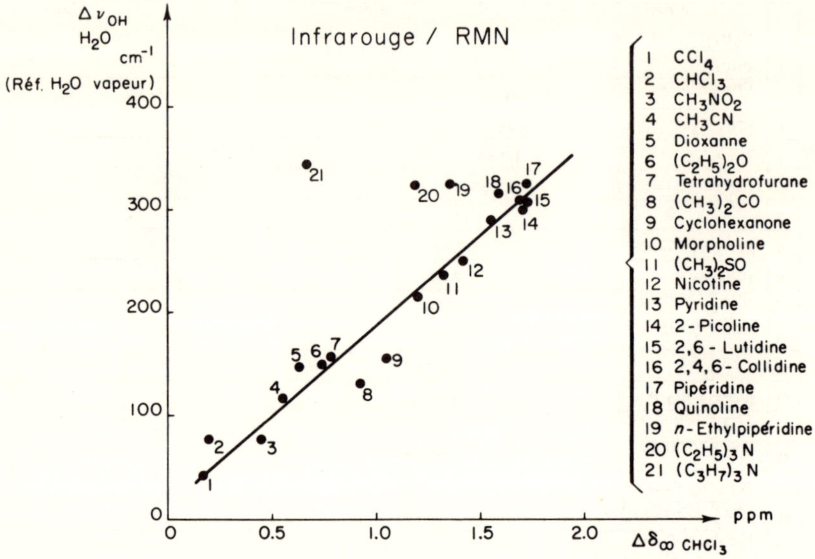
Δν OH
H2O
cm-1
(Réf. H2O vapeur)
Infrarouge / RMN
ppm
Δδ∞ CHCl3
1 CCl4
2 CHCl3
3 CH3NO2
4 CH3CN
5 Dioxanne
6 (C2H5)2O
7 Tetrahydrofurane
8 (CH3)2 CO
9 Cyclohexanone
10 Morpholine
11 (CH3)2SO
12 Nicotine
13 Pyridine
14 2-Picoline
15 2,6-Lutidine
16 2,4,6-Collidine
17 Pipéridine
18 Quinoline
19 n-Ethylpipéridine
20 (C2H5)3 N
21 (C3H7)3 N

FIG. 15.

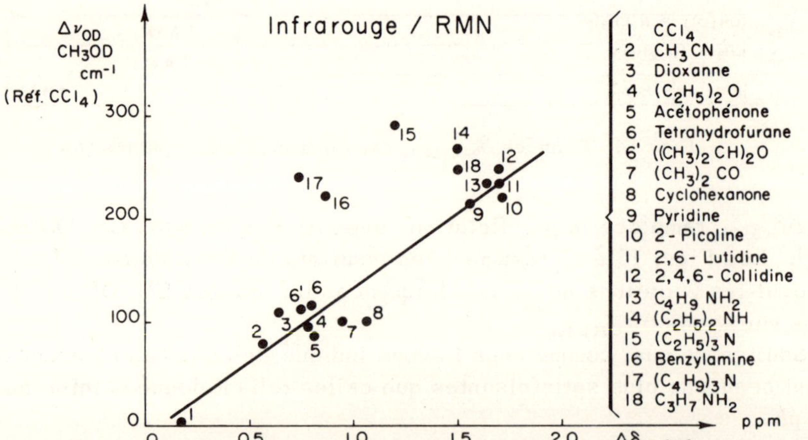
Δν OD
CH3OD
cm-1
(Réf. CCl4)
Infrarouge / RMN
ppm
Δδ∞ CHCl3
1 CCl4
2 CH3CN
3 Dioxanne
4 (C2H5)2 O
5 Acétophénone
6 Tetrahydrofurane
6' ((CH3)2 CH)2 O
7 (CH3)2 CO
8 Cyclohexanone
9 Pyridine
10 2-Picoline
11 2,6-Lutidine
12 2,4,6-Collidine
13 C4H9 NH2
14 (C2H5)2 NH
15 (C2H5)3 N
16 Benzylamine
17 (C4H9)3 N
18 C3H7 NH2

FIG. 16.

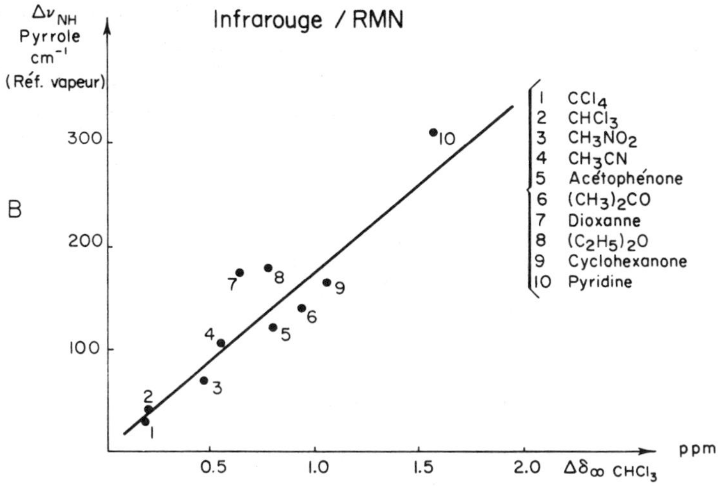

FIGS. 15–17. Résultats des travaux de Mme. Martin (*10*) reliant les variations de $\Delta\nu$ des fréquences infrarouges OH, OD ou NH aux variations $\Delta\delta_{\infty CHCl_3}$.

peuvent être comparées à la basicité définie indirectement par $\Delta\delta_{CHCl_3}$. C'est ce que montre la Fig. 18, due à Mme. Martin et G. J. Martin (*11*), qui ont étudié, d'une part, des composés saturés, d'autre part, des composés éthyléniques renfermant la liaison C—Br.

Ajoutons que Mme. et G. J. Martin (*11*) ont relié les variations $\Delta\nu$ de la fréquence infrarouge C—Br non seulement à $\Delta\delta_{CHCl_3}$ mais également aux déplacements $\Delta\delta$ des groupements CH voisins du brome.

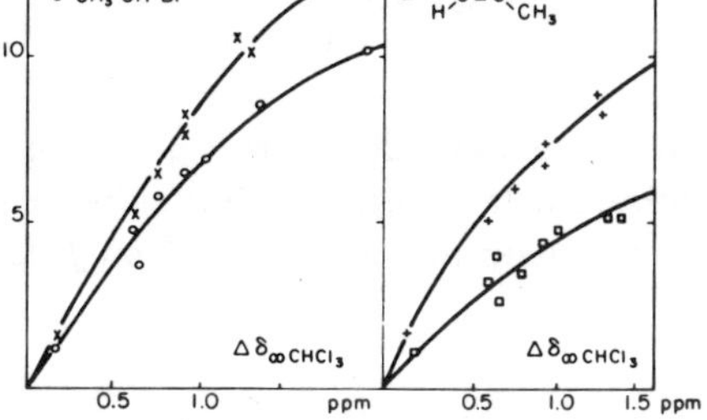

FIG. 18. Variations $\Delta\nu$ de la fréquence de vibration infrarouge C—Br comparées à la basicité définie $\Delta\delta_{\infty CHCl_3}$ (*11*).

$\Delta\nu_{H_2O}$ et $\Delta\delta_{H_2O}$. Rappelons que, dès 1960, G. Mavel (*12*) avait relié les variations des fréquences infrarouges de l'eau aux variations du déplacement chimique $\Delta\delta_{H_2O}$, pour les mêmes mélanges : les points s'alignent sur la courbe (Fig. 19), à l'exception de ceux qui correspondent à une très forte basicité.

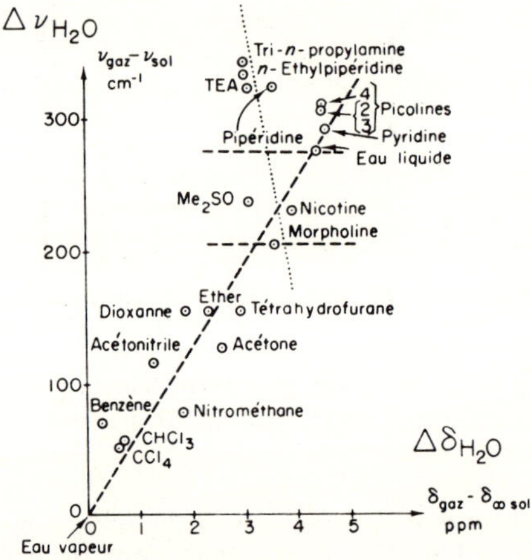

FIG. 19. Variations des fréquences infrarouges de l'eau reliées aux variations du déplacement chimique pour divers mélanges (*12*).

Relation entre $\Delta\delta$ et $L\ \Delta\nu$ des bandes infrarouges (de largeur L). Etudiant divers composés acétyléniques, et plus spécialement $C_6H_5C{\equiv}CH$ dans une trentaine de solutions, P. Jouve (*Compt. Rend.* **258,** 119, 1964) a essayé de relier les variations $\Delta\nu$ de la frequence ν ($\equiv$CH) aux déplacements correspondants $\Delta\delta$ en RMN. Il n'a pas obtenu de relation simple.

Par contre, en portant $L\Delta\nu$ (produit de la largeur L de la raie par le déplacement $\Delta\nu$) en fonction de $\Delta\delta$ RMN, il a obtenu les résultats que résume la Fig. 20.

En résumé, nous avons essayé de montrer que, pour les phénomènes intra- ou intermoléculaires, la position, l'intensité et la largeur des raies infrarouges peuvent être liées aux grandeurs δ ou J de la RMN.

Bibliographie

1. G. Amat, Thèse, Paris (1951).
2. G. M. Barrow et D. C. McKean, *Proc. Roy. Soc.* (*London*) **A213,** 27 (1952).

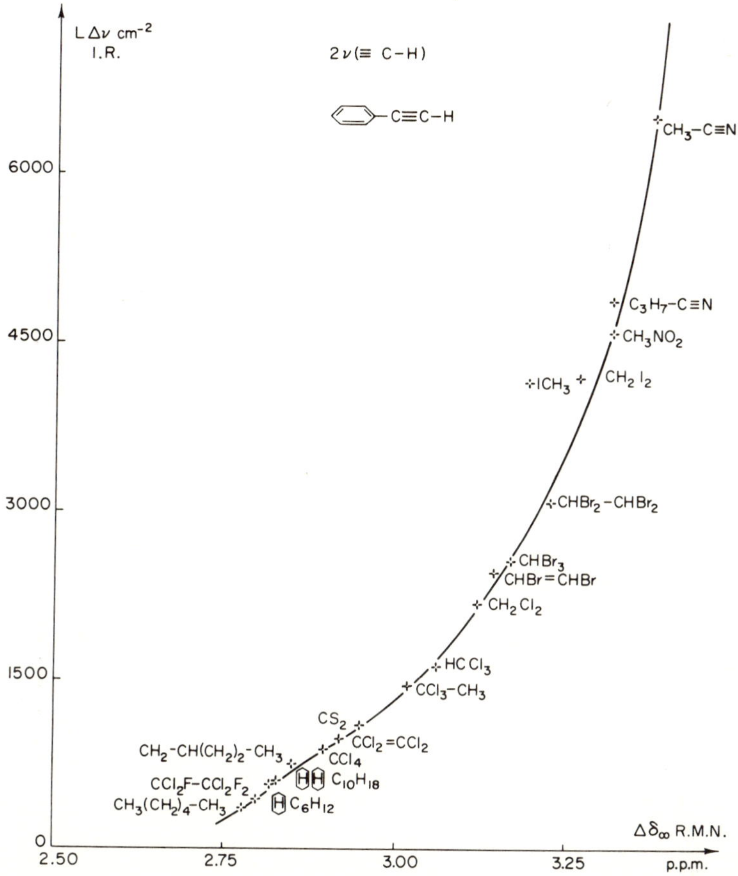

FIG. 20. Résultats des travaux de P. Jouve (*Compt. Rend.* **258,** 119, 1964) reliant $L\Delta\nu$ aux déplacements $\Delta\delta$ en RMN.

3. M. Chaigneau, E. Dayan et G. Widenlocher, *Compt. Rend.* **255,** 2597 (1962).
G. Widenlocher, E. Dayan et B. Vodar, *Compt. Rend.* **256,** 2584 (1963).
E. Dayan et G. Widenlocher, *Compt. Rend.* **257,** 883 (1963).
P. Jouve, G. Widenlocher et E. Dayan, *Compt. Rend.* **257,** 1079 (1963).
E. Dayan, G. Widenlocher et M. Chaigneau, *Compt. Rend.* **257,** 2455 (1963).
G. Widenlocher, E. Dayan et B. Vodar, *Compt. Rend.* **258,** 4242 (1964).
E. Dayan et G. Widenlocher, *Compt. Rend. 13ème Colloq. Ampère, Louvain* (1964) in press.

4. R. Freymann, B. Vodar et Yeou Ta, *J. Phys. Radium* **VII,** 282 (1938).

5. R. Freymann, *Ann. Phys.* (*Paris*) **20,** 243 (1933); *J. Chem. Phys.* **6,** 497 (1938).
R. Freymann et J. Gueron, *Compt. Rend.* **205,** 859 (1937).

R. Freymann, *Cahiers Phys.* **9**, 1 (1942); **14**, 19 (1943).
R. Freymann et M. Freymann, *Compt. Rend.* **219**, 515 (1944).

6. C. Giessner-Prettre, Thèse, Paris (1963).
7. P. Jouve, *Compt. Rend.* **256**, 1497; **257**, 121; **256**, 5120 (1963).
8. G. L. A. Martin et A. Besnard, *Compt. Rend.* **257**, 898, 2463 (1963).
9. G. Mavel et G. L. A. Martin, *J. Chim. Phys.* **59**, 762 (1962); *Compt. Rend.* **255**, 2095 (1962); *J. Phys. Radium* **24**, 108 (1963); *Compt. Rend.* **257**, 1703 (1963).
10. Mme. M. L. Martin, Thèse, Paris (1961); *Ann. Phys.* (*Paris*) **7**, 35 (1962); *J. Chim. Phys.* **59**, 736 (1962).
11. Mme. M. L. Martin et G. J. Martin, *Bull. Soc. Chim. France*, p. 1637 (1963); *Compt. Rend.* **256**, 5099; **256**, 403; **257**, 1048 (1963).
12. G. Mavel, *J. Chim. Phys.* **59**, 683; 511 (1962); **61**, 182 (1964).
13. S. Penner et D. Weber, *J. Chem. Phys.* **50**, 649, 654 (1953).
14. J. A. Pople, *Proc. Roy. Soc.* (*London*) **A239**, 541 (1957).
15. J. N. Shoolery et B. P. Dailey, *J. Am. Chem. Soc.* **77**, 3977 (1955).
16. L. Sobel, Thèse, Paris (1964); *Compt. Rend.* **218**, 315, 347, 834 (1944).
17. E. Taillandier et M. Taillandier, *Compt. Rend.* **257**, 1522 (1963).
18. Vu Hai, Thèse, Paris (1960).
19. R. Wolf, Thèse, Toulouse (1964).

The Chemical Shifts of Protons in Transition Metal Hydrides

A. D. BUCKINGHAM and P. J. STEPHENS

Inorganic Chemistry Laboratory, Oxford University, Oxford, England

Introduction

The large high-field shifts in the proton magnetic resonance spectra of transition-metal hydrides are now well known (*1, 2*) and used as tests of the presence of a metal–hydrogen (M–H) bond. In compounds so far prepared, τ lies in the range 10–50; relevant data are shown in Table I. The shifts were first attributed either to diamagnetic shielding by the metal (*3*), the proton being embedded in the metal atom, or to the H being hydride ionlike (*4*). However, the former model is not consistent with the H being a normal ligand in a fixed stereochemical position with an M–H bond length $\geqslant$ 2 au (1 au = 0.529 Å); also, τ for H^- is 5 (*5*) ($\tau = -1.7$ for the H atom and -19.5 for the free proton), so even orbital contraction could hardly explain the larger shifts in the H^- model. It was then suggested that paramagnetic shielding by the metal atom could be important (*6*), as in HI (*1*), where $\tau = 23.2$. Further, since B-group metal hydrides do not generally show large high-field shifts [τ = 5–8 in Ge and Sn hydrides (*7, 8*)] this paramagnetic shielding is probably associated with the nonbonding metal *d* electrons, which cause temperature-independent paramagnetism and large magnetic antishielding of the ^{59}Co nucleus in octahedral Co(III) complexes (*9*). This paper summarizes calculations of this effect in octahedral d^6 and square-planar d^8 complexes, using a ligand-field model (*10, 11*).

Theory

In the coordinate system of Fig. 1, a change of origin from H to M for the electronic orbital angular momentum l_x in the Ramsey formula for nuclear shielding (*12*) leads to the following expressions for the proton shielding (*10*):

$$\sigma = \sigma^p + \sigma^d = \tfrac{1}{3}(\sigma^p_{xx} + \sigma^p_{yy} + \sigma^p_{zz}) + \sigma^d,$$

TABLE I

CHEMICAL SHIFTS OF TRANSITION-METAL AND OTHER HYDRIDES

Ref.[a]		τ	Ref.[a]		τ
a	$Co(CN)_5H^{3-}$	22.1	*j*	*trans*-$PtH(NO_3)(PEt_3)_2$	33.8
a	$Rh(CN)_5H^{3-}$	20.3	*j*	*trans*-$PtHCl(PEt_3)_2$	26.9
b	*trans*-$FeHCl[C_2H_4(PEt_2)_2]_2$	43.8	*j*	*trans*-$PtHBr(PEt_3)_2$	25.6
b	*trans*-$FeHI[C_2H_4(PEt_2)_2]_2$	38.6	*j*	*trans*-$PtHI(PEt_3)_2$	22.7
b	*trans*-$FeHCl[o\text{-}C_6H_4(PEt_2)_2]_2$	40.7	*j*	*trans*-$PtHNO_2(PEt_3)_2$	29.7
b, c	trans-$FeH_2[o\text{-}C_6H_4(PEt_2)_2]_2$	23.1	*j*	*trans*-$PtHNCS(PEt_3)_2$	23.2, 27.9
d	$RuHCl[C_2H_4(PEt_2)_2]_2$	31.8	*j*	trans-$PtHCN(PEt_3)_2$	17.8
d	$RuHBr[C_2H_4(PEt_2)_2]_2$	31.0			
d	$RuHI[C_2H_4(PEt_2)_2]_2$	29.3			
e	$RuHMe[C_2H_4(PPh_2)_2]_2$	18.4	*k*	$Co(CO)_4H$	20
e	$RuHPh[C_2H_4(PMe_2)_2]_2$	23.3	*l*	$Fe(CO)_4H_2$	20
d	$OsHCl[C_2H_4(PEt_2)_2]_2$	36.0	*m*	K_2ReH_9	18
d	$OsHI[C_2H_4(PEt_2)_2]_2$	30.8			
f	*cis*-$RhtrienH_2^+$	27.1			
f	*cis*-$Rhen_2H_2^+$	31.0	*n, o*	CH_4	10.0
f	*cis*-$RhtrienHCl^+$	28.5	*n*	SiH_4	7.0
g	$RuHCl(CO)(PEt_2Ph)_3$	17.1	*n*	GeH_4	6.8
e	*cis*-$RuHMe[C_2H_4(PPh_2)_2]_2$	18.9	*n*	SnH_4	6.1
			o	HCl	10.4
			o	HBr	14.3
	(a) 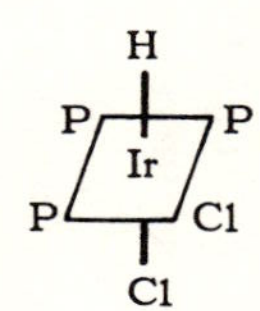	31.7	*o*	HI	23.2
g, h	$IrHCl_2(PEt_3)_3$				
	(b) 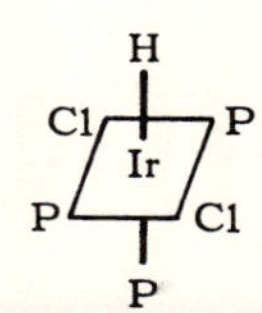	22.6			

i	$IrHCl_2(CO)(PEt_2Ph)_2$	(a) 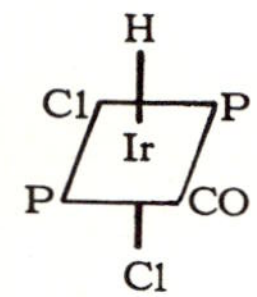	26.05	
		(b) 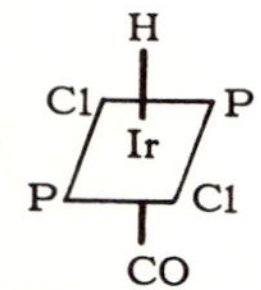	19.0	

[a]References:

a. W. P. Griffith and G. Wilkinson, *J. Chem. Soc.*, p. 2757 (1959).

b. J. Chatt and R. G. Hayter, *J. Chem. Soc.*, p. 5507 (1961).

c. J. Chatt, F. A. Hart, and D. T. Rosevear, *J. Chem. Soc.*, p. 5504 (1961).

d. J. Chatt and R. G. Hayter, *J. Chem. Soc.*, p. 2605 (1961).

e. J. Chatt and R. G. Hayter, *J. Chem. Soc.*, p. 6017 (1963).

f. R. D. Gillard and G. Wilkinson, *J. Chem. Soc.*, p. 3594 (1963).

g. J. Chatt and B. L. Shaw, *Chem. Ind. (London)*, p. 931 (1960).

h. J. Chatt, *Proc. Chem. Soc.*, p. 318 (1962).

i. J. Chatt, N. P. Johnson, and B. L. Shaw, *J. Chem. Soc.*, p. 1625 (1964).

j. J. Chatt and B. L. Shaw, *J. Chem. Soc.*, p. 5075 (1962).

k. R. A. Friedel, I. Wender, S. L. Shufler, and H. W. Sternberg, *J. Am. Chem. Soc.* **77**, 3951 (1955).

l. F. A. Cotton and G. Wilkinson, *Chem. Ind.*, p. 1305 (1956).

m. A. P. Ginsberg, J. M. Miller, J. R. Cavanaugh, and B. P. Dailey, *Nature* **185**, 528 (1960).

n. P. E. Potter, L. Pratt, and G. Wilkinson, *J. Chem. Soc.*, p. 524 (1964).

o. J. A. Pople, W. G. Schneider, and H. J. Bernstein, "High-Resolution Nuclear Magnetic Resonance," p. 90. McGraw-Hill, New York, 1959.

$$\sigma^{p}_{xx} = -\frac{e^2}{2m^2c^2}\sum_{n\neq 0}(E_n - E_0)^{-1}\left[<0\left|\sum_j l_{M_{xj}}\right|n>\times\right.$$
$$\left.<n\left|\sum_k r_{H_k}^{-3}\, l_{H_{xk}}\right|0> + <0\left|\sum_k r_{H_k}^{-3}\, l_{H_{xk}}\right|n><n\left|\sum_j l_{M_{xj}}\right|0>\right], \tag{1}$$

$$\sigma^{d} = \frac{e^2}{3mc^2}<0\left|\sum_j \frac{r_{H_j}^2 + Rz_{H_j}}{r_{H_j}^3}\right|0>,$$

where $|\,0>$ and $|\,n>$ are ground and excited states with energies E_0 and E_n; σ^p and σ^d are the paramagnetic and diamagnetic contributions to the shielding.

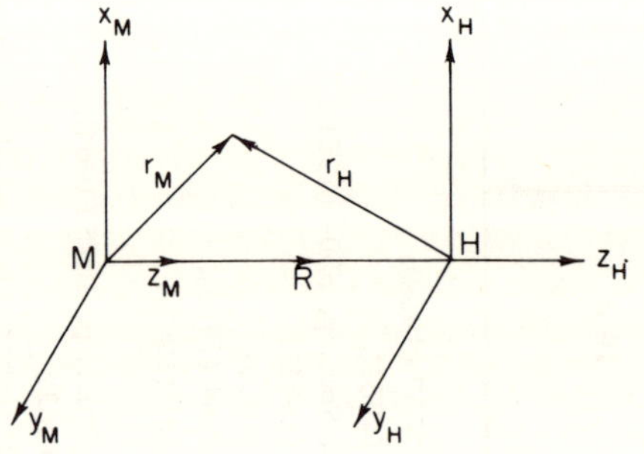

FIG. 1. The coordinate system.

In calculating the *d*-electron contribution to the proton shielding, we use the wave functions of the simple ligand-field theory and consider only those excited states that arise from excitation of *d*-electrons within the incomplete *d* shell; if spin-orbit interaction is neglected, only singlet states corresponding to the excitation of one electron contribute. Excited charge-transfer states, and those due to the excitation of other electrons, are not specifically considered, since they are also present in other hydrides and we wish to concentrate on the transition-metal anomaly.

Application to Octahedral Low-Spin d^6 Complexes

Consider complexes of O_h (e.g., MH_6), D_{4h} (*trans*-MX_4H_2), and C_{4v} (*trans*-MX_4HY) symmetries. In O_h, the ground state is $^1A_{1g}$ and the singlet states corresponding to the excitation of a t_{2g} electron to an e_g orbital are $^1T_{1g}$ and $^1T_{2g}$; since l_M transforms as T_{1g} in O_h, only the

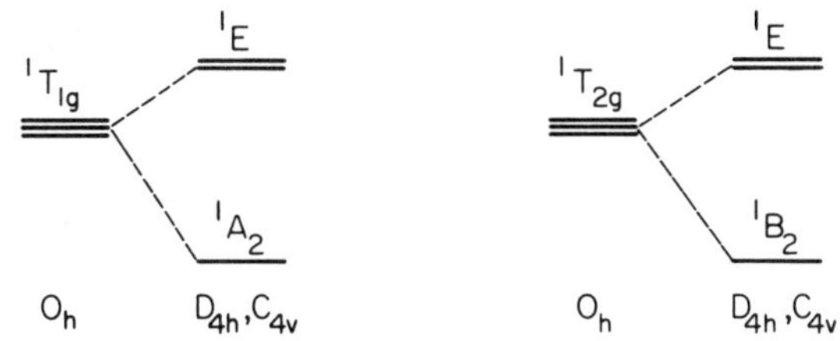

FIG. 2. Splitting of the triply degenerate excited states of the d^6 electrons due to deviations from octahedral symmetry.

former contributes to σ^p, as in the antishielding of M itself (*9*). In D_{4h} and C_{4v}, the states are split as shown in Fig. 2, and provided the perturbation is not too large the states arising from the T_{2g} functions still do not contribute to σ^p. Since l_{Mx} and l_{My} now transform as E, and l_{Mz} as A_2, the ^1E and $^1A_2$ states deriving from $^1T_{1g}$ contribute only to σ^p_{xx} ($=\sigma^p_{yy}$) and σ^p_{zz}, respectively.

Expressions for σ^p and σ^d are obtained by introducing explicit wave functions into Eq. (1); using one-parameter Slater orbitals with exponent k, results have been obtained (*10*) for $3d$ and $5d$ electrons and M–H bond lengths $R = 2$, 2.5, and 3 au. Those for $3d$ electrons and $R = 2$ are shown in Fig. 3; for larger R, the shielding is similar in form but smaller in magnitude. The integrals for σ were evaluated analytically, and the expressions reduce in the limit of $R = 0$ to the formula for the antishielding of the M nucleus (*9*) and for $k \longrightarrow \infty$ to the near-neighbor dipole re-

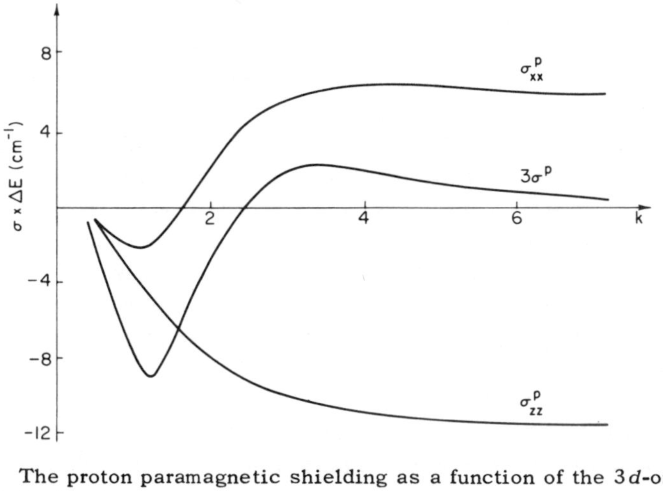

FIG. 3. The proton paramagnetic shielding as a function of the $3d$-orbital exponent k for $R = 2$ au.

sult producing shielding varying as R^{-3} and a vanishing σ^p (though not σ^p_{zz}) for an isotropic metal atom (*12, 13*); ΔE is to be interpreted as $E(^1T_1) - E(^1A_1)$ in O_h, and as $\Delta E_E = E(^1E) - E(^1A_1)$ in σ^p_{xx} and as $\Delta E_A = E(^1A_2) - E(^1A_1)$ in σ^p_{zz} in D_{4h} and C_{4v} symmetries.

The atomic wave-function calculations of Watson on the first-series transition metal atoms and ions (*14, 15*) indicate that it is reasonable to take $k = 3$ to 4 for the $3d$ shell, and σ is then positive and close to its maximum in an O_h complex. With $\Delta E = 25{,}000\ \text{cm}^{-1}$, $k = 3.5$ and $R = 2$, $\sigma^p = 25$ ppm and $\sigma^d = 3$ ppm; for the same ΔE but with $k = 2.5$, $R = 3$, $\sigma^p = 10$ ppm and $\sigma^d = 2$ ppm. Hence shifts of the required magnitude are obtained for reasonable values of ΔE and k provided R is close to 2 au. Early measurements of R in $FeH_2(CO)_4$ (*6*) and $MoH_2(C_5H_5)_2$ (*16*) yielded values close to 2 au, but recent determinations in K_2ReH_9 (*17*) and $RhH(CO)(PPh_3)_3$ (*18*) give about 3 au, in agreement with R in the transition-metal diatomic hydrides. Further bond-length measurements are required, but even if $R = 3$ should be confirmed, we think that although our equations would give rather small chemical shifts, the model is essentially valid, and that calculations with improved wave functions would give larger values of σ. Actual electron densities fall off much more slowly at these distances than a single-parameter Slater orbital indicates, and as σ^p depends on the inverse cube of the distance of the electron from the proton, an expansion of the orbitals in the vicinity of H increases σ.

Figure 3 shows that σ^p_{xx} is large and positive, while σ^p_{zz} is large and negative. This may be easily understood since σ^p_{zz} is caused by the mixing of the d_{xy} and $d_{x^2-y^2}$ orbitals by the magnetic field, while σ^p_{xx} arises from the mixing of d_{yz} with d_{z^2} and $d_{x^2-y^2}$; the d_{yz} and d_{z^2} orbitals are nearer to the H atom than the others, so the r^{-3} factor favors σ^p_{xx}. The signs of σ^p_{xx} and σ^p_{zz} correspond to those of the near-neighbor dipole model, but the geometry is such that σ^p_{xx} is larger, and σ^p_{zz} smaller, than the simple R^{-3} expressions, thereby making σ^p positive.

An important feature of the results is the sensitivity of σ^p to anisotropy in the complex. If $\Delta E_E \neq \Delta E_A$, as in a D_{4h} or C_{4v} complex, σ^p varies rapidly with energy changes, as shown in Fig. 4. Thus the large shift of 20 ppm in *trans*-FeHX(diphosphine)$_2$ when X is changed from Cl to H (*19, 20*) is probably due to a change in anisotropy. In complexes of the type *trans*-MX_4HY, increasing the ligand-field strength of Y increases the energy of the d_{z^2} orbital and therefore of the ^1E state, but not that of the $^1A_2$ since it is further from the d_{xy} and $d_{x^2-y^2}$ orbitals. Similarly, the major effect of increasing the strength of X is to increase ΔE_A. Thus τ tends to increase as the ligand-field strength of X is increased

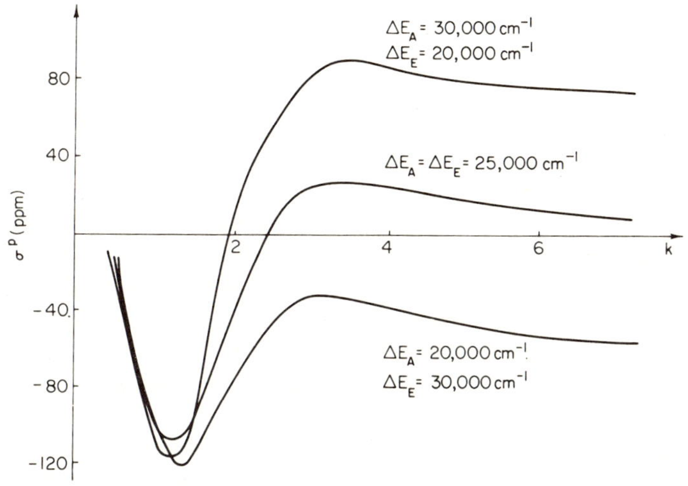

FIG. 4. The effect of anisotropy in the octahedral complex on the paramagnetic shielding as a function of k for $R = 2$ au.

and Y decreased. This result could be helpful in making stereochemical assignments.

One point that hinders detailed applications of the model is an uncertainty in the field strength of the H ligand. The spectrum of $Co(CN)_5H^{3-}$ apparently shows no splitting of the $^1T_{1g}$ band (*21*), suggesting that H and CN have similar strengths; this was also deduced (*22*) from the spectra of *trans*-RuXY(diphosphine)$_2$ complexes, but these spectra are difficult to interpret. Recent work (*23*) on Rh(III) complexes indicates a field strength intermediate between H_2O and NH_3. We shall assume a value at least as high as that of NH_3. The position in the spectrochemical series could vary with the metal atom.

Wherever there are large shifts in related complexes (see Table I), the theory satisfactorily explains the facts. Thus τ decreases in *trans*-MHX(diphosphine)$_2$ (M = Fe, Ru) from X = halogen to X = H, alkyl, or aryl (the ligand-field strenghts of alkyl and aryl are probably similar to those of phosphines), and τ also decreases from the (*a*) to (*b*) isomers of $IrHCl_2(PEt_3)_3$ and $IrHCl_2(CO)(PEt_2Ph)_2$. In $RuHCl(CO)(PEt_2Ph)_3$, τ is comparatively low, so the Cl ligand is probably *cis* to the H. Smaller variations in τ may be caused by changes in R, k, and σ^d; thus the decrease in τ in *trans*-MHX(diphosphine)$_2$ from X = Cl to X = I may be due to a decrease in k (that is, an expansion of the d orbitals) arising from a reduction in electronegativity.

Application to Square-Planar Low-Spin d^8 Complexes

Detailed calculations have been carried out on square-planar Pt(II) hydrides (*11*). These compounds differ from octahedral hydrides in that the Pt ion is anisotropic, even in PtH_4^{2-}, the paramagnetic susceptibility being greatest perpendicular to the plane. Hence even the near-neighbor dipole model predicts a high-field shift for a proton in the plane; for the same values of R, k, and ΔE, σ is greater in a d^8 than in a d^6 complex. Shifts of about 20 ppm are predicted by our equations at $R = 3$ au (*11*). Unlike the octahedral complexes, splitting of the excited states is unimportant in square-planar complexes, so major variations in τ may be due to changes in R associated with the "*trans* effect" of the ligand *trans* to H (*24*). In *trans*-$PtHX(PEt_3)_2$, τ tends to decrease with increasing *trans* effect of X.

If a proton were to be found above the plane and near the metal atom, then there should be a *low*-field shift. This arrangement might be achieved either by direct protonation or by using a chelate containing hydrogen atoms in its bridge connecting the atoms that are bonded to the metal, similarly to the position in 1,4-polymethylenebenzenes (*25*).

Conclusions

The paramagnetic shielding of the proton bonded to a transition metal arising from the mixing of the orbitals of the unfilled electron shell with the low-lying excited states through the application of the magnetic field can account for high-field shifts of 20–30 ppm. The magnitude of the shift is sensitive to the bond length R, to the detailed form of the d orbitals, and to the excitation energy, and changes in all of these variables are probably significant in influencing the chemical shift. Anisotropy in the complex is especially important in octahedral complexes.

A very large anisotropy in the chemical shift ($\sigma_{zz} - \sigma_{xx} = -500$ ppm) is predicted for octahedral complexes, since the paramagnetic shielding has opposite signs in the x and z directions (see Fig. 3), but in square-planar complexes it is about 10 times smaller.

A detailed calculation of the shielding in $CoH(CO)_4$ (*26*) and other molecules (*27*) has been carried out by Lipscomb *et al.* They included all electrons, whereas we have only evaluated the d-electron contributions, and they obtained large high-field shifts. However, their procedure involves approximations equivalent to the neglect of the paramagnetic term in Eq. (1), and we have found that this can be large when the metal atom may be considered to have an incomplete electron shell (or has low-lying excited states) as in all the complexes in Table I except the last 10, which were included for comparison purposes. Though some of these last 10 have large shifts, their τ values are much smaller than others in the table.

References

1. G. Wilkinson, *in* "Advances in the Chemistry of the Coordination Compounds," (S. Kirschner, ed.), p. 50. Macmillan, New York, 1961.
2. J. Chatt, *Proc. Chem. Soc.*, p. 318 (1962).
3. F. A. Cotton and G. Wilkinson, *Chem. Ind. (London)*, p. 1305 (1956).
4. R. A. Friedel, I. Wender, S. L. Shufler, and H. W. Sternberg, *J. Am. Chem. Soc.* **77**, 3951 (1955).
5. F. T. Ormand and F. A. Matsen, *J. Chem. Phys.* **30**, 368 (1959).
6. E. O. Bishop, J. L. Down, P. R. Emtage, R. E. Richards, and G. Wilkinson, *J. Chem. Soc.*, p. 2484 (1959).
7. J. E. Drake and W. L. Jolly, *J. Chem. Soc.*, p. 2807 (1962).
8. P. E. Potter, L. Pratt, and G. Wilkinson, *J. Chem. Soc.*, p. 524 (1964).
9. J. S. Griffith and L. E. Orgel, *Trans. Farad. Soc.* **53**, 601 (1957).
10. A. D. Buckingham and P. J. Stephens, *J. Chem. Soc.*, p. 2747 (1964).
11. A. D. Buckingham and P. J. Stephens, *J. Chem. Soc.*, p. 4583 (1964).
12. J. A. Pople, W. G. Schneider, and H. J. Bernstein, "High-Resolution Nuclear Magnetic Resonance," McGraw-Hill, New York, 1959.
13. J. A. Pople, *Discussions Farad. Soc.* **34**, 7 (1962).
14. R. E. Watson, *Phys. Rev.* **118**, 1036 (1960).
15. J. W. Richardson, N. C. Nieuwpoort, R. R. Powell, and W. F. Edgell, *J. Chem. Phys.* **36**, 1057 (1962).
16. M. J. Bennett, M. Gerloch, J. A. McCleverty, and R. Mason, *Proc. Chem. Soc.*, p. 357 (1962).
17. S. C. Abrahams, A. P. Ginsberg, and K. Knox, *Inorg. Chem.* **3**, 558 (1964).
18. S. J. LaPlaca and J. A. Ibers, *J. Am. Chem. Soc.* **85**, 3501 (1963).
19. J. Chatt and R. G. Hayter, *J. Chem. Soc.*, p. 5507 (1961).
20. J. Chatt, F. A. Hart, and D. T. Rosevear, *J. Chem. Soc.*, p. 5504 (1961).
21. J. M. Pratt, personal communication, May 1964.
22. J. Chatt and R. G. Hayter, *J. Chem. Soc.*, p. 772 (1961).
23. J. A. Osborn, R. D. Gillard, and G. Wilkinson, *J. Chem. Soc.*, p. 3168 (1964).
24. F. Basolo and R. G. Pearson, *Progr. Inorg. Chem.*, **4**, 381 (1962).
25. J. S. Waugh and R. W. Fessenden, *J. Am. Chem. Soc.* **79**, 846 (1957).
26. R. M. Stevens, C. W. Kern, and W. N. Lipscomb, *J. Chem. Phys.* **37**, 279 (1962).
27. L. L. Lohr and W. N. Lipscomb, *Inorg. Chem.* **3**, 22 (1964).

The Calculation of the Chemical Shift of Protons Due to >C═O, —OH, —Cl, and —C≡N Groups

R. F. ZÜRCHER

Physical Laboratories, Ciba Ltd., Basel, Switzerland

Introduction

The replacement of a $—CH_2—$ group in a molecule by a C═O group, or by a —CHX— group (X = —OH, —Cl, —C≡N), often has a marked influence on the chemical shift δ of nearby protons. It will be shown that in cases in which the inductive effect (through the bonds) and mesomeric effects (delocalization of electrons) can be neglected, the additional chemical shift $\Delta\delta$ of a proton due to the introduction of a substituent is brought about in high-resolution proton magnetic resonance spectroscopy by the combined action of (i) the electric dipole moment and (ii) the anisotropic magnetic susceptibility of this substituent. Rigid systems such as steroids and terpenoids, certain alkaloids, and bridged bicyclic compounds are best suited for such an investigation. It is well known that in steroids substituents separated by several C—C bonds from the protons of the tertiary methyl groups may cause appreciable additional chemical shifts of the latter (*1, 2*). In these cases inductive and mesomeric effects cannot play an important role and the total, experimentally determined additional chemical shift $\Delta\delta$ of a proton due to the introduction of a substituent may be expressed as

$$\Delta\delta = \Delta\delta_{el} + \Delta\delta_{magn}, \tag{1}$$

where $\Delta\delta_{el}$ is that part of the total additional chemical shift which is caused by the action of the electric dipole moment of the substituent and $\Delta\delta_{magn}$ the part caused by the anisotropy of the magnetic susceptibility of the substituent. A paramagnetic shift (to low magnetic field strength) is positive.

Buckingham (*3*) has shown that, "the electric field at a particular nucleus arising from polar groups in other parts of a molecule can lead to chemical shifts proportional to the first power of the field strength. The change $\Delta\sigma$ in the proton screening constant (or in other words the additional chemical shift $\Delta\delta_{el}$ of a proton) of an X—H bond when it is subjected to an electric field E is given by

$$\Delta\delta_{el} = -\Delta\sigma = k \cdot 10^{-12}\, E_z + 10^{-18}\, E^2, \tag{2}$$

where E_z is the component of E in the bond direction (and k is a proportionality factor). A field along the X—H bond draws the electrons in the enriched region between the nuclei away from the proton, thereby causing its resonance to occur at lower magnetic field strengths, while a a field in the H—X direction leads to resonance at higher fields. The electric field produced by the polarization of neighboring solvent molecules may also be important and lead to a solvent shift related to the dielectric constant of the solvent."

For our purposes Eq. (2) has been modified slightly. The term with E^2 was neglected, being of secondary importance as can be seen in an actual calculation and as Buckingham has pointed out; E_z has been replaced by e_z, the component of E in the X—H bond direction due to a substituent *with unit dipole moment* (1 debye). With these modifications Eq. (2) reads

$$\Delta\delta_{el} = k \cdot E_z \cdot 10^{-12} = k \cdot \mu \cdot e_z \cdot 10^{-12} = \kappa \cdot e_z \cdot 10^{-12}, \tag{3}$$

where

μ is the dipole moment of the substituent (in debyes), and
κ is the proportionality factor $10^{12} \cdot \Delta\delta_{el}/e_z$.

Here e_z shall be called a geometrical quantity and κ a physical quantity.

The influence due to the reaction field produced by the polarization of neighboring solvent molecules has also been neglected, mainly because the additional chemical shifts for a serious study should be known for solutions with widely different dielectric constants and because it is doubtful what should be chosen as the radius of the "sphere" of the solute molecule. Our results seem to indicate that the reaction field does not play an important role in chloroform or carbon tetrachloride solutions.

Pople (*4*) and McConnell (*5*) have shown that a substituent with an anisotropic magnetic susceptibility may give rise to a small additional magnetic field at surrounding sites, i.e., additional chemical shifts $\Delta\delta_{magn}$ of protons. In a tumbling molecule

$$\Delta\delta_{magn} = \frac{1}{3} R^{-3} \sum_{i=1,2,3} \chi_i (3\cos^2\gamma_i - 1), \tag{4}$$

where

R is the distance between substituent and proton,
χ_i is the principal magnetic susceptibility of the substituent in the i direction, and
γ_i is the angle between principal axis i of the substituent and R.

As the three angles γ_i are not independent of each other Eq. (4) may be written

$$\Delta\delta_{\text{magn}} = \frac{1}{3} R^{-3}[\Delta\chi_x(3\cos^2\gamma_x - 1) + \Delta\chi_y(3\cos^2\gamma_y - 1)] \qquad (5)$$

with the anisotropies of the magnetic susceptibility of the substituent $\Delta\chi_x = \chi_x - \chi_z$ and $\Delta\chi_y = \chi_y - \chi_z$ (physical quantities) and the geometrical factors $\frac{1}{3} R^{-3}$ $(3\cos^2\gamma - 1)$, the axes x, y, z coinciding with the principal axes i. For axially symmetric substituents $\chi_y \equiv \chi_z$ and $\Delta\chi_y \equiv 0$.

Calculations

If the relative spatial positions of the proton and the substituent, as well as the chemical shift caused by its introduction, are known, the geometrical factors of Eqs. (3) and (5) may be calculated, and the two (or three) unknown physical quantities κ and $\Delta\chi_x$ (and $\Delta\chi_y$) of the above equations may be determined. This was done by a least-squares treatment. The position of the center of the electric and induced magnetic dipole moments of the substituent, which is assumed to be identical, enters these calculations as a parameter. It turns out that the best fit with the experimental data is obtained if the center of the dipole moments is assumed to coincide roughly with the center of the multiple or C—X bonds (X = substituent).

Results

>C═O. The additional chemical shifts of the C-19 methyl protons in eight different ketosteroids (*2*) (as deuterochloroform solutions) have served as a basis for the least-squares calculations. It was not possible to obtain a fit with the experimental data when (*a*) only the magnetic anisotropy of the carbonyl group or (*b*) only its electric dipole moment was considered to be responsible for the additional chemical shift, and (c) when the methyl group was assumed to be freely rotating instead of being predominantly in a staggered position:

$\kappa = 12.2$; $\Delta\chi_x = 25.7 \cdot 10^{-30}$ cm^3; $\Delta\chi_y = 12.2 \cdot 10^{-30}$ cm^3 (or
$\Delta\chi_x = 15.4 \cdot 10^{-6}$ cm^3/mole; $\Delta\chi_y = 7.3 \cdot 10^{-6}$ cm^3/mole);

(x axis in C═O direction; y axis *in* and z axis *out* of C—CO—C plane);

center of electric and induced magnetic dipole moments: <u>0.60 Å</u> from C (halfway on C═O bond);

the measured additional chemical shifts $\Delta\delta$ which range from -0.03 to $+0.38$ ppm are represented with a standard deviation of ± 0.027 ppm.

The transition from propane (or *n*-hexane in $CDCl_3$) to acetone (or methylethylketone), both in $CDCl_3$, shifts the α-methyl group signal by 1.28

ppm. This is very nearly what one calculates using the above values ($\Delta\delta_{calc} = 1.21$ ppm). This success induced us to incorporate the respective ethyl derivatives in the following investigations. *A word of caution:* the additional chemical shifts of protons in α-position relative to a carbonyl group in *rigid* systems calculated in the outlined way are wrong. The same is true for the additional chemical shifts of protons in α or β positions in *rigid* compounds with a substituent X (X = —OH, —Cl, —C≡N).

C—Cl. The least-squares calculations were done with 12 different additional chemical shifts stemming from chlorosteroids (*2*) (as deuterochloroform solutions), bornyl chloride, and isobornyl chloride (in carbon tetrachloride) (*6*), and ethyl chloride (in deuterochloroform) (*7*). In this and the other investigations κ [cf. Eq. (3)] turned out to vary only slightly with a variation of the assumed center of the dipole moment, whereas the $\Delta\chi$'s often changed sensibly. Further, the inclusion of $\Delta\chi_x (\Delta\chi_y \equiv 0)$ in this and the following investigations did not improve the standard deviation from the measured additional chemical shifts markedly. It was, therefore, concluded that $\Delta\chi_x$ is of minor importance compared with κ and cannot be determined with reasonable accuracy for such a limited number of compounds.

$\kappa = 9.0$;
center of electric dipole moment: 0.90 Å from C on C—Cl axis;
the measured additional chemical shifts $\Delta\delta$ which range from 0.02 to 0.60 ppm are represented with a standard deviation of ±0.037 ppm.

C—OH. Twelve different values for additional chemical shifts of hydroxysteroids (*2*) and ethanol (as deuterochloroform solutions) were used. The hydroxyl group was treated as freely rotating and axially symmetric with respect to the C—O bond:

$\kappa = 5.5$;
center of electric dipole moment: 1.10 Å from C on C—O axis;
the measured additional chemical shifts $\Delta\delta$ which range from 0 to 0.33 ppm are represented with a standard deviation of ±0.017 ppm.

C—C≡N. Five different additional chemical shifts of cyanosteroids (*8, 9*) and 2-cyanopropane (*7*) were used:

$\kappa = 14.8$
center of electric dipole moment: 2.10 Å from C on C—C≡N axis, i.e., halfway on C≡N bond;
the measured additional chemical shifts $\Delta\delta$ which range from 0.05 to 0.45 ppm are represented with a standard deviation of ±0.039 ppm.

Discussion

The quantity κ is equal to the product $\mu \cdot k$, i.e., κ is proportional to the dipole moment μ of the substituent under consideration. $k (\equiv -10^{12} \times \Delta\sigma/\Delta E_z)$ measures the variation of the proton magnetic shielding σ with a change of E_z, the component of the electric field E in the C—H bond direction [cf. Eqs. (2) and (3)]. In this work k may be considered as constant, insofar as only protons belonging to methyl groups are dealt with. If κ is plotted against μ, the group dipole moments known from dipole moment measurements (*10, 11, 12*), a straight line passing through the origin should result. This is actually the case (Fig. 1). Only the C—OH group moment deviates slightly, a fact which may have its origin in the different solvents employed (deuterochloroform for NMR, benzene for dipole moment measurements), or in hindered rotations of the —OH group.

We consider this linear relationship as a proof that the electric field due to substituent dipole moments is of prime importance in causing chemical shifts of hydrogen nuclei.

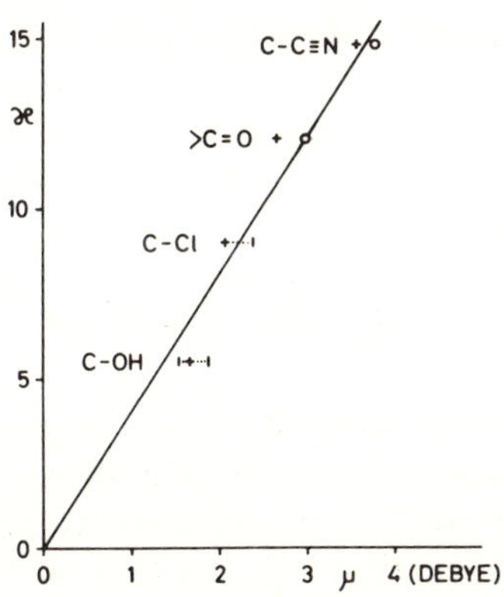

FIG. 1. The relationship between κ [cf. Eq. (3)] and the electric dipole moment μ of the substituents: + values from Ref. (*10*); o values for steroids from Ref. (*11, 12*).

Musher (*13*) has discussed the linear variation of proton magnetic shielding with electric field and has given a collection of theoretically and experimentally determined k values, ranging from 2.0 to 3.4. We find $k = 4.07$, which is remarkably higher.

In the previous treatment it was tacitly assumed that the C—H bond which is replaced by a C—X bond (X = substituent) has a negligible electric dipole moment and anisotropy of the magnetic susceptibility. Their consideration would scarcely affect the over-all picture.

The order of the principal susceptibilities of the carbonyl group, $\chi_x > \chi_y > \chi_z$ (χ_x most paramagnetic), corresponds to the order of the principal optical polarizabilities (*14*), $\alpha_x > \alpha_y > \alpha_z$, as determined by Kerr effect measurements (*15*).

Applications

Williams and associates (*16*) have reported an unusual chemical shift $\delta = 2.45$ ppm of the equatorial 1β-hydrogen atom in 5α-androstan-11-one. The signal of the axial 1α-hydrogen atom was located by a double resonance experiment at $\delta = 0.78$ ppm. As we do not know the respective chemical shifts of the unsubstituted steroid, we can only calculate the difference $\Delta\delta$(1α-H, 1β-H) and compare it with the measured value 1.67 ppm. We find

$$\begin{array}{rll} \Delta\delta(1\alpha\text{-H}) = & -0.27 \text{ ppm} & \text{additional chemical shift} \\ \Delta\delta(1\beta\text{-H}) = & +0.89 \text{ ppm} & \text{due to 11-keto group} \\ \hline \text{Difference} = & 1.16 \text{ ppm} & \end{array}$$

To this difference the usual, but in this case not precisely known, difference of the chemical shifts of an axial and equatorial hydrogen atom, amounting to 0.25 to 0.50 ppm, has to be added. Thus calculated, the difference $\Delta\delta$(1α-H, 1β-H) = 1.31 to 1.66 ppm compares favorably with the measured value 1.67 ppm. These calculations indicate that in unsubstituted 5α-androstane the signals of the 1α- and 1β-hydrogen atoms have to be expected at about δ(1α-H) = 1.05 ppm and δ(1β-H) = 1.56 ppm.

Applications along these lines should be helpful in the elucidation of unknown molecular configurations and conformations.

In cases where the anisotropy of the magnetic susceptibility may be neglected, it should be possible with the help of Eq. (6) [cf. Eq. (3)],

$$\kappa = 4.07 \cdot \mu \qquad (\mu \text{ in debyes}), \tag{6}$$

roughly to calculate additional chemical shifts due to the introduction of new substituents if their dipole moments are known.

A full account of this work will be given elsewhere.

References

1. J. N. Shoolery and M. T. Rogers, *J. Am. Chem. Soc.* **80,** 5121 (1958).

2. R. F. Zürcher, *Helv. Chim. Acta* **46,** 2054 (1963).

3. A. D. Buckingham, *Can. J. Chem.* **38,** 300 (1960).

4. J. A. Pople, *Proc. Roy. Soc.* (*London*) **A239,** 541, 550 (1957); J. A. Pople, W. G. Schneider, and H. J. Bernstein, "High Resolution Nuclear Magnetic Resonance," McGraw-Hill, New York, 1959, pp. 167–183.

5. H. M. McConnell, *J. Chem. Phys.* **27,** 226 (1957).
6. T. J. Flautt and W. F. Erman, *J. Am. Chem. Soc.* **85,** 3212 (1963).
7. N. S. Bhacca, L. F. Johnson, and J. N. Shoolery, "NMR Spectra Catalog," Vols. 1 and 2 (1962, 1963), Varian Associates, Palo Alto, California.
8. A. D. Cross and I. T. Harrison, *J. Am. Chem. Soc.* **85,** 3223 (1963).
9. J.-C. Jacquesy, J.-M. Lehn, and J. Levisalles, *Bull. Soc. Chim. France* **1961,** 2444.
10. K. B. Everard, L. Kumor, and L. E. Sutton, *J. Chem. Soc.* **1951,** 2807.
11. A. L. McClellan, "Tables of Experimental Dipole Moments," San Francisco, California, 1963.
12. J.-M. Lehn and G. Ourisson, *Bull. Soc. Chim. France* **1963,** 1113.
13. J. I. Musher, *J. Chem. Phys.* **37,** 34 (1962).
14. R. F. Zürcher, *J. Chem. Phys.* **37,** 2421 (1962).
15. C. G. LeFèvre, R. J. W. LeFèvre, and B. P. Rao, *J. Chem. Soc.* **1959,** 2340.
16. D. H. Williams, N. S. Bhacca, and C. Djerassi, *J. Am. Chem. Soc.* **85,** 2810 (1963).

Chemical Exchange Correlation Time Measurements by Proton NMR[1]

H. BENOIT and H. OTTAVI

Institut d'Electronique, Faculté des Sciences d'Orsay, France

Introduction

The proton NMR spectra of an aqueous solution of ammonium nitrate exhibits the triplet of the ammonium ion NH_4^+ (*1*) besides the water line. It is well known that this triplet appears only in acid solution (pH$<$1). If the pH is higher we have exchange between the protons of the NH_4^+ ions and NH_3 molecules, and the triplet disappears.

But there is another effect. Through chemical exchange, the indirect coupling $A = 53$ cps between N and H becomes time dependent, with the characteristic chemical exchange time τ. This gives a supplementary relaxation mechanism of the order of

$$\frac{4\pi^2 A^2 \tau}{1 + (\omega_I - \omega_S)^2 \tau^2},$$

where $A = 53$ cps is the scalar coupling, and ω_I and ω_S are the Larmor angular velocities of the proton and nitrogen nuclei.

In the case studied by Bloembergen (*2*), water doped with Mn^{2+} ion, we have $A = 2 \cdot 10^5$ cps and $\tau = 3 \cdot 10^{-9}$ sec. In our case A is considerably smaller, but the correlation time τ is of the order of 10^{-3} sec considerably longer. So if we are in fields low enough to satisfy the condition $(\omega_I - \omega_S)\tau << 1$, fields of a few gauss in our case, we shall still be able to see this effect.

The Exchange Mechanism

The ammonium ion is partly dissociated (*1*) into an NH_3 molecule and H^+ ion:

$$NH_4^+ \rightleftharpoons NH_3 + H^+.$$

So we have the relation

$$[NH_3] \cdot [H^+] = \text{constant}, \tag{1}$$

where $[NH_3]$ is the NH_3 concentration and $[NH_4^+]$ is assumed independent of the pH.

[1]This work is part of the thesis of H. Ottavi, to be published.

Protons are exchanged between NH_4^+ ions and NH_3 molecules,

$$NH_4^+ + NH_3 \longrightarrow NH_3 + NH_4^+,$$

so we may expect $\frac{1}{\tau}$ to be proportional to $[NH_3]$, and from Eq. (1) we have

$$\log \tau + \mathrm{pH} = \text{constant}. \tag{2}$$

Theory

The exact theory must take into account the proton I_z and nitrogen S_z nuclear polarizations. A set of coupled equations has been established by Solomon (*3, 4, 5*) in the HF molecule case[2]:

$$\begin{aligned} \frac{dI_z}{dt} &= -\left(\rho + \frac{1}{T_1'}\right)(I_z - I_0) - 4\sigma(S_z - S_0), \\ \frac{dS_z}{dt} &= -\left[4\rho'(1-P) + \frac{P}{T_1''}\right](S_z - S_0) - \sigma'(I_z - I_0), \end{aligned} \tag{3}$$

where

$$\rho = S(S+1)\left(\tfrac{4}{3}\,\delta + Y\right);$$
$$\sigma = I(I+1)\left(\tfrac{2}{3}\,\delta - Y\right);$$
$$\rho' = I(I+1)\left(\tfrac{4}{3}\,\delta + Y\right);$$
$$\sigma' = S(S+1)\left(\tfrac{2}{3}\,\delta - Y\right);$$
$$I = \tfrac{1}{2}; \qquad s = 1;$$

δ is a term due to the proton-nitrogen nucleus dipolar coupling and

$$Y = \frac{2}{3} \cdot \frac{4\pi^2 A^2 \tau}{1 + (\omega_I - \omega_S)^2 \tau^2}; \tag{4}$$

A is the indirect coupling;

τ is the chemical exchange correlation time;

$\omega_I = \gamma_I H_R$ and $\omega_S = \gamma_S H_R$ are the proton and nitrogen nuclear angular velocities in the field H_R in which we study the relaxation;

P is the probability of finding a nitrogen nucleus in an NH_3 molecule;

T_1' and T_2'' are small corrections which take into account proton dipolar interaction in the NH_4^+ ion and quadrupole relaxation in the NH_3 molecule.

In Eq. (3) we know or may estimate *a priori* all the factors except the correlation time τ which appears in Y [Eq. (4)]. Moreover τ appears in

[2]These equations are slightly modified by a factor 4 before σ and ρ' because of the four protons of this NH_4^+ ion.

the only field-dependent factor Y [Eq. (4)] and Y depends quite critically on the value of the field H_R.

This critical value H_C is given by

$$(\omega_I - \omega_S)\tau = (\gamma_I - \gamma_S)H_C \, \tau = 1. \tag{5}$$

For $H_R \geqslant 3H_C$, Y is negligible, and for $H_R \leqslant \frac{1}{3} H_C$, Y has nearly attained its full value $\frac{2}{3} A^2\tau$.

So even if we know the other parameters to only a crude approximation we may obtain a good measurement of τ by studying the field dependence of the relaxation.

Experimental

To measure the relaxation time in low field the liquid is first polarized in the 20 kG field of an electromagnet. Then it stays for some 3.2 sec in a low adjustable field H_R where the magnetization decreases according to the rate equations (4), and finally the magnetization is measured with a standard NMR spectrometer.

This is accomplished in an easy way: we use a permanent flow of liquid (6). There are three containers (Fig. 1): C_1 in the gap of the magnet, C_2 in the magnetic field H_R, and C_3 in the middle of a set of Bloch

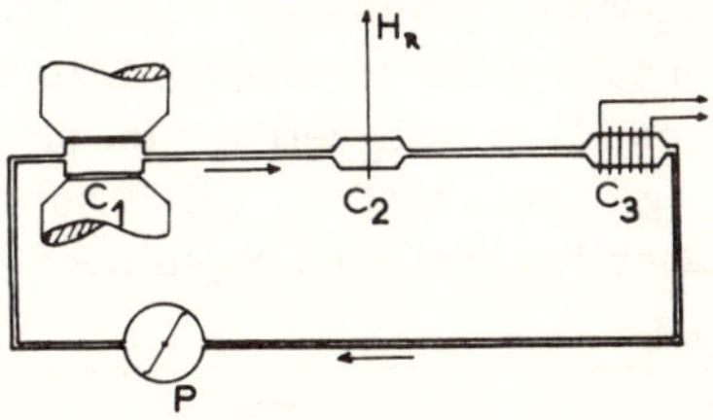

FIG. 1. Schematic diagram of the experiment. With the help of the pump P the liquid flows through the containers C_1, C_2, and C_3. C_1 is in the gap of an electromagnet, C_2 is in the adjustable low field H_R where the relaxation takes place, and C_3 in the middle of the Bloch coils with which we measure the proton polarisation I_z.

coils. These containers are connected with small-bore tubing in order to avoid spurious relaxation during the transport time.

The electromagnet is designed to have a very small stray field, so the C_2 container is placed at 1 meter from this magnet and the Bloch spectrometer is 2 meters away. The NMR spectrometer works in a low field of around 3 G, and the liquid is spun in C_3 in order to reduce the linewidth (7).

In Fig. 2 we see typical records of the proton NMR for two values of the field H_R.

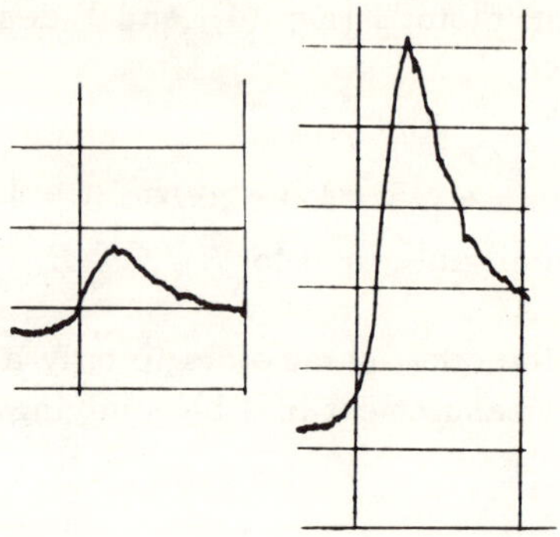

FIG. 2. Records of the NMR proton signal, I_z, for two values of the relaxation field, $H_R = 2$ G (left) and $H_R = 120$ G (right), for an ammonium nitrate solution • pH 8.2.

Results and Conclusion

In Fig. 3 are shown, on a logarithmic scale, relative values of the proton nuclear magnetization I_z *versus* the field H_R, for different values of the pH. In Eqs. (4) we have to adjust the values of τ in order that the solution fits the experimental curves. We have found reasonably good agreement and the corresponding results are shown in Fig. 4.

We have been able to measure the correlation time between $2 \cdot 10^{-4}$ and $6 \cdot 10^{-7}$ sec. Equation (2) is well verified for pH between 3 and 5.5. For pH > 6 it is interesting to note that Eq. (2) is not verified: τ becomes constant. This discrepancy has not been explained; probably the ex-

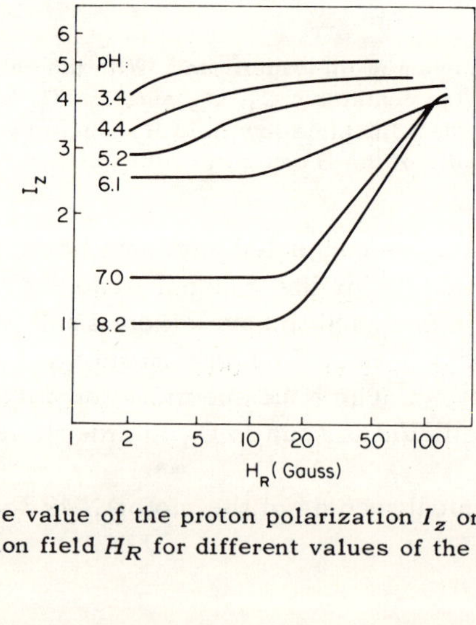

FIG. 3. Relative value of the proton polarization I_z on a logarithmic scale *versus* the relaxation field H_R for different values of the pH.

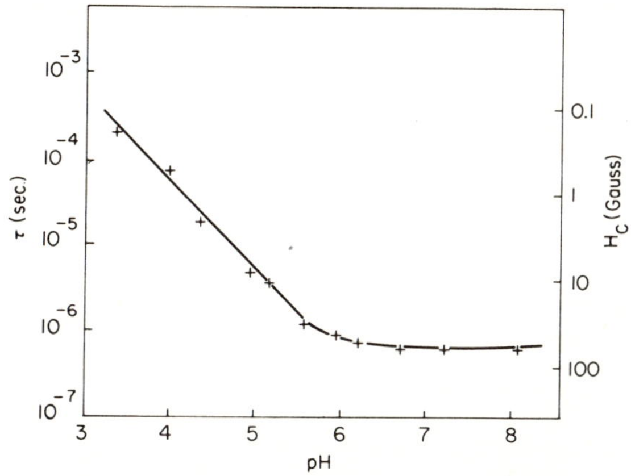

FIG. 4. Chemical exchange correlation time τ in seconds on a logarithmic scale *versus* the pH. On the right are the corresponding critical fields H_C of Eq. (5).

change mechanism is more complicated than expected and the protons of the water solution take part in the chemical exchange.

This method is useful whenever chemical exchange modifies a hyperfine interaction. In low fields it is possible to measure exchange times of the order of 10^{-3} sec even with a weak hyperfine interaction.

There is another possibility, that of studying the relaxation, not in a steady magnetic field, but in a rotating one (*8, 9*). With this modification it is possible to extend the preceding results toward longer correlation time.

References

1. R. A. Ogg, *Discussions Farad. Soc.* **17**, 215 (1954).
2. N. Bloembergen, *J. Chem. Phys.* **27**, 572 (1957).
3. I. Solomon, *Phys. Rev.* **99**, 559 (1955).
4. I. Solomon and N. Bloembergen, *J. Chem. Phys.* **25**, 261 (1956).
5. A. Abragam, "Nuclear Magnetism," p. 333. Oxford Univ. Press, London, 1961.
6. H. Benoit, *Arch. Sci.* (*Geneva*), **11**, fasc. spec., 282 (1958).
7. J. Hennequin, *Ann. Phys.* (*Paris*), p. 949 (1961).
8. I. Solomon, *Compt. Rend.* **249**, 1631 (1959).
9. H. Benoit, H. Ottavi, and J. Pommier, *Compt. Rend.*, **256**, 399 (1963).

Nuclear Relaxation by Inter- and Intramolecular Interactions in Two-Spin Systems

E. L. MACKOR and C. MacLEAN

Koninklijke/Shell-Laboratorium, Amsterdam (Shell Research N. V.) Amsterdam, The Netherlands

Contributions to relaxation in molecules such as HF, $CHFCl_2$ with two nonidentical spins may be made by intra- and intermolecular interactions between identical and nonidentical spins. It is possible to devise an experiment which allows the contributions from the various interactions to be determined. The nuclear spins (S and I) are supposed to be coupled through a small scalar interaction $J\mathbf{S}\cdot\mathbf{I}$. For $S = I = \frac{1}{2}$ four lines are observed corresponding to four magnetizations. These magnetizations are proportional to differences in occupation numbers of the energy levels, e.g., $S_+ = \text{const.}\ (N_{++} - N_{-+})$, etc. Transition probabilities ω_0, ω_2, ω_1, and ω_1' are defined according to Solomon (*1*).

The equations of motion of the total S and I magnetizations are (*1*)

$$\frac{dS}{dt} = \frac{d}{dt}(S_+ + S_-) = -(\omega_0 + 2\omega_1' + \omega_2)(S - S^0) - (\omega_2 - \omega_0)(I - I^0), \tag{1a}$$

$$\frac{dI}{dt} = \frac{d}{dt}(I_+ + I_-) = -(\omega_2 - \omega_0)(S - S^0) - (\omega_0 + 2\omega_1 + \omega_2)(I - I^0), \tag{1b}$$

where S^0 $(= S_+^0 + S_-^0)$ is the equilibrium value of the S magnetization.

Shimizu and Fujiwara (2) also considered the decay of the difference between the magnetizations of each spin; this decay is purely exponential:

$$\frac{d}{dt}(S_+ - S_-) = -2(\omega_1 + \omega_1')(S_+ - S_-), \tag{1c}$$

$$\frac{d}{dt}(I_+ - I_-) = -2(\omega_1 + \omega_1')(I_+ - I_-). \tag{1d}$$

The relaxation equations for intramolecular interaction have to be modified for intermolecular relaxation. One then has to consider relaxation transitions within a collision complex of two molecules. This four-spin system has available 16 energy levels between which ω_2^*, ω_0^*, ω_1^* and $\omega_1^{*\prime}$ transitions are possible. The transition probabilities are starred to indicate that they originate from translation.

We distinguish between *SI*, *SS*, and *II* relaxation processes. In the first type an ω_2^* process means a simultaneous transition of two spins due to interaction of an *S* and an *I* spin *on different molecules*, e.g., from the state $(++)_A(++)_B$ to $(+-)_A(-+)_B$, etc. for molecules *A* and *B*.

The equations of motion, when *SI* interactions are considered, analogous to (1), become:

$$\frac{dS}{dt} = -(\omega_2^* + 2\omega_1^{*\prime} + \omega_0^*)(S - S^0) - (\omega_2^* - \omega_0^*)(I - I^0), \tag{2a}$$

$$\frac{dI}{dt} = -(\omega_2^* - \omega_0^*)(S - S^0) - (\omega_2^* + 2\omega_1^* + \omega_0^*)(I - I^0), \tag{2b}$$

$$\frac{d}{dt}(S_+ - S_-) = -(2\omega_1^{*\prime} + 2\omega_1^* + 2\omega_2^* + 2\omega_0^*)(S_+ - S_-), \tag{2c}$$

$$\frac{d}{dt}(I_+ - I_-) = -(2\omega_1^{*\prime} + 2\omega_1^* + 2\omega_2^* + 2\omega_0^*)(I_+ - I_-). \tag{2d}$$

Thus the time constant of $(S_+ - S_-)$ [or $(I_+ - I_-)$] depends on whether the relaxation is intra- or intermolecular. The question whether the relaxation is due to interactions between like or unlike spins can be answered when the magnitude of the Overhauser effect is known.

The experiment has been performed at 56.4 MHz on liquid $CHFCl_2$ at $-135\,^\circ C$, where the main relaxation mechanism is the dipole-dipole interaction (*3*). In the steady-state Overhauser experiments an enhancement of 0.40 has been measured for both spins, showing that fluorine is relaxed by hydrogen and vice versa.

To study the total and difference magnetizations, S_+ is inverted by an adiabatic passage and the decays S_+ and S_- are simultaneously recorded; $(S_+ + S_-)$ and $(S_+ - S_-)$ can then be computed as a function of time. The same can be done for the *I* spins. For a dipolar relaxation mechanism (*1*):

$$\omega_2^* : \omega_0^* : \omega_1^{*\prime} : \omega_1^* = \omega_2 : \omega_0 : \omega_1' : \omega_1 = 12 : 2 : 3 : 3.$$

From this relation and the experimental decay of $(S_+ + S_-)$ the slope of $(S_+ - S_-)$ can be predicted for the two limiting cases of intra- and intermolecular relaxation.

It follows from the experimental data that at $-135\,^\circ C$ $CHFCl_2$ relaxes predominantly by interactions between unlike spins. The intramolecular contribution to the relaxation is about twice as large as the intermolecular one.

References

1. I. Solomon, *Phys. Rev.* **99**, 559 (1955).
2. H. Shimizu and S. Fujiwara, *J. Chem. Phys.* **34**, 1501 (1961).
3. R. J. C. Brown, H. S. Gutowsky, and K. Shimomura, *J. Chem. Phys.* **38**, 76 (1962).

Proton Chemical Shifts and π-Electron Charge Distributions of Anthracene and Perinaphthene Aromatic Ions

W. G. SCHNEIDER

Division of Pure Chemistry, National Research Council, Ottawa, Canada

It has been established previously (*1, 2, 3*) that, in simple aromatic systems, the proton chemical shift is directly proportional to the excess π-electron density on the carbon atom to which the proton is bonded. Empirically this may be expressed by the simple linear relationship

$$\delta_H = k\Delta q, \tag{1}$$

where δ_H is the proton shift, Δq the excess charge density, and k a constant. In practice it is convenient to use benzene as a reference for chemical shifts and hence Δq is equal to $(1 - q)$, where q is the total π-electron density. Experimentally the constant k is found to have the value 10 to 11 ppm per electron (*1, 2, 3*). Previous work (*3*) has shown that proton resonance chemical shifts provide reliable information about charge density distributions provided that no interfering effects are present. Such effects are, for example, contributions to the proton chemical shift arising from large magnetic anisotropies of neighboring atoms or substituent groups.

Aromatic ions are of particular interest since the excess charges are substantially larger than in neutral molecules. Attempts were made to obtain the proton resonance spectrum of the anthracene dinegative ion, which has two extra electrons paired in the lowest antibonding orbital. Accordingly a fairly symmetrical proton resonance spectrum may be anticipated, somewhat like anthracene, but because of the negative charges in the ion, the entire spectrum would be displaced upfield (i.e., increased proton screening). Attempts to prepare the dinegative ion, by treating anthracene in dimethoxyethane solvent with excess lithium metal, yielded an unsymmetrical proton spectrum of the ABCD type. By further experiments, and also by employing 9,10-dideuteroanthracene and 1,4,5,8-tetradeuteroanthracene, it was revealed that the stable ionic species being formed was not the dinegative ion, but a protonated (or methylated) mononegative ion. The proton and/or methyl group must derive from the solvent. A more convenient way of preparing the protonated ion, suggested by the work of Velthorst (*4*), consists in treating an equimolar mixture of anthracene and 9.10-dihydroanthracene with lithium metal in the ether solvent, according to the reaction

$$A + AH_2 \xrightarrow{Li} 2\ AH^- . \tag{2}$$

Under the present experimental conditions, it must be concluded that the dinegative ion when formed is rather reactive, and, in the absence of proton donor species, will extract a proton or methyl group from the solvent.

Figure 1 shows the observed proton resonance spectrum of the protonated negative ion, together with that of neutral anthracene, on a common scale relative to benzene. The proton assignments of both spectra

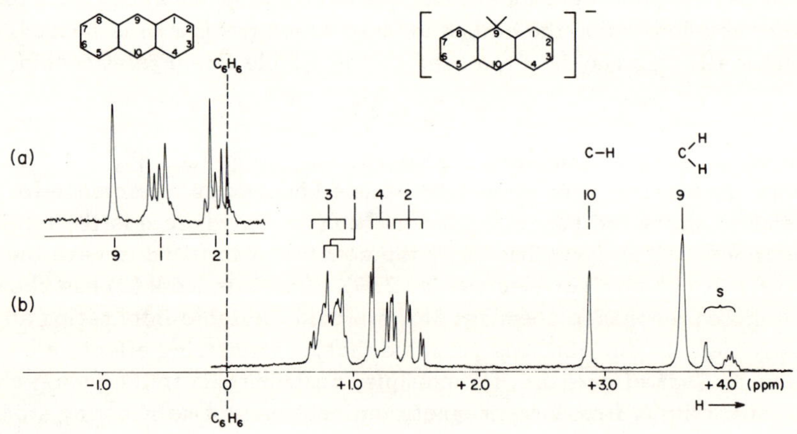

FIG. 1. Proton resonance spectrum at 60 Mc/s of (a) anthracene (2 mole % solution in dimethoxyethane), and (b) anthracene negative ion (2 mole % solution in deuterated dimethoxyethane). The position of internal reference benzene is also shown.

are shown. The spectrum of the negative ion was recorded by employing deuterated dimethoxyethane (1,2-dimethoxyethane) as solvent so as not to obscure the resonance peak of the CH_2 group at the 9 position. The residual protons in the deuterated solvent give rise to the weak signals labeled S in Fig. 1.

In evaluating the charge density distribution by Eq. (1), it is assumed that in the ion the aromatic ring current is absent in the center ring. Accordingly no corrections for the ring current of neighbor rings need be applied for the observed proton shifts of the two outer ring protons. But a correction is required to the observed shift of proton 10. This consists of two parts:

(*i*) Contribution of the ring currents in the two outer rings, +1.04 ppm (calculated).

(*ii*) Contribution due to lack of ring current in center ring, −1.50 ppm (taken as the difference in proton shift between benzene and cyclohexadiene (*3*). Applying these corrections, we can obtain the charge density distribution shown in Fig. 2. Also shown for comparison is the density

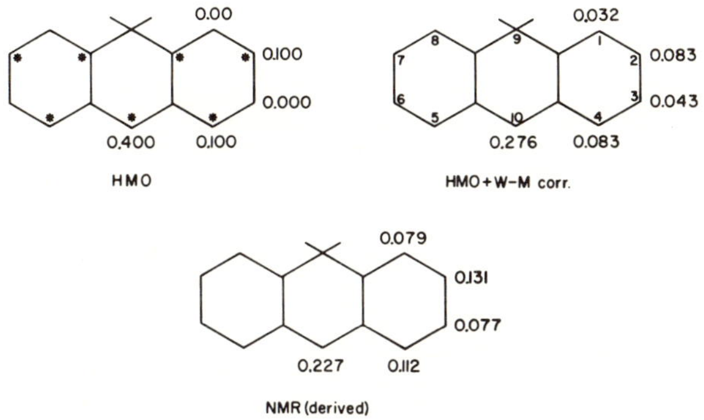

FIG. 2. Protonated anthracene negative ion.

distribution calculated on the basis of simple Hückel MO theory, as well as the results after applying the Wheland-Mann correction, as suggested by MacLean and Mackor (*5*). The NMR-derived charges are slightly too high (their sum being somewhat in excess of unity), but the relative charge distribution around the ring may be expected to be quite reliable. This distribution is considerably more uniform than that calculated on the basis of Hückel MO theory, and accords more satisfactorily with the distribution as modified by the Wheland–Mann correction.

The aromatic ions of perinaphthene are of particular interest since it proved possible to prepare both the positive and the negative ion. The positive ion was prepared by a method due to Reid and Sutherland (*6*) and was measured in a solution of nitromethane. The negative ion was prepared by treating the parent perinaphthene hydrocarbon with lithium metal in a suitable ether solvent. The proton resonance spectrum of the negative ion is reproduced in Fig. 3a, that of the positive ion in Fig. 3b. (Note that the sweep rate in Fig. 3a is somewhat greater than that in Fig. 3b.) The two spectra are mirror images of each other and are of the expected simple AB_2 type. The actual chemical shifts of the B protons relative to internal benzene are nearly equal, but in opposite sense. The shift of the A protons in the negative ion is somewhat greater than that in the positive ion.

Since in the negative ion (as compared with the positive ion) two extra electrons are contained in the nonbonding orbital, it is reasonable to assume these electrons will not contribute to the aromatic ring current.

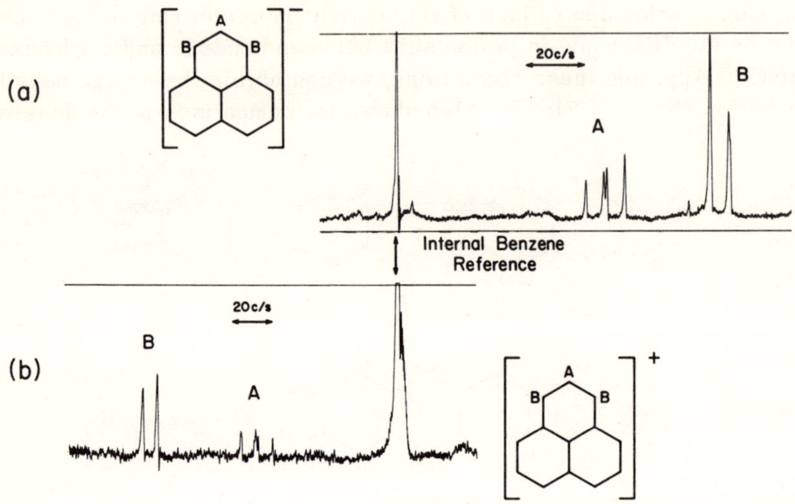

FIG. 3. Proton resonance spectrum at 60 Mc/s of (a) perinaphthene negative ion, and (b) perinaphthene positive ion. As indicated the sweep rates for the two spectra are not identical.

Thus the aromatic ring current will be similar in the two ions. On applying Eq. (1) to evaluate the local charge densities, the latter turn out to be much too large. The reason for this is not clear, but it may be that for structures of this type the use of benzene as a reference point is quite inappropriate. The ring current circulation in these ions will be almost exclusively around the periphery, so that the structures cannot be treated simply as made up of three benzene rings.

In order to evaluate the electron densities it is possible to make use of a normalization condition, $\Sigma\delta_H = 2$, that is, the arithmetic sum of all the proton shifts between the two ions must correspond to a total charge of two electrons. This neglects the charges on the four inner carbon atoms without protons, but the charges here are undoubtedly small and will introduce little error. In this way the value of the constant k [Eq. (1)] is found to be 16.4 ppm per electron, considerably larger than the value for simple benzenoid systems (~11 ppm per electron). With this value the resulting charge densities for the positive and negative ions are shown in Fig. 4, together with those calculated by Hückel MO theory. As in the anthracene ion, the charge distributions derived from the proton resonance shifts exhibit a more uniform distribution than that obtained from simple MO calculations.

The proton resonance shifts are assumed to arise from σ-π interaction. The excess charge in the carbon p orbital polarizes the σ core, which includes the C—H bond, and hence alters the proton screening. In odd-

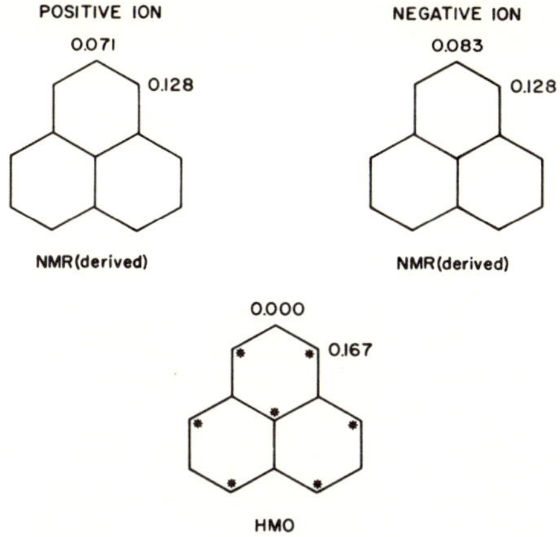

FIG. 4.

alternant systems, such as those being considered here, MO theory predicts the unstarred carbon atoms (Fig. 4) to have zero density in the nonbonding orbital. On the other hand, the NMR results indicate a density about 60% as large as the density on the neighboring starred atoms. The proton screening is altered primarily by polarization of the C—H sigma bond. If all the charge were assumed to reside only in the *p* orbitals of the neighboring starred carbon atoms, polarization of the σ core could extend to the intermediate unstarred carbon and to the proton bonded to this carbon. However, this polarization may be expected to become greatly attenuated through a distance of two sigma bonds, and it appears difficult to account for the large shifts of the protons bonded to the unstarred carbon atoms unless the latter also possess an appreciable π-electron density. However, more quantitative considerations of this question require a very detailed study of σ-π interaction.

References

1. G. Fraenkel, R. E. Carter, A. McLachlan, and J. H. Richards, *J. Am. Chem. Soc.* **82**, 5846 (1960).
2. H. Spiesecke and W. G. Schneider, *Tetrahedron Letters*, No. 14, 468 (1961).
3. T. Schaefer and W. G. Schneider, *Can. J. Chem.* **41**, 966 (1963).
4. N. H. Velthorst, Thesis, Univ. of Amsterdam, 1963.
5. C. MacLean and E. L. Mackor, *Mol. Phys.* **4**, 241 (1961).
6. D. H. Reid and R. G. Sutherland, *J. Chem. Soc.*, p. 3295 (1963).

High-Resolution Nuclear Magnetic Resonance Studies of Some Substituted Phenanthrenes

K. D. BARTLE and J. A. S. SMITH[1]

School of Chemistry, The University of Leeds, Leeds, England

Introduction

The high-resolution NMR spectra of phenanthrene and its substituted derivatives can be expected to show many features of interest; apart from the values of chemical shifts and spin-spin splittings within the ring systems 1,2,3,4 and 5,6,7,8, (the numbering used throughout this paper is given in Fig. 1), strong steric effects can be expected from substituents in the 9,10 or 4,5 position, with the additional possibility of transmission effects through the short C(9)–C(10) bond.

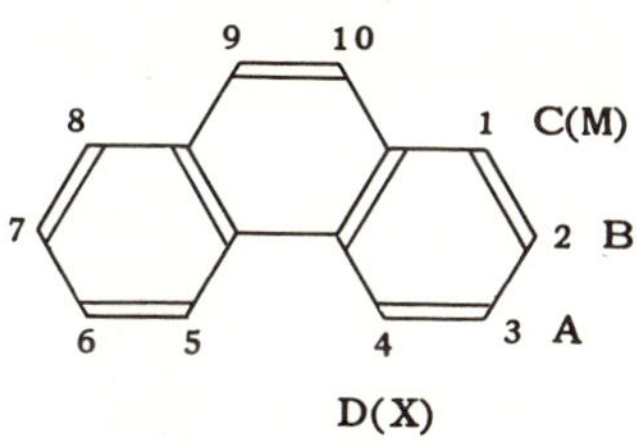

FIGURE 1.

Some of these effects have already been noted in the high-resolution spectrum of phenanthrene itself (*1*) and of several unsymmetrically substituted phenanthrenes (*2*). This paper is principally concerned with a description of the spectral features of 9,10-disubstituted phenanthrenes, which have been solved as ABMX systems, and some of their unsymmetrically substituted derivatives. The practical interest of these studies is illustrated by the identification of the substitution position (hitherto unknown) of an ethyl-9,10-dimethylphenanthrene.

Experimental

The spectra were recorded at 60 Mc/s on a Varian A-60 spectrometer at the Bradford Institute of Technology. All the 9,10-disubstituted phe-

[1] *Present address*: School of Molecular Sciences, The University of Warwick, Coventry, England.

nanthrenes were studied in at least two different solutions in CS_2; the line positions are quoted with respect to tetramethylsilane as internal standard and in most cases were obtained by extrapolation to infinite dilution. 9,10-Dimethoxycarbonylphenanthrene was also studied in $CDCl_3$. The chemical shifts are considered to be reliable to ±0.005 ppm.

Discussion

The basis of the discussion is the analysis of the NMR spectrum of phenanthrene given by Batterham *et al.* (*1*). They assumed, as in previous work (*3*), that there are significant low-field steric shifts for the 1,4 protons, and so derived starting parameters by treating the system as ABMX (as in Fig. 1); there is evidence from deuterium-substitution and related studies that the 9,10 protons do not interfere (*4*). The analysis was then refined in a symmetrical ABCD system (viz., with $J_{12} = J_{34}$, $J_{13} = J_{24}$) by an iterative computer technique. In the spectrum of 9,10-dimethylphenanthrene (Fig. 2) an additional steric shift operates on H(1,8), which move downfield from the AB group [H(2,3) and H(6,7)] to

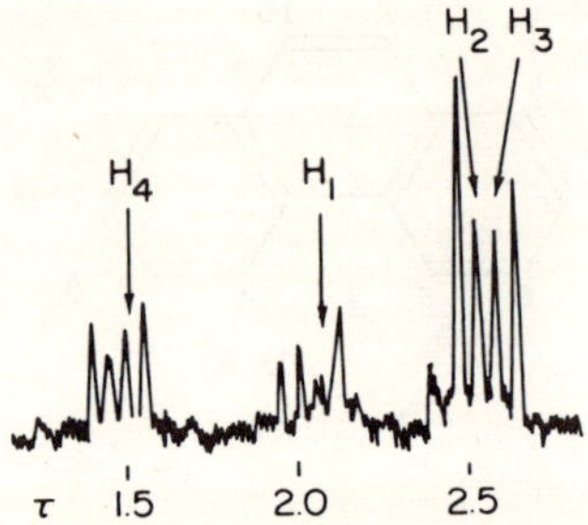

FIG. 2. Ring spectrum of 9, 10-dimethylphenanthrene in CS_2 at 60 Mc/s.

fall approximately midway between the latter and H(4,5). *Assuming* these assignments, the parameters can be deduced both in ABMX and in symmetrical ABCD (in which we acknowledge the assistance of Dr. T. J. Batterham). The two sets of data compare well with respect to line positions; consequently, the ABMX analysis is used here for all the 9,10-disubstituted derivatives. It should be noted that this is an approximation, for the relative intensities are only poorly reproduced, but it has the possible advantage that we do *not* assume that $J_{12} = J_{34}$ and $J_{13} = J_{24}$, and in fact in the analyses these equalities are not maintained.

Chemical shifts and spin splittings are given in Table I for four 9,10-disubstituted phenanthrenes analyzed as ABMX systems, together with the analyses of Batterham *et al.* (*1*) for phenanthrene and phenanthrene-

TABLE I

CHEMICAL SHIFTS (τ-VALUES IN PPM) AND SPIN-SPIN SPLITTING CONSTANTS (C/S) FOR PHENANTHRENE AND SOME 9,10-DISUBSTITUTED DERIVATIVES

	Phenanthrene, sym. ABCD(1), in $CDCl_3$ (10%)	9,10-Dimethyl-phenanthrene		9,10-Diethyl-phenanthrene ABMX, in CS_2 (infinite dilution)	9,10-Dibromo-phenanthrene, ABMX, in CS_2 (6.4% w/w)	9,10-Dimethoxy-carbonyl-phenanthrene, ABMX, in CS_2 (infinite dilution)	Phenanthrene-9,10-diquinone, sym. ABCD (1), in $CDCl_3$ (10%)
		sym. ABCD, in CS_2 (infinite dilution)	ABMX, in CS_2 (infinite dilution)				
ν_1	1.87_5	2.03_5	2.02_5	1.96	1.63_5	1.93_5	1.99_8
ν_2	2.17_5	2.52_5	2.53_5	2.47	2.44_5	2.44_5	2.28_5
ν_3	2.12	2.54	2.55	2.49	2.41_5	2.36	2.54_3
ν_4	1.07	1.45_5	1.45	1.39	1.49_5	1.38	1.83_0
J_{12}	8.4	8.5	9.0	8.9	8.1	8.4	8.3
J_{13}	1.6	1.4	0.9	1.1	1.7	1.4	1.4
J_{14}	0.5	0.5	0.5	0.5	0.5	0.5	0.5
J_{23}	7.3	6.9	6.9	6.8	6.9	6.9	7.4
J_{24}	1.6	1.4	1.7	1.6	1.0	1.9	1.4
J_{34}	8.4	8.5	8.0	7.8	8.8	8.1	8.3
$\nu_3 - \nu_2$(c/s)	−3.3	+0.9	+0.9	+1.2	−1.8	−5.1	+15.5

9,10-diquinone (in 10% solutions in $CDCl_3$). The assignment quoted for 9,10-dibromo- and 9,10-diquinonephenanthrene is one of two possibilities; the other is derived by interchanging proton 1 with 4 and proton 2 with 3. The ambiguity partly arises because the shift of proton 1, presumably caused by the 9,10 substituents, becomes so large that the signals from H(1) and H(4) overlap considerably and an absolute distinction between the two is impossible; the present assignment is tentative and is based largely on the solvent shifts (Table II), and the assumption that H(4) is still the lowest-field signal.

Assuming these assignments, therefore, Table I shows clear evidence of steric effects on the chemical shift. For ν_4, there is a rough dependence on the size of the substituent. For ν_1, the steric shift seems to depend to some extent on magnetic anisotropy, as well as size, of the group attached to position 10. A low-field shift is found with the C–Br bond as neighbor, as has been observed in some bromo-substituted chelate complexes (*5*); on the other hand, the magnetic anisotropy of the >C=O group in the 9,10-diquinoyl derivative might be expected to produce a low-field shift of ν_1 (*6*) whereas a small high-field shift (relative to phenanthrene) is observed with the assignment given in Table I.

Except for 9,10-dibromophenanthrene, the spin-spin splittings maintain a similar order; it is noticeable that in the 9,10-dialkyl-substituted derivatives and 9,10-dimethoxycarbonylphenanthrene, the ABMX analysis (where the symmetry restrictions on the J's have been removed) gives the order $J_{12} > J_{34} > J_{23}$ for the spin-spin splitting constants, the inverse order ($d_{12} < d_{34} < d_{23}$) of the C–C distances in phenanthrene (*7*) suggesting that double-bond character is having an important effect.

An interesting feature of these spectra is the difference in solvent shifts between the various peaks; this has already been noted in monosubstituted phenanthrenes (*2*). Table II shows results for four 9,10-

TABLE II

SOLVENT SHIFTS[a] (C/S PER 1% W/W) FOR SOME 9,10-DISUBSTITUTED PHENANTHRENES

	9,10-Dimethylphenanthrene in CS_2	9,10-Diethylphenanthrene in CS_2	9,10-Dimethoxycarbonylphenanthrene in		9,10-Dibromophenanthrene in CS_2
			CS_2	$CDCl_3$	
H(2,3) (AB)	0.40	0.62	0.54	0.50	0.54
H(1)(M)	0.45	0.45	0.33	0.19	0.41
H(4)(X)	0.60	0.79	0.74	0.58	0.79

[a]Mean estimated standard deviation = 0.05 c/s per one percent weight/weight.

disubstituted phenanthrenes; each figure represents the slope of a plot of line position in cycles per second (with respect to tetramethylsilane) against percent weight/weight, averaged over the stronger lines of *one* of the three groups in an ABMX analysis. Provided that the assignments in 9,10-dibromo and 9,10-dimethoxycarbonylphenanthrene are correct, the table supports the contention that H(4) shows a considerably stronger concentration dependence than H(1). A similar pattern occurs in 3-formylphenanthrene (*2*) but is not necessarily general for all positions of substitution. At least, in the 9,10-disubstituted phenanthrenes, the larger high-field shift at increasing solute concentration may be diagnostic of H(4), and is used as such in some of the assignments made in Table I.

The high-field shifts could be explained by association of the solute molecules in parallel planes; the ring currents of one then produce increased shielding at the proton positions of the other. Furthermore, if the solute molecules pack in a similar orientation, H(4), being much closer to the ring 5,6,7,8 of a neighboring molecule than H(1) and the same distance from the other two, will experience a greater shielding effect than H(1), in agreement with experiment. At the present, this explanation must be regarded as tentative, since other arrangements would also produce a similar order.

Identification of Positions of Substitution

The practical importance of these studies is illustrated by the assignment of the position of substitution (hitherto unknown) of an ethyl-9,10-dimethylphenanthrene. The unknown compound was obtained by reduction of an acetyl-9,10-dimethylphenanthrene prepared by acetylation of the parent hydrocarbon; the former was tentatively identified as 3-acetyl-9,10-dimethylphenanthrene on the basis of the behavior of the derived amine (*8*).

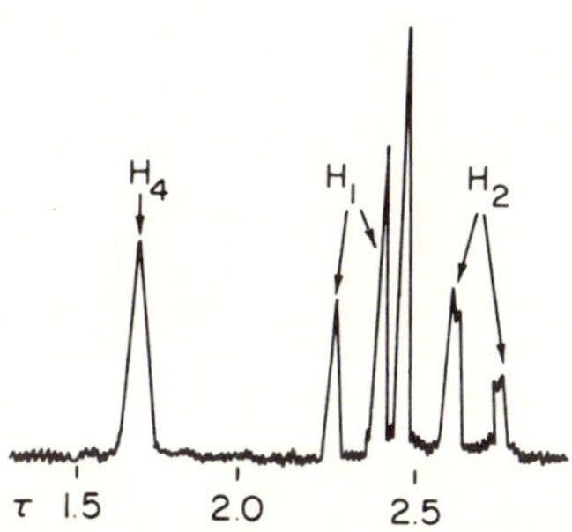

FIG. 3. Ring spectrum of 3,6-dimethylphenanthrene in CS_2 at 60 Mc/s.

TABLE III
CHEMICAL SHIFTS (τ-VALUES IN PPM) AND SPIN-SPIN SPLITTING CONSTANTS (C/S) IN SOME SUBSTITUTED PHENANTHRENES

	3,6-Dimethyl-phenanthrene, ABX, in CS_2 (infinite dilution)	3-Methyl-phenanthrene, ABX, in CCl_4 (infinite dilution)	3-Ethyl-phenanthrene, ABX, in CCl_4 (infinite dilution)	3-Ethyl-9,10-dimethyl-phenanthrene, ABX, A′B′M′X′, in CS_2 (infinite dilution)
ν_1	2.38	2.32	2.27_5	2.11
ν_2	2.71_5	2.67_5	2.64_5	2.65
ν_4	1.68_5	1.61	1.58	1.66_5
ν_5	1.68_5	1.40	1.36	1.45_5
ν_6	—	—[a]	—[a]	2.54
ν_7	2.71_5	—[a]	—[a]	2.54
ν_8	2.38	—[a]	—[a]	2.04
J_{12}	8.2	8.2	8.0	8.4
J_{14}	0.3	0.4	0.4	0.3
J_{24}	1.8	1.6	1.6	1.8
J_{56}	—	—[a]	—[a]	8.5
J_{57}	1.8	—[a]	—[a]	1.2
J_{58}	0.3	—[a]	—[a]	0.4
J_{67}	—	—[a]	—[a]	6.9
J_{68}	—	—[a]	—[a]	1.3
J_{78}	8.2	—[a]	—[a]	8.6

[a]Not measured.

Before an assignment could be made, it was necessary to study the effect of alkyl substituents at positions other than 9,10. The analyses are given in Table III. The spectrum of 3,6-dimethylphenanthrene (Fig. 3) is solved by an ABX treatment (A = H(1), B = H(2), X = H(4)) almost exactly, since $\nu_X - \nu_B/J_{BX}$ is about 30. In the spectra of 3-methyl- (Fig. 4) and 3-ethylphenanthrene, the resonances of H(2) and H(4) in the substituted

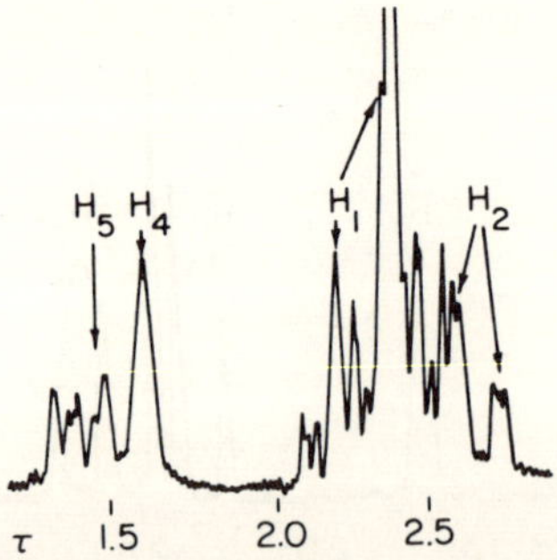

FIG. 4. Ring spectrum of 3-methylphenanthrene in CCl_4 at 60 Mc/s.

ring are shifted to a slightly higher field than those of H(7) and H(5) (respectively) in the unsubstituted ring. The H(4) signal takes the form of a broad singlet (width ~ 5 c/s), presumably because of weak and unresolved coupling to H(1) and H(2). Thus the ABX system H(1), H(2), and H(4) may be distinguished from H(5), H(6), H(7), and H(8) and an analysis made.

The ethyl-substituted 9,10-dimethylphenanthrene gave the spectrum reproduced in Fig. 5 and the analysis quoted in Table III. By analogy with 9,10-dimethylphenanthrene, the lines of the ABMX pattern of H(5,6,7,8) can be readily picked out, since the separation of the outermost and strongest lines of the AB, M, and X groups will be very nearly the same.

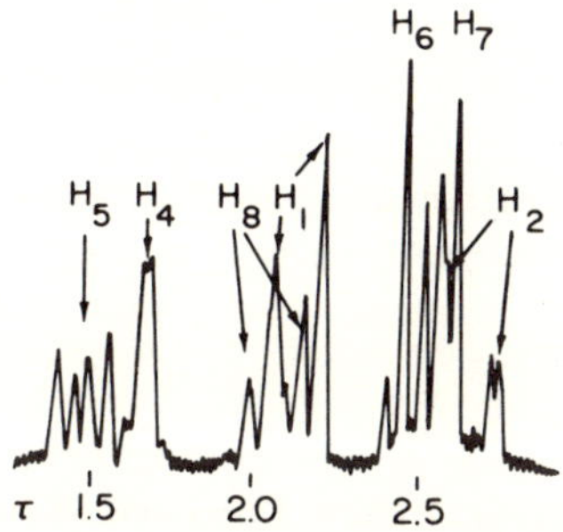

FIG. 5. Ring spectrum of an ethyl-substituted 9,10-dimethylphenanthrene in CS_2 at 60 Mc/s.

As for the rest, the broad singlet 0.2 ppm to the high-field side of the center of H(5) (X, at the lowest field) has an intensity equivalent to one proton, and is in the predicted position for H(4) with small *meta* and *para* couplings. On the high-field side of this, H(8) (M) is superimposed on two slightly broadened lines separated by 8.4 c/s, which we attribute to H(1) coupled to another proton with $J = 8.4$ c/s. The magnitude of this coupling suggests that the latter must be an *ortho* proton; in confirmation, we find in the highest-field group two doublets, whose centers are 8.4 c/s apart, which we assign as H(2) lines. We conclude that the spectrum is explained if the ethyl group is in the 3 position.

In final confirmation of these assignments, we may consider the spectra we would predict were substitution to have occurred in other positions. (i) In 4-ethyl-9,10-dimethylphenanthrene, the signals at $\tau < 1.8$ ppm would correspond to one proton [H(5)] rather than two in the observed spectrum. Furthermore, we could expect an extra downfield shift of about 0.3 ppm, by analogy with the shift of 0.28 ppm observed for H(5) of 4-methylphenanthrene relative to H(4,5) of phenanthrene (*2*). (ii) In 2-ethyl-

9,10-dimethylphenanthrene, we expect a reasonably sharp doublet for H(4) as in the spectrum of 2,7-dimethylphenanthrene (Fig. 6) rather than the broad singlet observed. (iii) Finally, in 1-ethyl-9,10-dimethylphenanthrene we expect the H(4) lines to consist of a complex pattern superimposed on H(5) and resonance from only one proton [H(8)] would be found near $\tau = 2.1$.

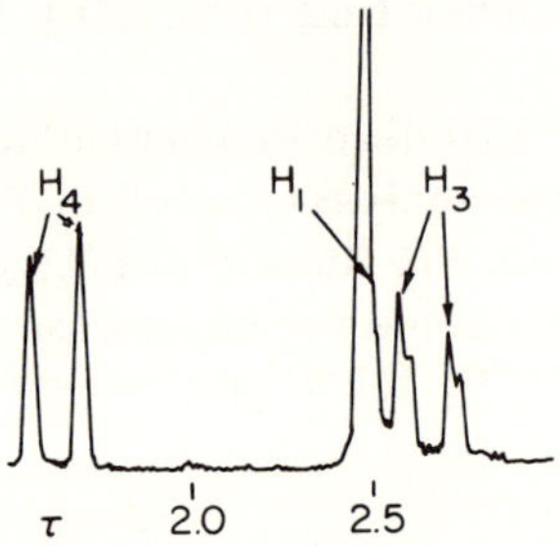

FIG. 6. Ring spectrum of 2,7-dimethylphenanthrene in CS_2 at 60 Mc/s.

Acknowledgments

We thank Dr. P. M. G. Bavin of Smith Kline and French Laboratories Ltd., and Dr. I. T. Millar of Keele University for samples, Drs. E. G. Edwards and D. W. Jones of the Bradford Institute of Technology for permission to use the A-60 spectrometer, and the Coal Tar Research Association for financial assistance and samples.

References

1. T. J. Batterham, L. Tsai, and H. Ziffer, *Australian J. Chem.* **17**, 163 (1964).
2. R. H. Martin, N. Defay, F. Geerts-Evrard, and H. Figeys, *Bull. Soc. Chim. Belges* **73**, 199 (1964).
3. N. Jonathan, S. Gordon, and B. P. Dailey, *J. Chem. Phys.* **36**, 2443 (1962).
4. J. A. Pople, W. G. Schneider, and H. J. Bernstein, "High-Resolution Nuclear Magnetic Resonance," p. 250, McGraw-Hill, New York, 1959.
5. J. A. S. Smith and J. D. Thwaites, *Discussions Farad. Soc.* **34**, 143 (1962).
6. J. A. Pople, *Discussions Farad. Soc.* **34**, 7 (1962).
7. J. Trotter, *Acta Cryst.* **16**, 605 (1963).
8. P. M. G. Bavin, *Can. J. Chem.* **38**, 911 (1960).

Subspectral Analysis of NMR Spectra

P. DIEHL

Physics Department, University of Basel, Basel, Switzerland

Regularities within the observed NMR spectra have always formed the basis for all methods that try to break down the complex problem into simpler ones. To start with, the sorting out of repeated spacings may be mentioned. Figure 1 presents a typical example.

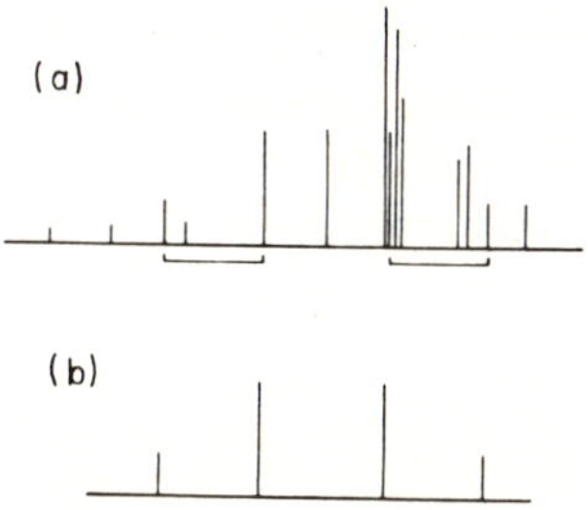

FIG. 1. Calculated spectra of AB_3^* system for $J/\delta = 0.5$. (a) Superposition of the ($I_A = 1/2$, $I_B = 3/2$) and the ($I_A = 1/2$, $I_B = 1/2$) subspectra. (b) The ab subspectrum ($I_A = 1/2$, $I_B = 1/2$).

The figure shows the calculated spectrum of an AB_3^* system like methyl mercaptan. Abraham *et al.* (*1*), who first discussed the system in 1958, pointed out that two pairs of lines showed a repeated spacing equal to the coupling constant J_{AB}, as indicated in Fig. 1a. They did not mention, however, that these four lines are the four transitions of an AB system such as shown in Fig. 1b. Actually this AB-type subspectrum contains all the information required to characterize the whole AB_3^* system, i.e., the coupling constant J_{AB} and the chemical shift δ_{AB}.

Waugh and Dobbs (*2*) and somewhat later Whitman *et al.* (*3*) demonstrated that NMR spectra of all systems of magnetically equivalent nuclei $A_m^* B_n^* \ldots$ can be broken down into simpler subspectra. The breakdown can be explained by the fact that within a group of magnetically equivalent nuclei there are no transitions that involve a change in the total spin quantum number I_T (maximum component of the spin vector in the z direction). The breakdown follows also from the rule that transitions between states of different symmetry are forbidden.

In the AB_3^* case of Fig. 1, $I_T(B_3^*)$ can have the values 3/2 and 1/2, and $I_T(A)$ is 1/2, so that we have two systems as shown. The alternative explanation is given in terms of group theory (*4*). The wave functions of the nuclei B_3^* will be either A_1 or E species so that there are no transitions between the two systems. In general the simplest subspectra of the $A_m^* B_n^*$ spectrum are

$$\begin{array}{ll} ab & \text{for } m \text{ odd, } n \text{ odd,} \\ ab_2^* & \text{for } m \text{ odd, } n \text{ even,} \\ a_2^* b_2^* & \text{for } m \text{ even, } n \text{ even} \end{array}$$

(lower-case letters are used to designate nuclei in subspectra). Once the simplest subspectrum has been identified, all the parameters of the problem are directly obtainable.

As a second example a typical $A_2^* B_3^*$ spectrum of an ethyl group is shown in Fig. 2. The spectrum of Fig. 2b can be sorted out with the help of a graphical method using a plot of the relative positions of the eight $A_2^* B$ transitions as a function of J/δ. The analysis of this simple $A_2^* B$ type of subspectrum again provides the two parameters of the whole $A_2^* B_3^*$ problem.

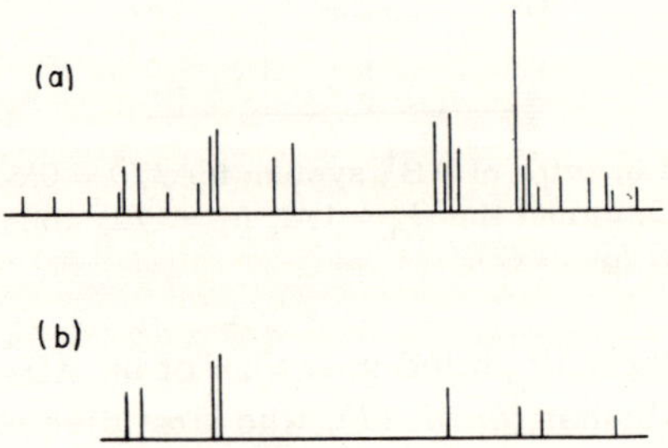

FIG. 2. Calculated spectra of $A_2^* B_3^*$ system for $J/\delta = 0.25$. (a) Superposition of the ($I_A = 1$, $I_B = 3/2$), the ($I_A = 1$, $I_B = 1/2$), the ($I_A = 0$, $I_B = 3/2$), and the ($I_A = 0$, $I_B = 1/2$) subspectra. (b) The $a_2^* b$ subspectrum ($I_A = 1$, $I_B = 1/2$) and ($I_A = 0$, $I_B = 1/2$).

This method of breakdown has been called the "composite particle model" because the simplification of the problem is actually achieved by treating each group of n magnetically equivalent nuclei of spin 1/2 as one particle with spin $n/2$.

Another scheme of subspectral analysis is the "effective Larmor frequency method" (*5*). As an example, paradeuterofluorobenzene is shown in Fig. 3.

The condition for the applicability of this method is that the molecule contains one group of strongly coupled nuclei (A_2B_2) which are weakly coupled to other groups of magnetically equivalent nuclei (D and F). The latter nuclei again have to be weakly coupled to each other. For

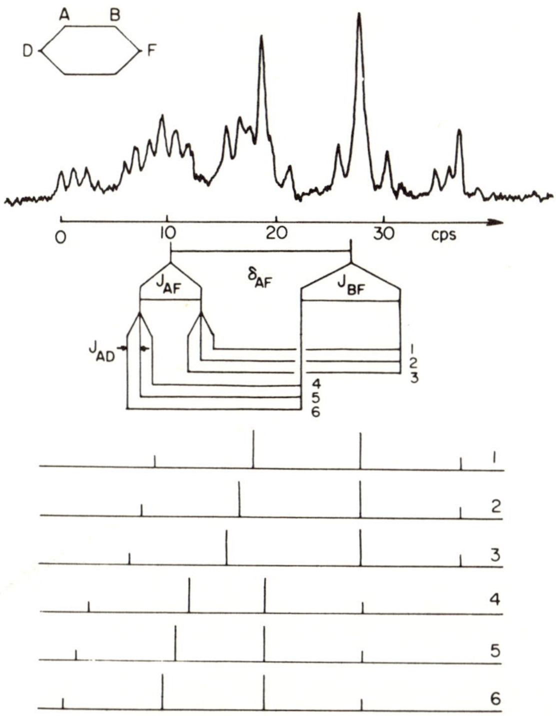

FIG. 3. The effective Larmor frequency subspectra of paradeuterofluorobenzene (proton spectrum at 60 Mc/sec). Not all of the a_2b_2 subspectra transitions are shown. Effective chemical shifts: (1) $\delta_{AB} + 1/2(J_{BF} - J_{AF}) - J_{AD}$; (2) $\delta_{AB} + 1/2(J_{BF} - J_{AF})$; (3) $\delta_{AB} + 1/2(J_{BF} - J_{AF}) + J_{AD}$; (4) $\delta_{AB} - 1/2(J_{BF} - J_{AF}) - J_{AD}$; (5) $\delta_{AB} - 1/2(J_{BF} - J_{AF})$; (6) $\delta_{AB} - 1/2(J_{BF} - J_{AF}) + J_{AD}$.

such systems it can be shown that the total Hamiltonian effectively depends on the weakly coupled nuclei only through the z components of their spins. It is possible to think of A_2B_2 states that are associated with each of the values of $F_z(\text{D})$ and $F_z(\text{F})$. The effect of these nuclei is equivalent to replacing the Larmor frequencies ν_A and ν_B by the effective Larmor frequencies:

$$\begin{aligned} \nu_A^* &= \nu_A \pm 1/2\, J_{AF} \pm J_{AD} && \text{for} \quad F_z(\text{F}) = \pm 1/2, \quad F_z(\text{D}) = \pm 1, \\ &= \nu_A \pm 1/2\, J_{AF} && \text{for} \quad F_z(\text{F}) = \pm 1/2, \quad F_z(\text{D}) = 0, \\ \nu_B^* &= \nu_B \pm 1/2\, J_{BF} && \text{for} \quad F_z(\text{F}) = \pm 1/2, \quad J_{BD} \approx 0. \end{aligned}$$

The six possible effective chemical shifts are now defined by the relations given in the caption of Fig. 3. The subspectra can be sorted out and analyzed.

The fact that we have six different effective chemical shifts in the same molecule could equally well be interpreted in terms of an effective external magnetic field. The spectrum of Fig. 3 is identical with a superposition of six spectra of the same hypothetical molecule A_2B_2 taken at six different external magnetic fields.

In principle the effective chemical shift can become very small, for instance, when $1/2(J_{BF} - J_{AF}) \approx \delta_{AB}$. The corresponding subspectrum will then degenerate into a single line. As a consequence we may observe deceptively simple spectra (6). A typical example is parafluorotoluene the proton spectrum (at 30 Mc/sec) of which consists of two lines corresponding to two degenerate a_2b_2 subspectra. Inspection of Fig. 3 shows that a degeneration of the two subspectra (numbers 2 and

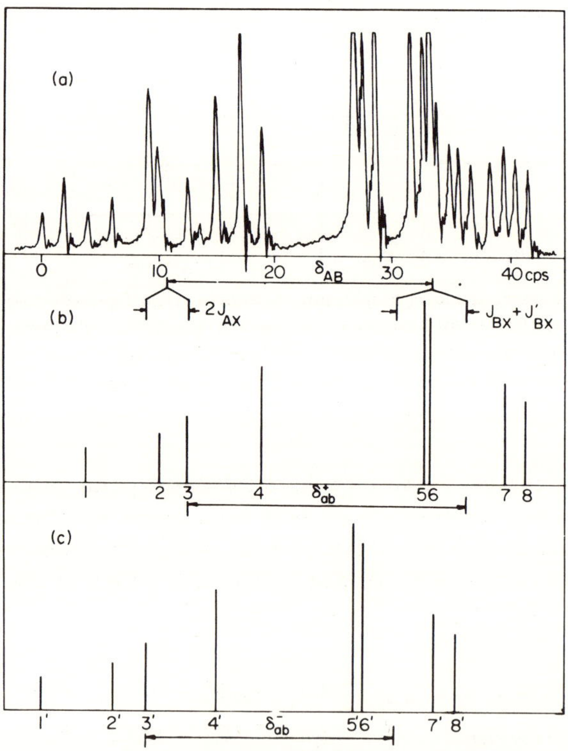

FIG. 4. The proton NMR spectrum of pyridine at 60 Mc/sec. (a) The AB_2 part of the AB_2X_2 spectrum. (b) The ab_2^* subspectrum corresponding to $F_z(X_2) = +1$, $J/\delta = J_{AB}/\delta^+_{ab} = J_{AB}/[\delta_{AB} + 1/2(J_{BX} + J'_{BX} - 2J_{AX})]$. (c) The ab_2^* subspectrum corresponding to $F_z(X_2) = -1$, $J/\delta = J_{AB}/\delta^-_{ab} = J_{AB}/[\delta_{AB} - 1/2(J_{BX} + J'_{BX} - 2J_{AX})]$.

5) is only possible if simultaneously $\delta_{AB} \approx 0$ and $J_{BF} \approx J_{AF}$. The doublet will then show a separation of $1/2(J_{AF} + J_{BF})$.

As pointed out above the effective Larmor frequency method cannot be used for a complete subspectral breakdown of more complex systems like the AB_2X_2 spectrum of pyridine. It is possible, however, to obtain a partial subspectral analysis in terms of effective Larmor frequencies for all NMR systems containing at least one weakly coupled nucleus (7).

This originates in the unique spin configurations of the X_2 group where all the spins are "up" (+1/2, +1/2) or all "down" (−1/2, −1/2). The subpart of the Hamiltonian corresponding to these X states again depends upon the X nuclei through $F_z(X_2)$ only. Similar statements for the X part can be made when $F_z(AB_2) = \pm 3/2$. For other values of $F_z(X_2)$ and $F_z(AB_2)$ this is no longer true, so that there are no corresponding effective Larmor frequency subspectra.

As an example the two ab_2^* subspectra of the AB_2X_2 system of pyridine, corresponding to the two values $F_z(X_2) = \pm 1$, are shown in Fig. 4. The analysis of these subspectra, which is easily accomplished by means of graphical methods, will yield the following parameters: J_{AB}, δ_{AB}, J_{AX}, and $J_{BX} + J'_{BX}$. Furthermore δ_{AX} can be obtained by identification of the X transitions corresponding to $F_z(AB_2) = \pm 3/2$.

The example shows that without a complete analysis of the problem, still all the parameters of the strongly coupled group as well as sums of coupling constants between the groups of nuclei can be obtained. This method will be of particular interest in studies of solvent effects of complex molecules where now the chemical shift can be measured simply by picking out the subspectra.

Generally speaking subspectral methods will simplify the analysis of NMR spectra and will indicate whether and to what extent the help of a computer is necessary.

References

1. R. J. Abraham, J. A. Pople, and H. J. Bernstein, *Can. J. Chem.* **36**, 1302 (1958).
2. J. S. Waugh and F. Dobbs, *J. Chem. Phys.* **31**, 1235 (1959).
3. D. R. Whitman, L. Onsager, M. Saunders, and H. T. Dubbs, *J. Chem. Phys.* **32**, 67 (1960).
4. H. M. McConnell, A. D. McLean, and C. A. Reilly, *J. Chem. Phys.* **23**, 1152 (1955).
5. J. A. Pople and T. Schaefer, *Mol. Phys.* **3**, 547 (1961); P. Diehl and J. A. Pople, *Mol. Phys.* **3**, 557 (1961).
6. P. Diehl, *Helv. Chim. Acta* **47**, 1 (1964).
7. P. Diehl, R. G. Jones, and H. J. Bernstein, *Can. J. Chem.* **43**, 81 (1965).

NMR Studies on Geometrical Isomers of Paraldehyde and Other Substituted Trioxanes

J. L. JUNGNICKEL and C. A. REILLY

Shell Development Company, Emeryville, California

Introduction

Cyclic trimers of aldehydes having the general formula of a substituted 1,3,5-trioxane

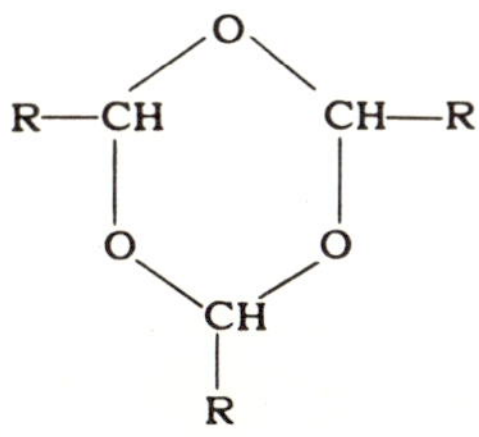

can, in principle, exist in either a boat or a chair form. However, it has been found experimentally that, for this type of isomerism, the chair form is generally the more stable one. For example, 1,3,5-trioxane itself has been shown to exist in the chair form by x-ray diffraction (*1*) and by infrared (*2*) measurements. Also, both the α and β isomers of parachloral have been shown by dipole moment, infrared, and NMR measurements to have the chair form (*3*). As a further example, the chair form of cyclohexane is more stable than the boat form by about 5 or 6 kcal/mole (*4, 5*), the free energy for the interconversion chair ⟷ chair being estimated as 10 or 11 kcal/mole at −67°C (*6*). Therefore, it will be assumed that the *sym*-trioxanes studied here exist predominantly in the chair form. The NMR spectral results will be examined for inconsistencies with this assumption.

Four geometrical isomers (see Fig. 1) of the chair form of *sym*-trioxanes are possible. In the two *cis-cis* isomers the three R groups can be either equatorial (A) or axial (B). Form A is interconvertible with form B by flipping through a "boat" conformation. However, it is to be expected that the population of form A will greatly exceed that of form B because of steric hindrance of the R groups in the latter form. In the two *cis-trans* isomers, two of the three R groups can either be equatorial (C) or axial (D). The AB pair cannot be converted to the CD pair without breaking bonds. In paraldehyde, the cyclic trimer of acetaldehyde ($R = CH_3$),

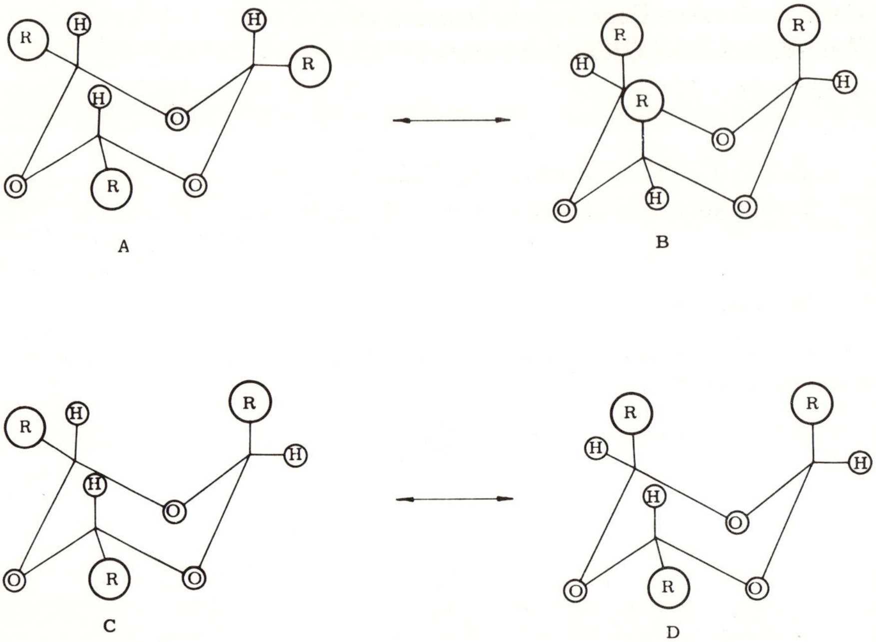

FIG. 1. Possible stereoisomers of the chair form of *sym*-trioxanes.

only one isomer has been positively known until recently. Diehl and Freeman (*7*) concluded from the simplicity of its high-resolution NMR spectrum that this isomer had all three methyl groups equatorial (Fig. 1A). The structure shown in Fig. 1B was ruled out solely on steric grounds. Recently, Craven *et al.* (*8*) identified a second isomer with one of the methyl groups axial. This isomer and the ordinary one with all three methyl groups equatorial were identified by means of their infrared spectra.

Prior to the publication by Craven *et al.*, E. R. Bell of this laboratory isolated two forms of the cyclic trimer of acetaldehyde by gas chromatography of the products from the radiolysis of polyacetaldehyde. Chemical shift data obtained from the high-resolution NMR spectra of these two forms were used to determine their structures. The present paper describes this identification together with NMR studies on the cyclic trimers of six other aldehydes.

Experimental

Unless otherwise indicated, the NMR spectra were obtained with a Varian Model A-60 analytical NMR spectrometer. Most samples were examined as dilute solutions in CCl_4. However, because of some interesting solvent effects, acrolein cyclic trimer was examined in other solvents. The cyclic trimer of β-cyanopropionaldehyde was dissolved in $CDCl_3$. Tetramethylsilane (TMS) was used as an internal reference in all cases. The samples were not degassed.

When necessary, the spectra were analyzed completely with the aid of the Swalen–Reilly iterative method (*9*); otherwise a conventional approximate analysis was made. Chemical shifts (δ) were expressed as parts per million (ppm) from TMS and spin-spin coupling constants (*J*) were expressed in cycles per second (cps). Spinning sidebands (labeled SSB) and impurity lines (labeled Imp) have been indicated on the spectra. In some cases integrals were obtained and are so indicated.

Results and Discussion

Acetaldehyde Cyclic Trimers

The NMR spectrum of one of the forms (mp 12.4 °C) of acetaldehyde cyclic trimer, 2,4,6-trimethyl-1,3,5-trioxane (I),

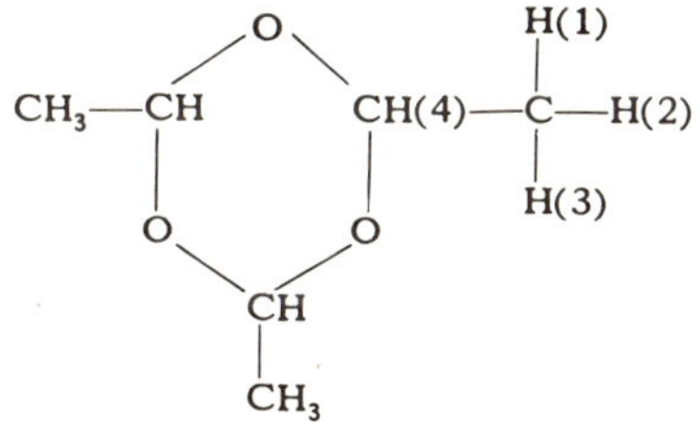

(I)

is shown in Fig. 2. The spectrum contains a doublet due to three equivalent methyl groups and a quartet due to three equivalent O—CH—O hydrogens. Since all three methyl groups are equivalent and all three O—CH—O hydrogens are equivalent, this must be a *cis-cis* isomer in which all three methyl groups are either axial or equatorial. A sample of ordinary commercial paraldehyde was found to have an identical NMR spectrum. The NMR spectra reported by Freymann *et al.* (*10*) and by Martin and Cantacuzene (*11*) for acetaldehyde cyclic trimer are similar to the one shown in Fig. 2.

The NMR spectrum of the other form (mp 14.7 °C) of acetaldehyde cyclic trimer is shown in Fig. 3. The methyl and O—CH—O regions are shown in more detail as inserts. This spectrum contains a doublet due to two equivalent methyl groups, another doublet half as intense due to the other methyl group, a quartet due to two equivalent O—CH—O hy-

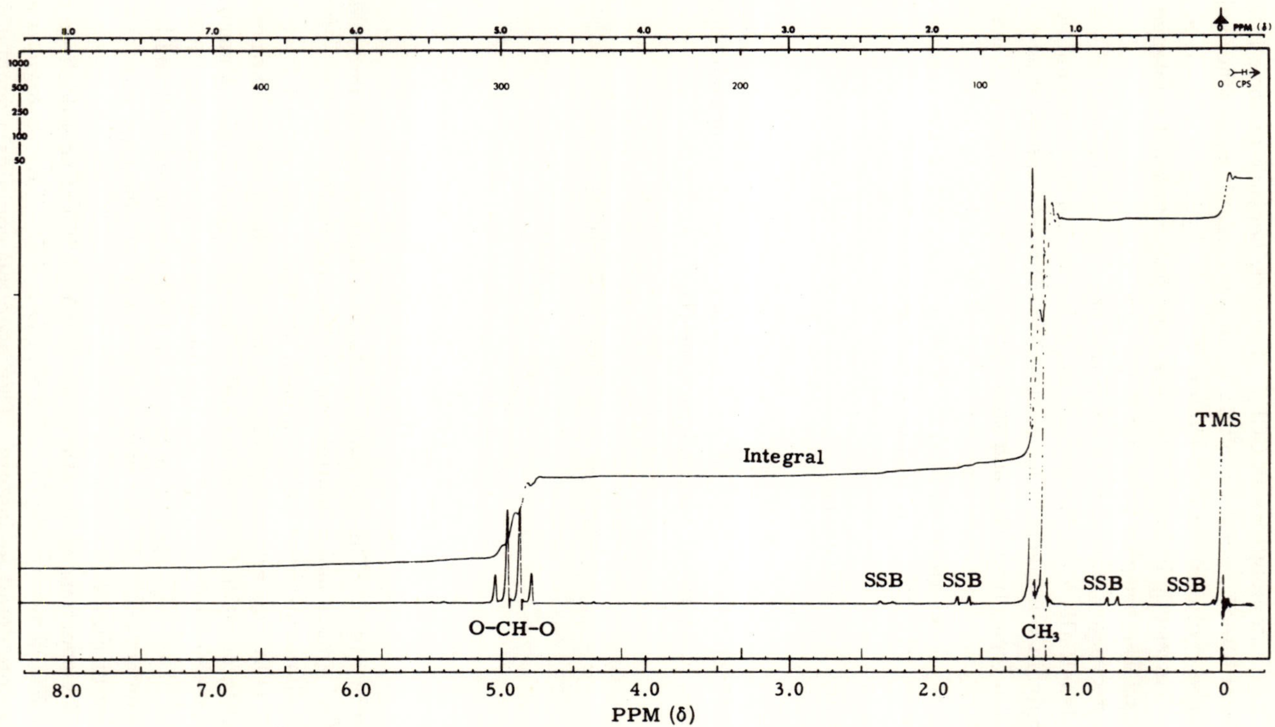

FIG. 2. NMR spectrum of acetaldehyde cyclic trimer (*cis-cis* isomer).

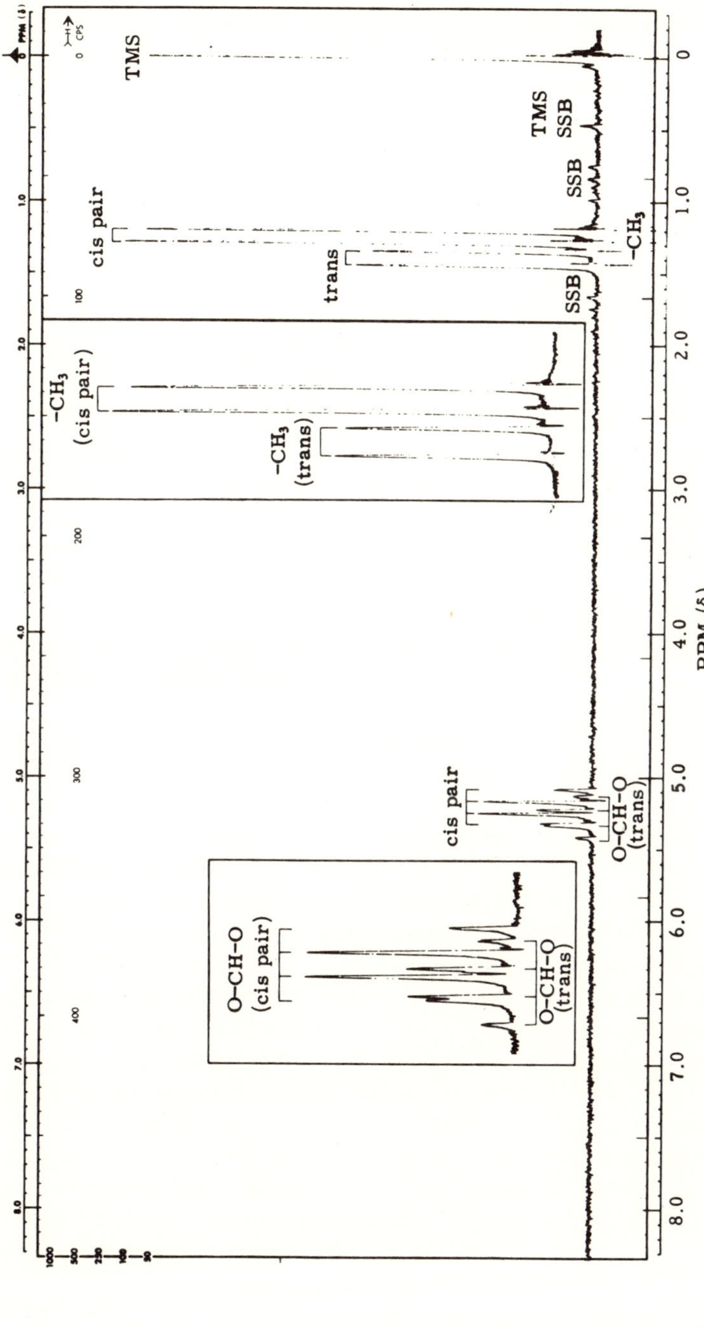

FIG. 3. NMR spectrum of acetaldehyde cyclic trimer (*cis-trans* isomer).

drogens, and another quartet half as intense due to the other O—CH—O hydrogens. This clearly must be a *cis-trans* isomer, either with one methyl group axial and the other two equatorial, or vice versa.

Chemical shifts and spin-spin coupling constants, obtained by the first first-order AX_3 approximation from the spectra at high dispersion, are as shown in the accompanying tabulation.

	δ ppm				$J_{H—CH_3}$ (cps)	
	H(1) = H(2) = H(3)		H(4)		$J_{14} = J_{24} = J_{34}$	
Isomer	Axial	Equat.	Axial	Equat.	Axial H	Equat. H
cis-trans	1.39	1.24	5.21	5.28	5.0	5.8
cis-cis		1.26	4.92		5.1	

It is noted that in the *cis-trans* isomer, the equivalent pair of protons, H(4), is more shielded (smaller chemical shift from TMS) by 0.07 ppm than is the unique proton. There is a considerable amount of evidence (*12*) in the literature to show that axial protons are more shielded than equatorial ones. Hence, it is concluded that the equivalent pair of protons in the *cis-trans* isomer is axial, i.e., that the molecule has the structure shown in Fig. 1C. In this case, the spin-spin coupling constant between an axial proton and the equatorial methyl group bonded to the same C atom is 5.0 cps and the corresponding coupling constant between an equatorial proton and an axial methyl group is 5.8 cps. A comparison of the chemical shifts of the methyl groups in the *cis-trans* and the *cis-cis* isomers together with a comparison of the H to CH_3 spin-spin coupling constants now identified the *cis-cis* isomer (paraldehyde) as the one with the methyl groups all equatorial (Fig. 1A). This was the conclusion reached also by Diehl and Freeman (*7*) on a less firm basis.

It is a little surprising to find such a relatively large chemical shift difference between the axial and equatorial methyl group protons. This difference is larger than that for the axial and equatorial H atoms and is in the opposite direction. This fact in itself could raise some doubts as to the structural assignment made in the last paragraph. On the other hand, the axial H of the *cis-cis* isomer is more shielded by 0.29 ppm than are the axial H's in the *cis-trans* isomer. This is consistent with the corresponding difference of 0.43 ppm observed for the axial H's of the α- and β-parachloral isomers (*13*). The internal consistency of all the data available together with the steric argument against three axial methyl groups can leave no doubt that the correct assignment has been made. By analogy with the corresponding α and β isomers of parachloral, it is proposed that ordinary paraldehyde be called β-paraldehyde and the other isomer be called α-paraldehyde.

Dichloroacetaldehyde Cyclic Trimer

The NMR spectrum of the cyclic trimer of dichloroacetaldehyde, 2,4,6-trisdichloromethyl-1,3,5-trioxane (II) is shown in Fig. 4. The spec-

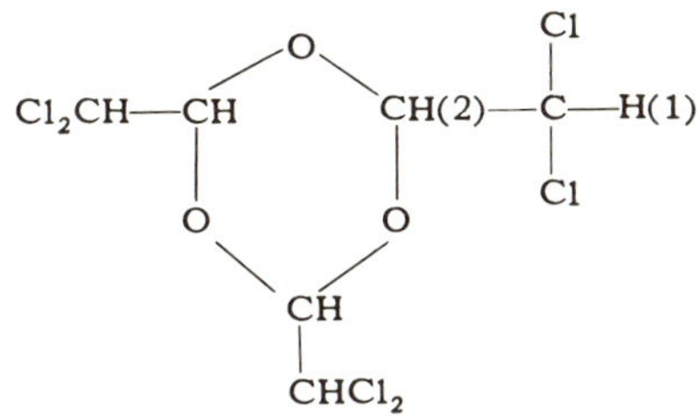

(II)

trum shows only a simple AB quartet pattern characteristic of two hydrogens on adjacent carbon atoms. Since the spectrum shows only one such pattern, it follows that the three —$CHCl_2$ hydrogens are equivalent and that the three hydrogens attached to the ring are also equivalent. This sample therefore is composed entirely of a *cis-cis* isomer and undoubtedly has the structure A of Fig. 1 because the R groups are even larger than in the corresponding acetaldehyde trimer. The same steric repulsion argument against axial R groups applies.

The AB pattern was analyzed according to the usual treatment (*14*) to calculate the chemical shifts and spin-spin coupling constants:

$$\delta_1 = 5.16 \text{ ppm},$$
$$\delta_2 = 5.64 \text{ ppm},$$
$$J_{12} = 4.8 \pm 0.1 \text{ cps}.$$

The assignment of the pairs of doublets shown in Fig. 4 is based on the following arguments. First, it is noted that the lines of the upper field doublet show slightly smaller peak intensities than do the corresponding lines in the other doublet, i.e., these lines are slightly broader. This broadening is assumed to be due to relaxation by the nearby Cl nuclei. Furthermore, it might be expected that the effect of two chlorine atoms in the neighboring R group would be to increase δ for the ring hydrogen considerably over that observed for the corresponding ring hydrogen atoms in the acetaldehyde cyclic trimers.

Chloroacetaldehyde Cyclic Trimer

The NMR spectrum of the chloroacetaldehyde cyclic trimer, 2,4,6-trischloromethyl-1,3,5-trioxane (III) is shown in Fig. 5. It was analyzed

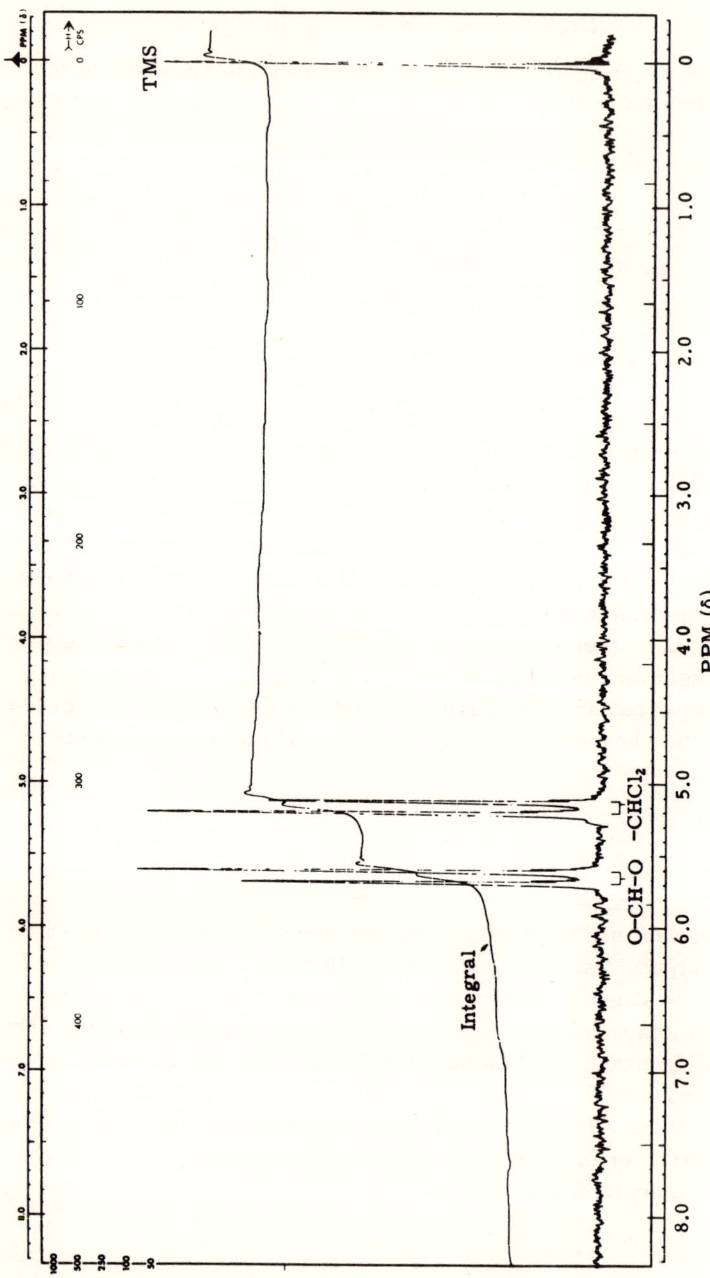

FIG. 4. NMR spectrum of dichloroacetaldehyde cyclic trimer.

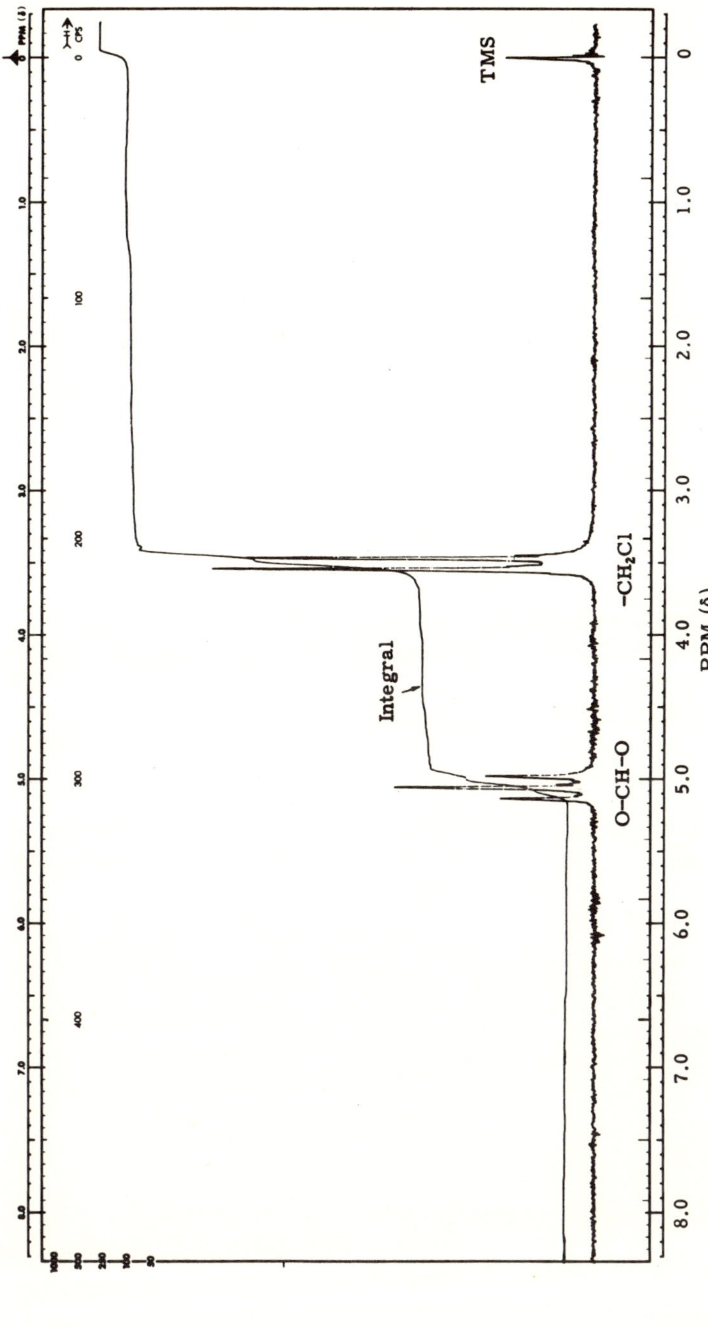

FIG. 5. NMR spectrum of chloroacetaldehyde cyclic trimer.

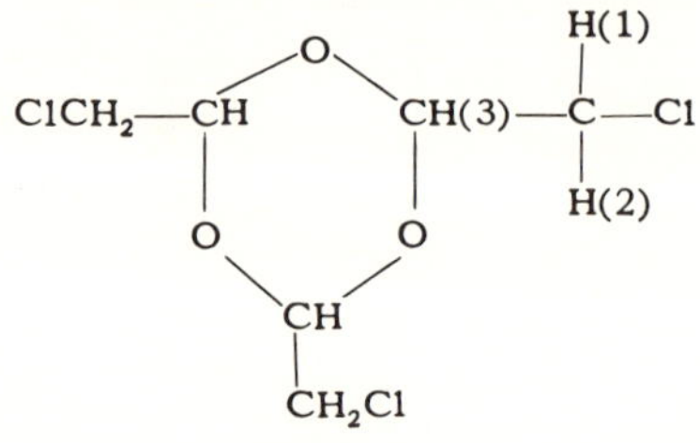

(III)

both by means of the trace relations and with the aid of the iterative method (*9*). Details of the analysis by means of the trace relations are given in the Appendix. The chemical shifts and spin-spin coupling constants so determined are as follows:

$$\delta_1 = \delta_2 = 3.52 \text{ ppm},$$
$$\delta_3 = 5.05 \text{ ppm},$$
$$J_{12} \text{ is indeterminate},$$
$$J_{13} = 4.77 \pm 0.14 \text{ cps}.$$

The spectrum is a simple A_2B pattern. It follows that the three $—CH_2Cl$ groups are equivalent and the three ring hydrogens also are equivalent. This then must be a *cis-cis* isomer and is doubtless the one with three equatorial $—CH_2Cl$ groups.

Propionaldehyde Cyclic Trimer

The NMR spectrum of 2,4,6-triethyl-1,3,5-trioxane (IV) is shown in Fig. 6A. Because the spectrum shows considerable higher-order split-

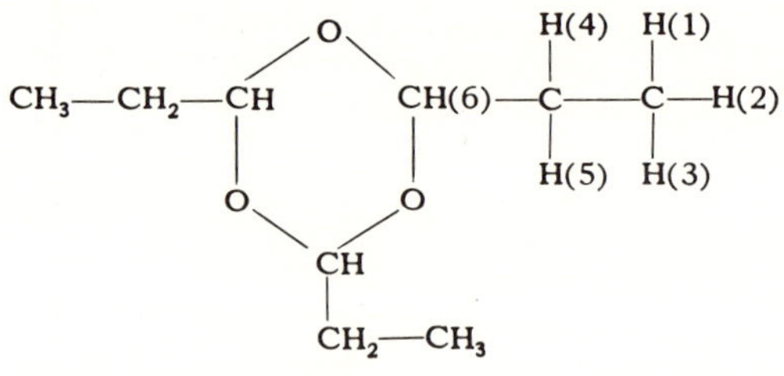

(IV)

ting, it was analyzed completely by the iterative method (*9*). The spectrum was treated as a six-spin system, i.e., the three C_3H_6 groups were assumed to be equivalent and completely isolated from one another. The computed chemical shifts and spin-spin coupling constants are as follows:

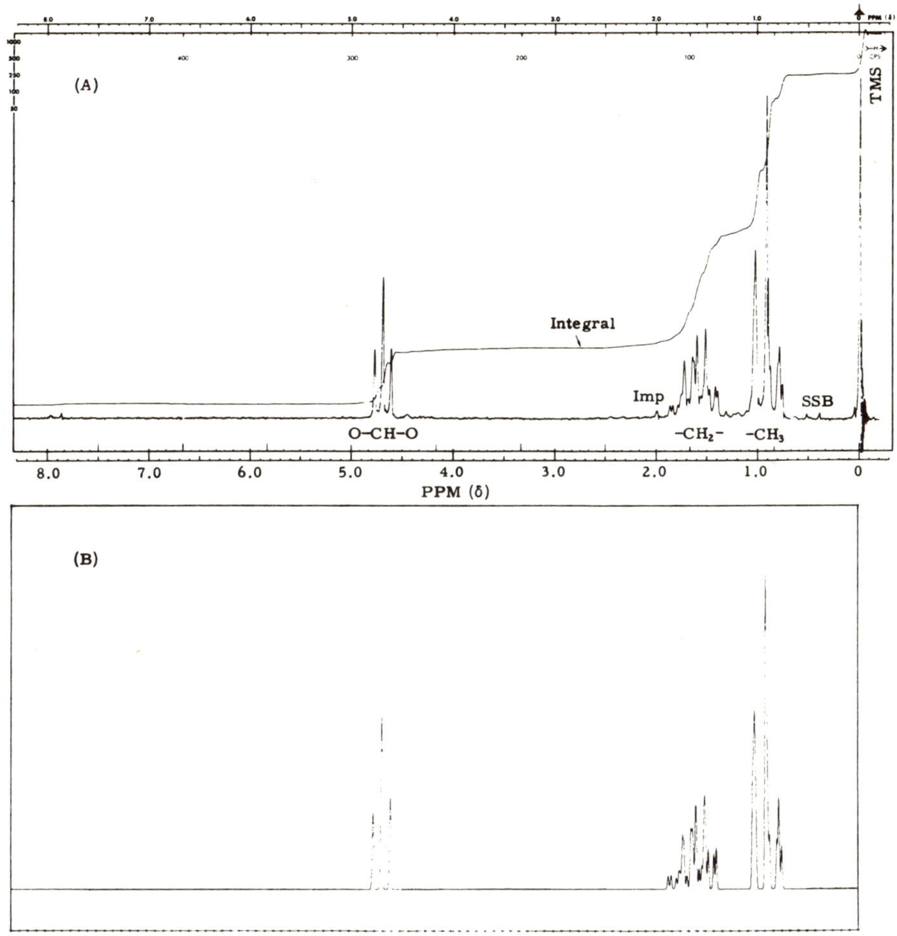

FIG. 6. NMR spectrum of propionaldehyde cyclic trimer. (A) Observed; (B) calculated.

$$\delta_1 = \delta_2 = \delta_3 = 0.91 \text{ ppm},$$
$$\delta_4 = \delta_5 = 1.61 \text{ ppm},$$
$$\delta_6 = 4.68 \text{ ppm},$$
$$J_{14} = J_{15} = J_{24} = J_{25} = J_{34} = J_{35} = 7.55 \pm 0.27 \text{ cps},$$
$$J_{16} = J_{26} = J_{36} = -0.05 \pm 0.21 \text{ cps},$$
$$J_{46} = J_{56} = 5.15 \pm 0.28 \text{ cps}.$$

The rather large estimated errors in the coupling constants are a consequence of the large number of overlapping lines in the observed spectrum. The computed spectrum assuming a Gaussian line shape is shown in Fig. 6B. Because this computed spectrum is in such excellent agreement with the observed spectrum (Fig. 6A), the three ethyl groups must be equivalent, as assumed. Therefore, this sample consists solely of a *cis-cis* isomer and also has the structure of Fig. 1A. A more complex spectrum would be expected for the *cis-trans* isomer.

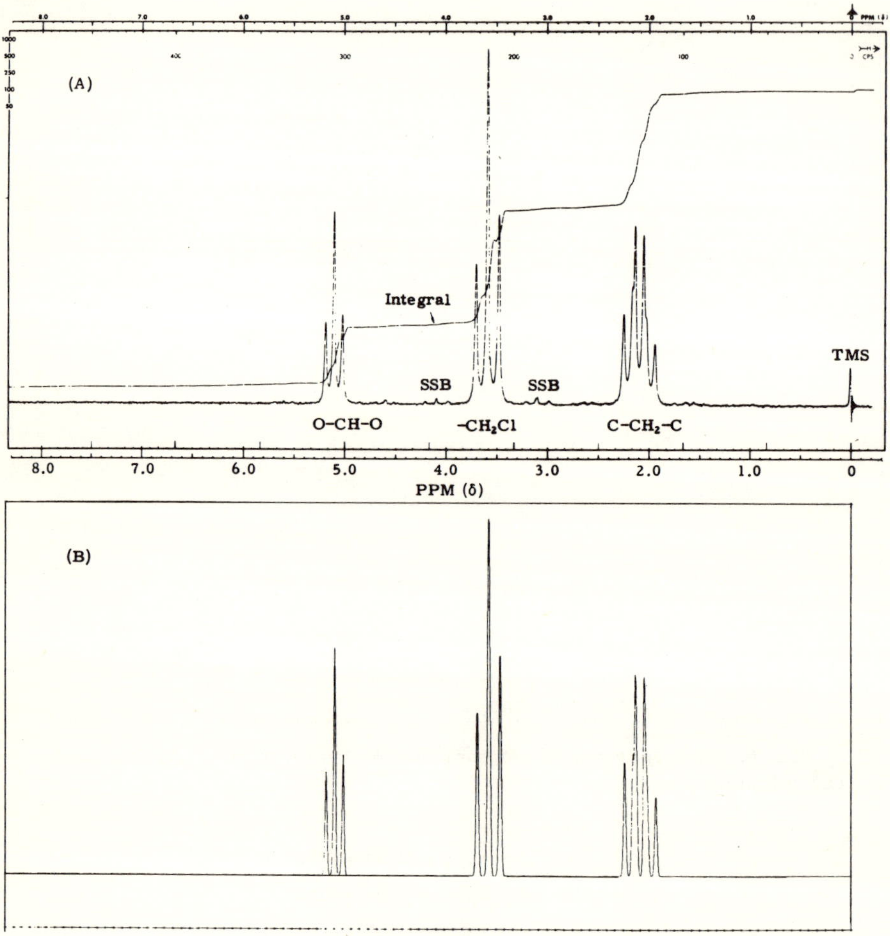

FIG. 7. NMR spectrum of β-chloropropionaldehyde cyclic trimer. (A) Observed; (B) calculated.

β-Chloropropionaldehyde Cyclic Trimer

The NMR spectrum of 2,4,6-tris-2-chloroethyl-1,3,5-trioxane (V) is shown in Fig. 7A. This spectrum was analyzed as that of a five-spin

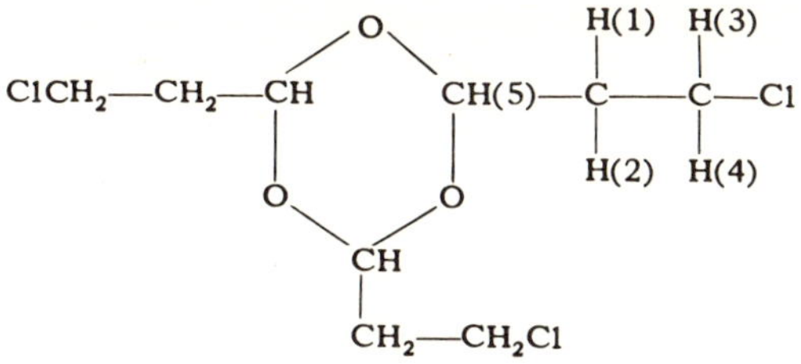

(V)

system by the iterative method, assuming all three of the C_3H_5Cl groups to be equivalent. The computed chemical shifts and coupling constants are as follows:

$$\delta_1 = \delta_2 = 2.09 \text{ ppm},$$
$$\delta_3 = \delta_4 = 3.57 \text{ ppm},$$
$$\delta_5 = 5.08 \text{ ppm},$$
$$J_{13} = J_{14} = J_{23} = J_{24} = 6.72 \pm 0.06 \text{ cps},$$
$$J_{15} = J_{25} = 5.00 \pm 0.09 \text{ cps},$$
$$J_{35} = J_{45} = -0.02 \pm 0.14 \text{ cps}.$$

The computed spectrum is shown in Fig. 7B. As this is also in excellent agreement with the observed spectrum (Fig. 7A), the three $-CH_2-CH_2Cl$ groups must all be equivalent and undoubtedly are equatorial.

β-Cyanopropionaldehyde Cyclic Trimer

The NMR spectrum of 2,4,6-tris-2-cyanoethyl-1,3,5-trioxane (VI) is shown in Fig. 8A. This spectrum was analyzed as a five-spin system by

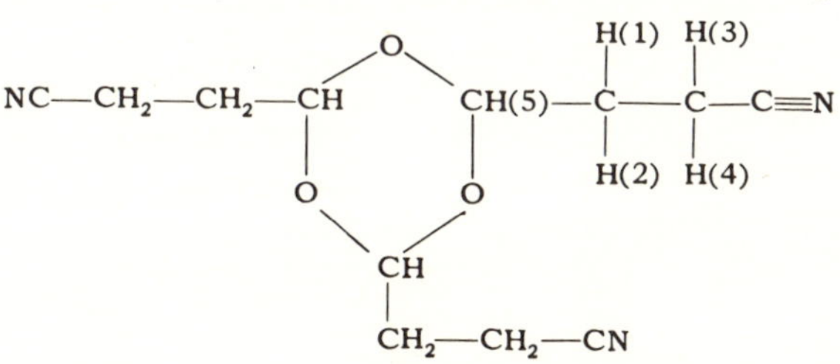

(VI)

the iterative method, assuming all three of the C_3H_5CN groups to be equivalent. The computed chemical shifts and coupling constants are as follows:

$$\delta_1 = \delta_2 = 2.05 \text{ ppm},$$
$$\delta_3 = \delta_4 = 2.54 \text{ ppm},$$
$$\delta_5 = 5.15 \text{ ppm},$$

$$J_{13} = J_{14} = J_{23} = J_{24} = 7.27 \pm 0.05 \text{ cps},$$
$$J_{15} = J_{25} = 4.46 \pm 0.07 \text{ cps},$$
$$J_{35} = J_{45} = -0.13 \pm 0.10 \text{ cps}.$$

The computed spectrum is shown in Fig. 8B. This is in very good agreement with the observed spectrum (Fig. 8A), so that the assumption that the $-CH_2-CH_2CN$ groups are all equivalent is verified. The sample is therefore a *cis-cis* isomer with equatorial $-CH_2-CH_2CN$ groups, as would be expected since it was prepared from the *cis-cis* isomer of β-chloropropionaldehyde cyclic trimer (V).

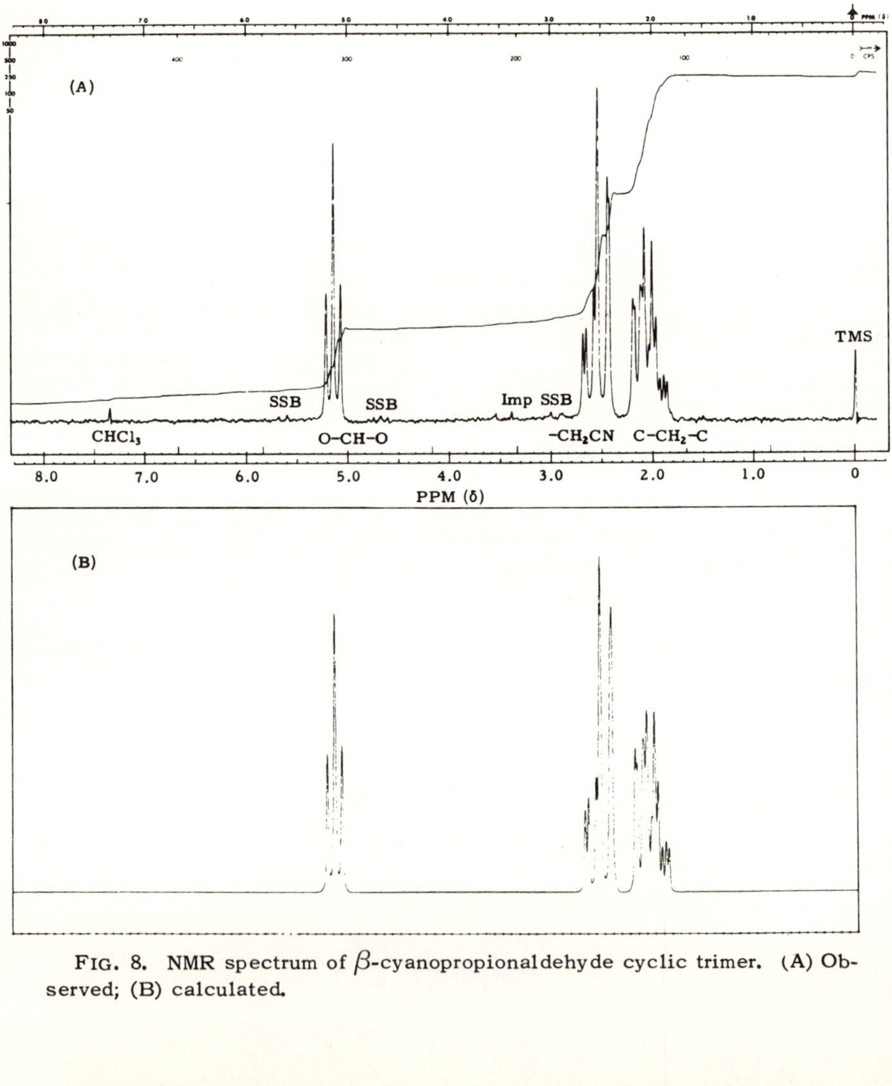

FIG. 8. NMR spectrum of β-cyanopropionaldehyde cyclic trimer. (A) Observed; (B) calculated.

Metacrolein (Acrolein Cyclic Trimer)

The NMR spectra of 2,4,6-trivinyl-1,3,5-trioxane (VII) dissolved in pyridine, in CCl_4, and in $CDCl_3$ are shown in Figs. 9, 10, and 11, respec-

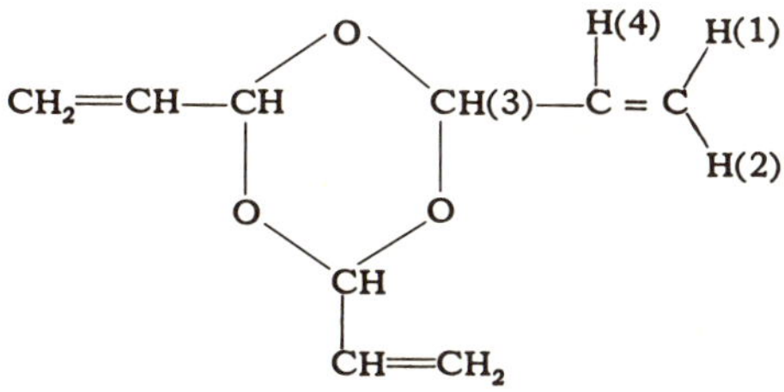

(VII)

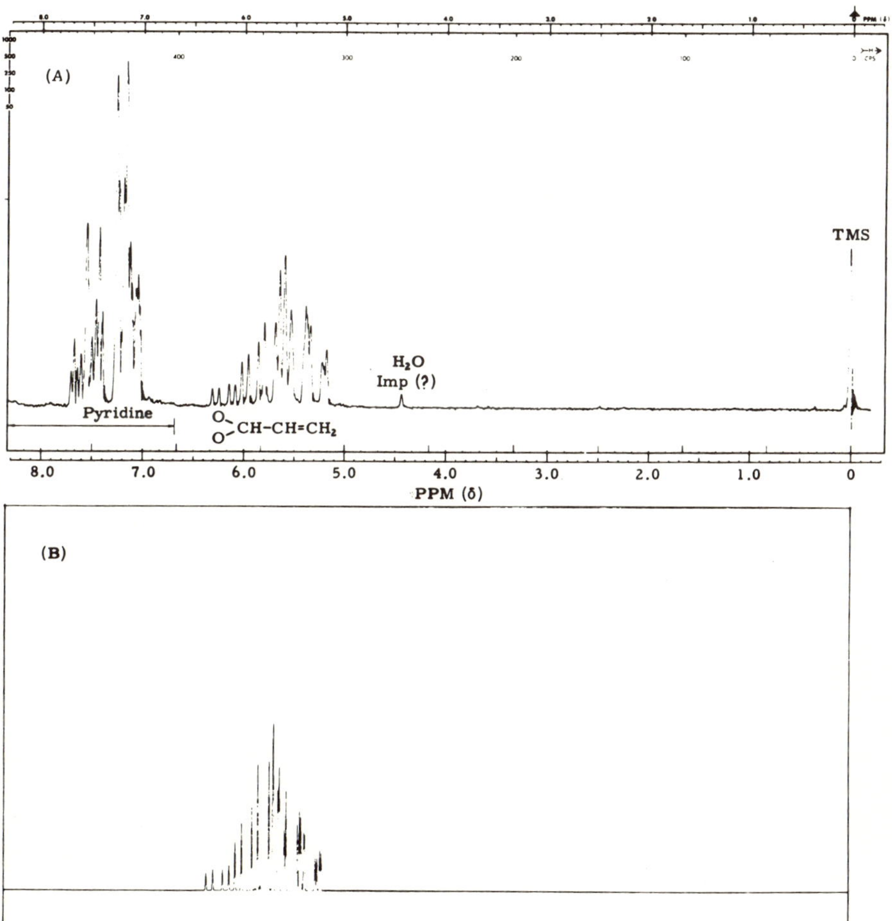

FIG. 9. NMR spectrum of metacrolein (acrolein cyclic trimer) dissolved in pyridine.

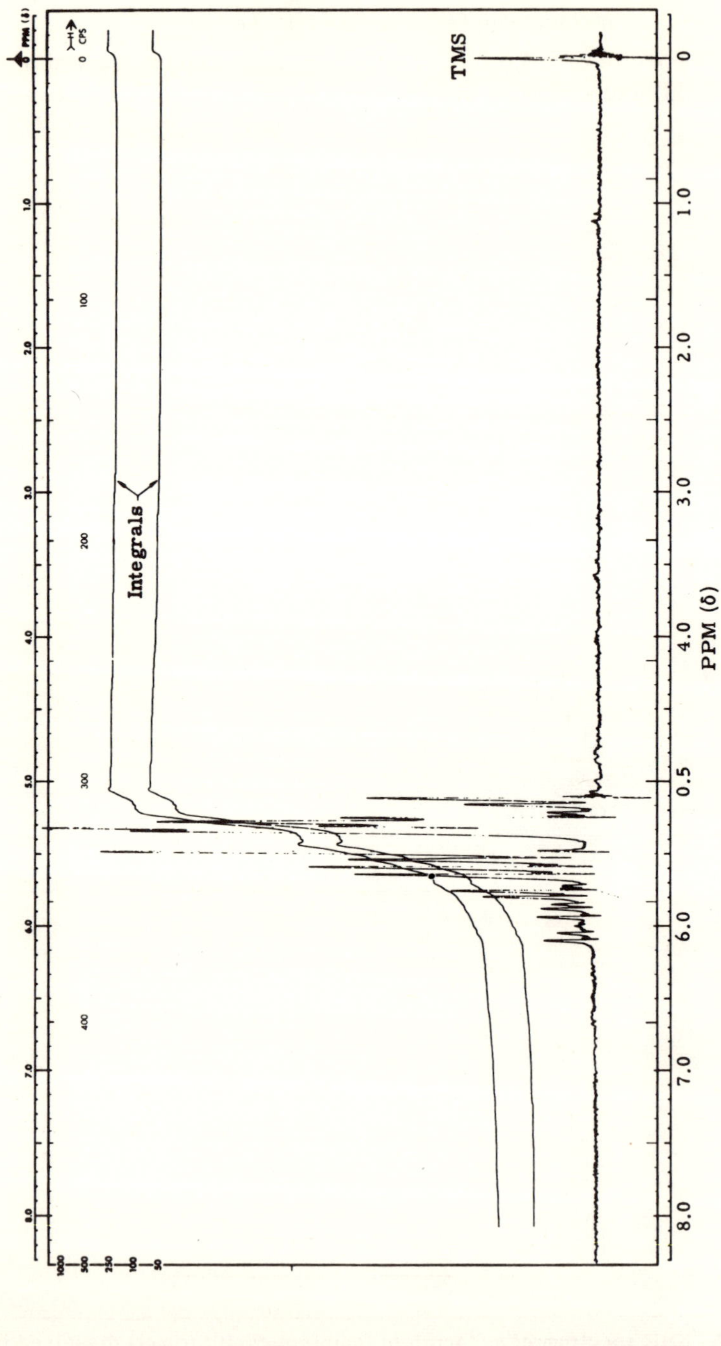

FIG. 10. NMR spectrum of metacrolein (acrolein cyclic trimer) dissolved in carbon tetrachloride.

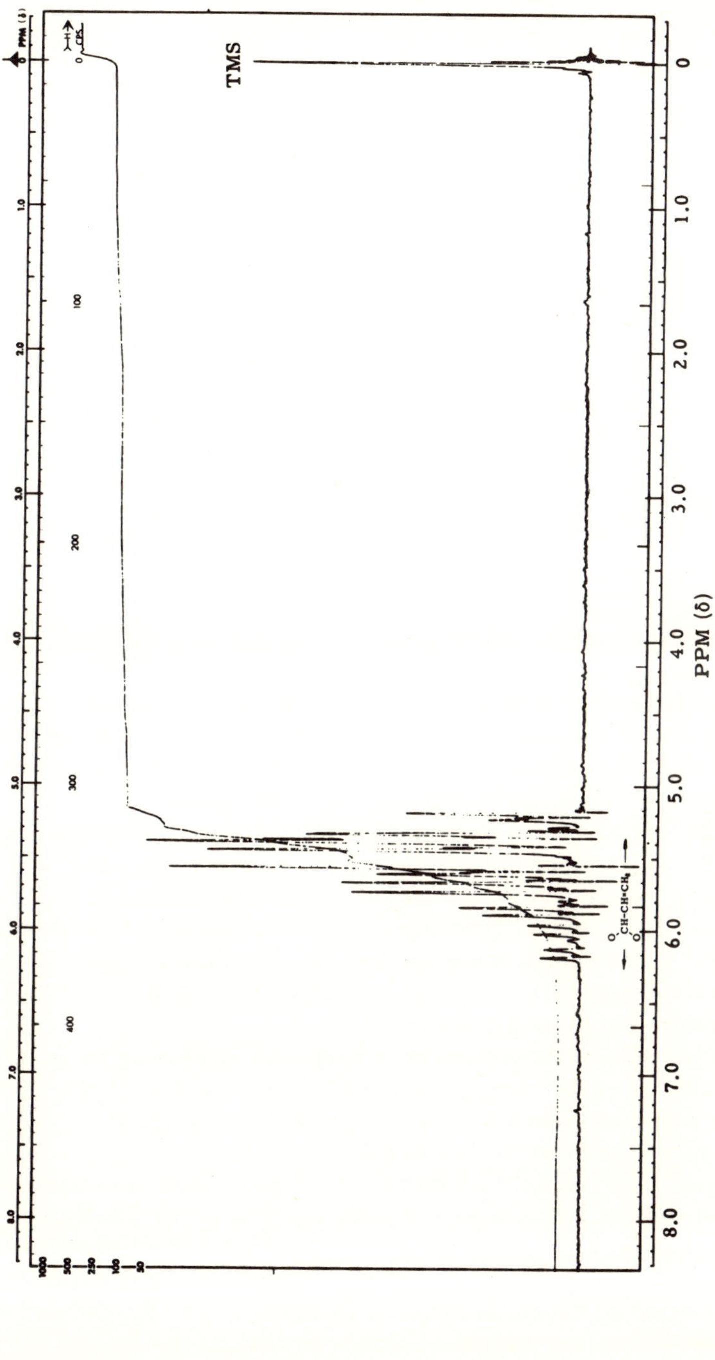

FIG. 11. NMR spectrum of metacrolein (acrolein cyclic trimer) dissolved in deuterochloroform.

tively. These spectra are quite complex and do not yield readily to analysis. Many attempts were made, with only partial success, to analyze the spectrum obtained in pyridine solution since this is the simplest one. However, the 100-Mc/sec spectrum (*15*) of this solution shown in Fig. 12A can be analyzed readily. Two sets of parameters, I and II (Table I), were obtained by the iterative technique, depending on whether

TABLE I

	Set I	Set II
δ_1	= 5.28 ppm	5.28 ppm
δ_2	= 5.56 ppm	5.56 ppm
δ_3	= 5.57 ppm	5.57 ppm
δ_4	= 6.04 ppm	6.04 ppm
J_{12}	= 1.55 ± 0.11 cps	−2.40 ± 0.12 cps
J_{13}	= −1.06 ± 0.14 cps	−1.06 ± 0.13 cps
J_{14}	= 10.79 ± 0.15 cps	10.63 ± 0.15 cps
J_{23}	= −1.18 ± 0.15 cps	−1.19 ± 0.15 cps
J_{24}	= 17.67 ± 0.24 cps	17.90 ± 0.22 cps
J_{34}	= 4.35 ± 0.24 cps	4.35 ± 0.21 cps

the geminal coupling J_{12} was assumed to be positive or negative, respectively. Although set II fits the observed frequencies as well as set I does, the intensities in the computed spectrum (Figs. 12B and 12C) obtained with set I are in better agreement with those observed (note the intensity of the lines indicated by arrows in Fig. 12). Hence, it is concluded that set I with J_{gem} ($\equiv J_{12}$) positive is the correct set.

Incidentally, the above analysis illustrates that in order to determine whether the sign of a coupling constant is positive or negative, it is *not* sufficient to merely reverse the sign of the constant in question and to recalculate the spectrum. The spectrum must be reassigned and iterations again carried out to reach the last set of parameters corresponding to the changed sign. The importance of doing this seems not to be fully appreciated in the literature (*16*).

These two sets of parameters were next used as starting parameters for the analysis of the corresponding 60-Mc/sec spectrum (Figs. 9 and 13A). In each case, the analysis now proceeded without any difficulty to give the sets of parameters in Table II.

The spectra calculated with these sets of parameters are shown in Figs. 13B and 13C. It is even more obvious than in the 100-Mc/sec spectrum that set I, i.e., with J_{gem} ($\equiv J_{12}$) positive is the correct set for this molecule.

A comparison of the parameter sets determined at 60 Mc/sec and at 100 Mc/sec shows some apparent discrepancies. In the case of set II at

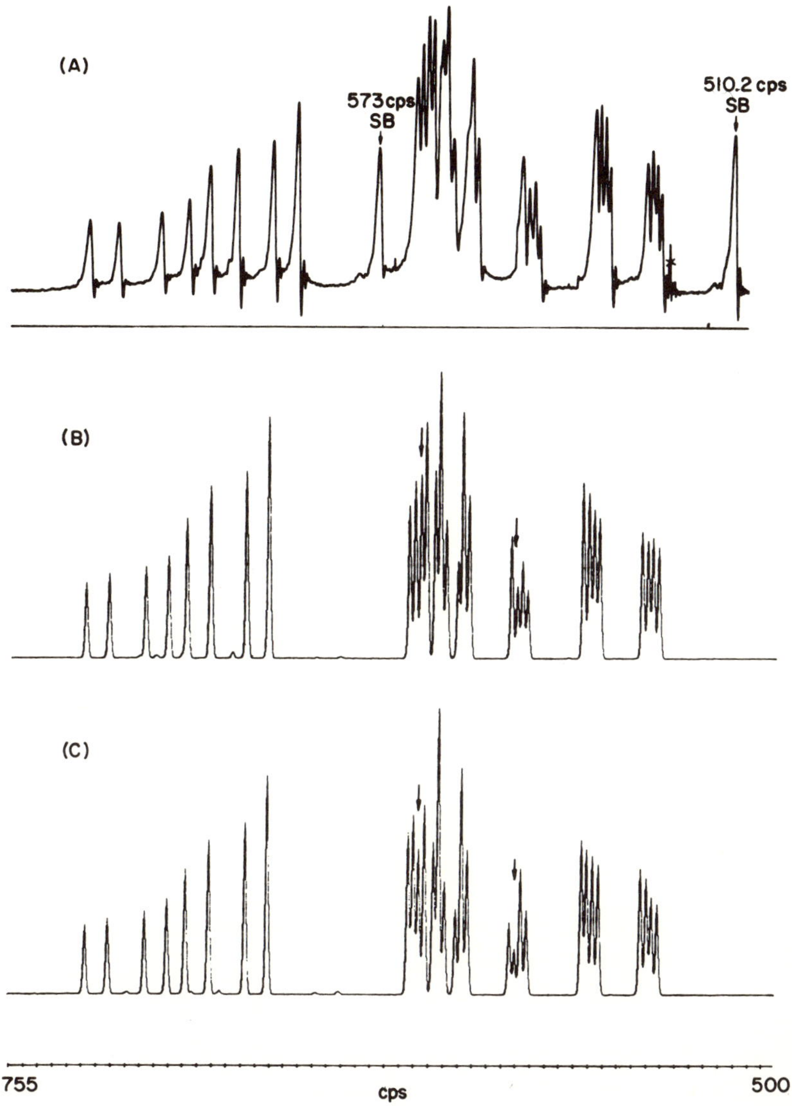

FIG. 12. 100-Mc/sec NMR spectrum of metacrolein (acrolein cyclic trimer) dissolved in pyridine. (A) Observed; (B) calculated with J_{gem} positive; (C) calculated with J_{gem} negative.

least some of the discrepancy can be accounted for by the fact that it is an incorrect set. For the correct set I, the discrepancies in the chemical shifts are certainly due to the fact that the measurements were made on two different solutions and probably at different concentrations. The

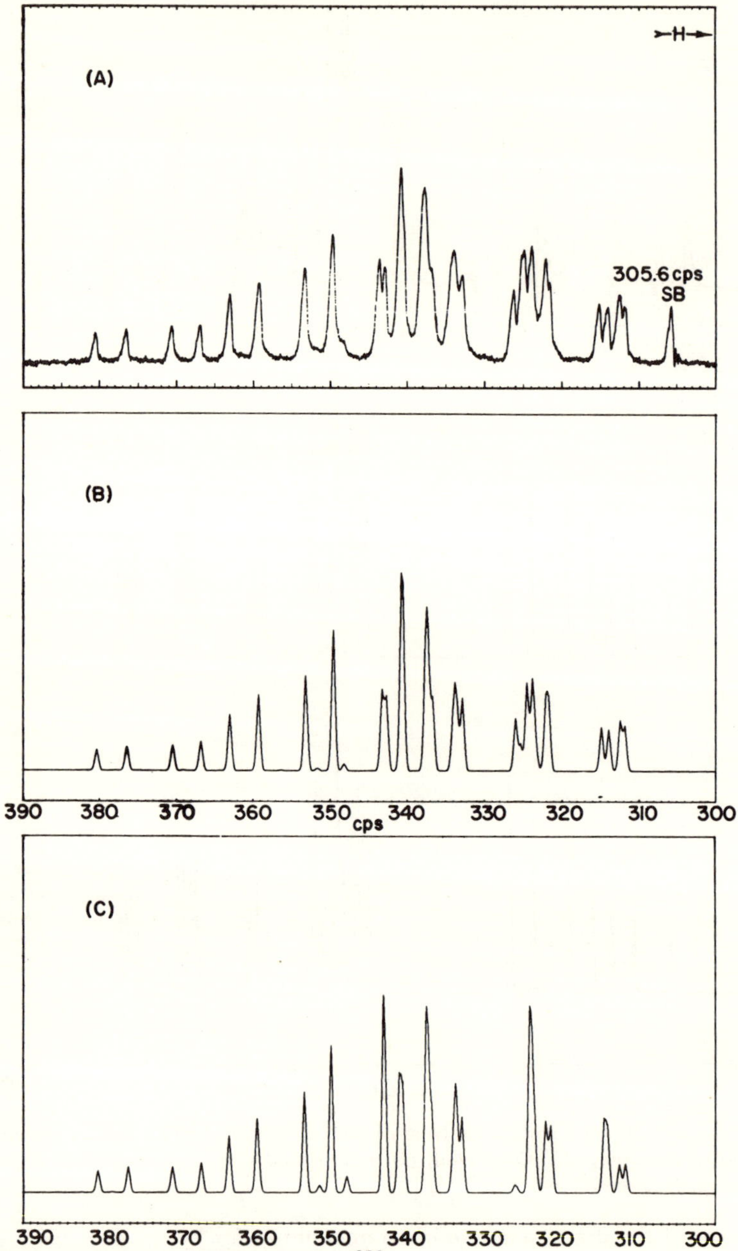

FIG. 13. 60-Mc/sec NMR spectrum of metacrolein (acrolein cyclic trimer) on an expanded scale. (A) Observed; (B) calculated with J_{gem} positive; (C) calculated with J_{gem} negative.

TABLE II

Set I	Set II
$\delta_1 = 5.32$ ppm	5.32 ppm
$\delta_2 = 5.60$ ppm	5.60 ppm
$\delta_3 = 5.59$ ppm	5.59 ppm
$\delta_4 = 6.02$ ppm	6.02 ppm
$J_{12} = 1.21 \pm 0.13$ cps	-3.41 ± 0.19 cps
$J_{13} = -0.85 \pm 0.11$ cps	-0.74 ± 0.18 cps
$J_{14} = 10.70 \pm 0.16$ cps	9.86 ± 0.22 cps
$J_{23} = -0.89 \pm 0.12$ cps	-0.86 ± 0.20 cps
$J_{24} = 17.34 \pm 0.23$ cps	17.94 ± 0.35 cps
$J_{34} = 4.42 \pm 0.22$ cps	4.42 ± 0.33 cps

agreement for the coupling constants is probably within their rather large standard (rms) errors. Hence, the following will be taken as the best values for these constants.

$$J_{12} = 1.4 \pm 0.1 \text{ cps}$$
$$J_{13} = -1.0 \pm 0.1 \text{ cps}$$
$$J_{14} = 10.7 \pm 0.1 \text{ cps}$$
$$J_{23} = -1.0 \pm 0.1 \text{ cps}$$
$$J_{24} = 17.5 \pm 0.2 \text{ cps}$$
$$J_{34} = 4.4 \pm 0.2 \text{ cps}$$

These results show that the three vinyl groups must be equivalent. Hence, this molecule is also a *cis-cis* isomer with equatorial $-CH{=}CH_2$ groups, as expected from its preparation from the *cis-cis* isomer of β-chloropropionaldehyde cyclic trimer.

Comparison of Chemical Shifts and Spin-Spin Coupling Constants

With the general formula

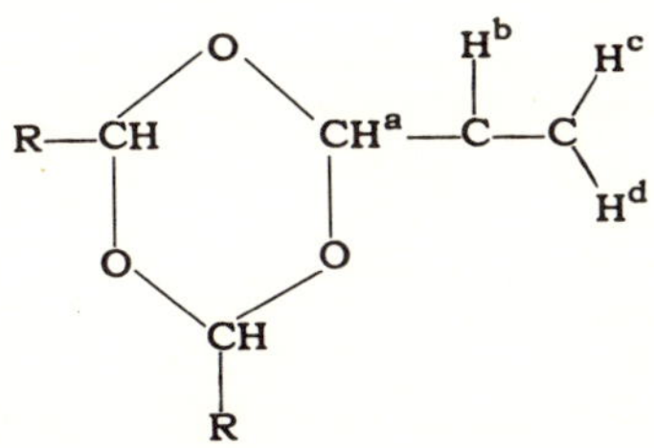

as reference, the chemical shifts and coupling constants of the seven compounds studied can be compared. Table III shows the chemical shifts of the various types of hydrogen. An interpretation of these data

TABLE III

CHEMICAL SHIFTS OF PROTONS IN SUBSTITUTED TRIOXANES[a]

R Group	Geometrical isomer	δ, ppm from TMS			
		H^a	H^b	H^c	H^d
$—CH_3$	*cis-cis*	4.92 (axial)	1.26 (equat.)	–	–
	cis-trans				
	Axial	5.21	1.39	–	–
	Equat.	5.28	1.24	–	–
$—CHCl_2$	*cis-cis*	5.64	5.16	–	–
$—CH_2Cl$	*cis-cis*	5.05	3.52	–	–
$—CH_2—CH_3$	*cis-cis*	4.68	1.61	0.91	0.91
$—CH_2—CH_2Cl$	*cis-cis*	5.08	2.09	3.57	3.57
$—CH_2—CH_2CN$	*cis-cis*	5.15[b]	2.05[b]	2.54[b]	2.54[b]
$—CH{=}CH_2$	*cis-cis*	5.59[c]	6.02[c]	5.32[c]	5.60[c]

[a]See structural formula in text for identification of protons. All samples dissolved in CCl_4 except where noted.

[b]Solvent = $CDCl_3$.

[c]Solvent = pyridine.

has been given, when significant, in previous sections of this paper. A summary of the spin-spin coupling constants is given in Table IV. It should be noted that, in the data presented here, the sign of each vicinal coupling constant, J_{ab}, has been taken to be positive so as to be in accord with the most recent theoretical and experimental work (*17*). The remaining constants then have the signs shown in the table.

All the vicinal coupling constants to the O—CH—O proton of the ring (J_{ab}) are somewhat smaller in magnitude than those normally found (6 to

TABLE IV

SPIN-SPIN COUPLING CONSTANTS OF PROTONS IN SUBSTITUTED TRIOXANES[a]

R Group	Geometrical isomer	Spin-spin coupling constant, cps					
		J_{ab}	J_{ac}	J_{ad}	J_{bc}	J_{bd}	J_{cd}
$—CH_3$	*cis-cis*	5.1	–	–	–	–	–
	cis-trans	{5.0	–	–	–	–	–
		5.8	–	–	–	–	–
$—CHCl_2$	*cis-cis*	4.8	–	–	–	–	–
$—CH_2Cl$	*cis-cis*	4.77	–	–	–	–	–
$—CH_2—CH_3$	*cis-cis*	5.15	−0.05	−0.05	7.55	7.55	–
$—CH_2—CH_2Cl$	*cis-cis*	5.00	−0.02	−0.02	6.72	6.72	–
$—CH_2—CH_2CN$	*cis-cis*	4.42	−0.10	−0.10	7.37	7.37	–
$—CH{=}CH_2$	*cis-cis*	4.4	−1.0	−1.0	10.7 (*cis*)	17.5 (*trans*)	1.4

[a]See structural formula in text for identification of protons.

8 cps) when both protons occur in groups of a chain (e.g., compare J_{ab} with J_{bc} and J_{bd} in Table II). The relative magnitudes of the various J_{ab} couplings substantiate the earlier conclusion that our *cis-trans* isomer of paraldehyde is the one with one axial CH_3 group. The J_{ab} values for equatorial R groups and axial ring hydrogens range from about 4.4 to 5.2 cps, whereas $J_{ab} = 5.8$ cps for the axial CH_3 group and the equatorial ring hydrogen. The signs and magnitudes of the remaining coupling constants are within the range usually found for such constants.

Appendix. The Analysis of the NMR Spectra of A_2B-Type Molecules by Means of Trace Relations

The Hamiltonian matrix for an A_2B type of molecules is given in Table 6-6 of Ref. (*14*). Because of the molecular symmetry this matrix is composed of two independent ones corresponding to the symmetric and antisymmetric operations describing this symmetry. The diagonal elements of the various submatrices together with their traces are given in Table V.

TABLE V[a]

F_z	*Hnn*	*Traces*
Symmetric matrix		
$+\frac{3}{2}$	$2h_A + h_B + \frac{1}{4}J_{AB} + \frac{1}{4}J'$	$2h_A + h_B + \frac{1}{4}J_{AB} + \frac{1}{4}J'$
$+\frac{1}{2}$	$\left\{\begin{array}{l} h_B + \frac{1}{4}J' \\ 2h_A - h_B - \frac{1}{4}J_{AB} + \frac{1}{4}J' \end{array}\right\}$	$2h_A - \frac{1}{4}J_{AB} + \frac{1}{2}J'$
$-\frac{1}{2}$	$\left\{\begin{array}{l} -2h_A + h_B - \frac{1}{4}J_{AB} + \frac{1}{4}J' \\ -h_B + \frac{1}{4}J' \end{array}\right\}$	$-2h_A - \frac{1}{4}J_{AB} + \frac{1}{2}J'$
$-\frac{3}{2}$	$-2h_A - 2h_B + \frac{1}{4}J_{AB} + \frac{1}{4}J'$	$-2h_A - 2h_B + \frac{1}{4}J_{AB} + \frac{1}{4}J'$
Antisymmetric matrix		
$+\frac{1}{2}$	$h_B - \frac{3}{4}J'$	$h_B - \frac{3}{4}J'$
$-\frac{1}{2}$	$-h_B - \frac{3}{4}J'$	$-h_B - \frac{3}{4}J'$

[a]In this table $h_i = (\gamma_i H_0/4\pi)(1 - \delta_i)$ and $J' = J_{AA}$.

A transition between symmetric spin states and antisymmetric ones is strictly forbidden. Otherwise, the selection rule that normally applies is $\Delta F_z = \pm 1$. It will be noted that the coupling constant J' between the two magnetically equivalent A nuclei will not appear in any transition between spin states and hence is experimentally inaccessible. Hence, we set $J' = 0$ for simplicity. In this case the traces become those shown in Table VI.

The nuclear spin energy levels are obtained as the eigenvalues of the Hamiltonian matrix, and these in turn are obtained by diagonalization of

TABLE VI

	Traces	
F_z	Symmetric matrix	Antisymmetric matrix
$\frac{3}{2}$	$2h_A + h_B + \frac{1}{4} J_{AB}$	
$\frac{1}{2}$	$2h_A \quad - \frac{1}{4} J_{AB}$	h_B
$-\frac{1}{2}$	$-2h_A \quad - \frac{1}{4} J_{AB}$	$-h_B$
$-\frac{3}{2}$	$-2h_A - h_B + \frac{1}{4} J_{AB}$	
	Trace = 0	Trace = 0

this matrix. In such a transformation, it is well known from matrix algebra that the trace of a matrix remains invariant. Therefore, we can equate the traces of the various submatrices given above to the sums of corresponding energy levels. Numerical values for the latter are obtained from the assigned spectrum.

The energy level scheme for an A_2B-type molecule together with the labeling of the eight possible transitions is represented conveniently by Fig. 14 [see Ref. (*14*)]. The order of the numbers of the transitions correspond, in a first-order spectrum ($J_{AB} \ll h_B - h_A$), to increasing fre-

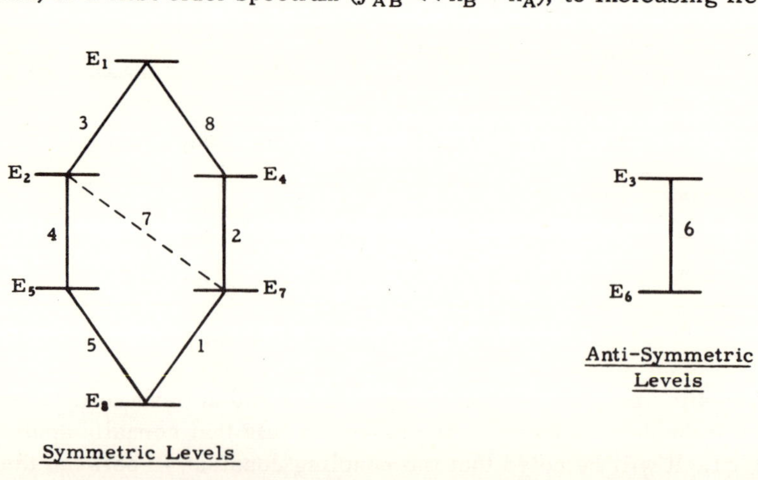

FIG. 14. Labeling of energy levels and transitions for an A_2B-type molecule.

TABLE VII.

F_z	Symmetric matrix	Antisymmetric matrix
$\frac{3}{2}$	$2h_A + h_B + \frac{1}{4} J_{AB} = E_1$	
$\frac{1}{2}$	$2h_A \quad - \frac{1}{4} J_{AB} = E_2 + E_4$	$h_B = E_3$
$-\frac{1}{2}$	$-2h_A \quad - \frac{1}{4} J_{AB} = E_5 + E_7$	$-h_B = E_6$
$-\frac{3}{2}$	$-2h_A - h_B + \frac{1}{4} J_{AB} = E_8$	

quency (i.e., energy) of the transition. Hence, we write the trace relations as Table VII.

The chemical shift $2h_B$ for the B nucleus is obtained directly as $E_3 - E_6$. The remaining relations can be combined in an obvious way to yield the other parameters.

The application of these relations to the analysis of the NMR spectrum of the cyclic trimer of chloroacetaldehyde follows.

It is first necessary to assign the observed lines to the pairs of energy levels from which they arise. This assignment will be different for a positive sign of J_{AB} from what it is for a negative sign. The correct assignment can be made unequivocably by means of either the "tickling" form or the nuclear Overhauser form of double resonance. In the absence of facilities for performing such experiments we have assumed that J_{AB} is positive in agreement with the findings for the signs of vicinal coupling constants in other molecules (*17*). The assignment is shown in Table VIII. From this assignment, one obtains immediately that $E_3 = +151.5$ and $E_6 = -151.5$. The remaining seven transitions together with

TABLE VIII

Transition no.	E Level origin	Obs. frequency
1 2	7-8 4-7	209.0 cps
3 4	1-2 2-5	213.7 cps
5	5-8	298.5 cps
6 7	3-6 2-7	303.0 cps
8	1-4	308.1 cps

the additional requirement that $E_1 + E_2 + E_4 + E_5 + E_7 + E_8 = 0$ are more than sufficient to determine the remaining six levels. This is done conveniently with the computer program NMREN2; but can also be done by an equivalent method without the aid of a computer (*9*). The values found with NMREN2, together with their standard errors are as follows:

$$\begin{array}{ll} E_1 & 365.35 \pm 0.10 \text{ cps}, \\ E_2 & 151.56 \pm 0.08 \text{ cps}, \\ E_4 & 57.35 \pm 0.10 \text{ cps}, \\ E_5 & -62.11 \pm 0.10 \text{ cps}, \\ E_7 & -151.56 \pm 0.08 \text{ cps}, \\ E_8 & -360.59 \pm 0.10 \text{ cps}. \end{array}$$

The trace relations given earlier now yield the following values for the parameters:

$$2h_A = 211.36 \text{ cps}, \qquad \delta_A = 3.523 \text{ ppm},$$
$$2h_B = 303.18 \text{ cps}, \qquad \delta_B = 5.053 \text{ ppm},$$
$$J_{AB} = 4.77 \text{ cps}.$$

These values are in good agreement with those determined by the iterative method and quoted in the body of this paper. The errors in the parameters were determined with the ERROR subroutine of the NMRIT computer program (*9, 16*). For this purpose the errors in E_3 and E_6 were assumed to be ± 0.10 cps and the errors quoted above were used for the remaining levels.

Acknowledgments

The authors are greatly indebted to E. R. Bell, E. Bergman, L. H. Gale, and G. A. Kurhajec of this laboratory for supplying most of the compounds studied.

References

1. N. R. Moerman, *Rec. Trav. Chim.* **56**, 161 (1937).
2. D. A. Ramsay, *Trans. Farad. Soc.* **44**, 289 (1948).
3. A. Novak and E. Whalley, *Can. J. Chem.* **36**, 1116 (1958).
4. N. L. Allinger and L. A. Freiberg, *J. Am. Chem. Soc.* **82**, 2393 (1960).
5. W. S. Johnson, J. Bauer, J. L. Margrave, M. Frisch, L. Dreger, and W. Hubbard, *J. Am. Chem. Soc.* **83**, 606 (1961).
6. F. R. Jensen, D. S. Noyce, C. Sederholm, and A. Berlin, *J. Am. Chem. Soc.* **82**, 1256 (1960); R. H. Harris and N. Sheppard, *Proc. Chem. Soc.*, p. 418 (1961).
7. P. Diehl and R. Freeman, *Mol. Phys.* **4**, 39 (1961).
8. E. C. Craven, H. Jowitt, and W. R. Ward, *J. Appl. Chem.* (*London*) **12**, 526 (1962).
9. J. D. Swalen and C. A. Reilly, *J. Chem. Phys.* **37**, 21 (1962).
10. M. R. Freymann, M. Freymann, M. Koechlin, M. Martin, and M. G. Mavel, *Arch. Sci.* (*Geneva*) **12**, Fasc. Spec., 207 (1959).

11. M. Martin and J. Cantacuzene, *Compt. Rend.* **249**, 1216 (1959).
12. L. M. Jackman, "Applications of Nuclear Magnetic Resonance Spectroscopy in Organic Chemistry," p. 116. Pergamon, New York, 1959.
13. R. U. Lemieux, R. K. Kullnig, H. J. Bernstein, and W. G. Schneider, *J. Am. Chem. Soc.* **80**, 6098 (1958).
14. J. A. Pople, W. G. Schneider, and H. J. Bernstein, "High-Resolution Nuclear Magnetic Resonance," p. 119. McGraw-Hill, New York, 1959.
15. N. S. Bhacca, Varian Associates, Palo Alto, California.
16. R. Freeman, N. S. Bhacca, and C. A. Reilly, *J. Chem. Phys.* **38**, 293 (1963).
17. P. C. Lauterbur and R. J. Kurland, *J. Am. Chem. Soc.* **84**, 3405 (1962).

The Determination of the Coupling Constants of the Rotational Isomers of 1-Chloro-2-bromoethane from the Solvent Dependence of the Spectrum

R. J. ABRAHAM and L. CAVALLI[1]

The Robert Robinson Laboratories, The University of Liverpool, Liverpool, England,

and

K. G. R. PACHLER

The Chemical Physics Group, South African Council for Scientific and Industrial Research, Pretoria, Republic of South Africa

Introduction

The determination of the coupling constants of the individual rotational isomers of a substituted ethane has been of considerable importance in the development of NMR spectroscopy, both as a test of the theory of coupling constants (*1, 2*), and from the application of the results to more complex (e.g., cyclic) systems (*3*). In some compounds these couplings can be obtained directly from the spectra at such low temperatures that the rate of interconversion of the rotamers is negligibly small (*4*). Usually it is not feasible to achieve sufficiently low temperatures, and thus these couplings have been determined in most cases by analysis of the J versus temperature curve at high temperatures. The observed coupling is now the weighted mean of the couplings of the rotamers and from the analysis of the curve the unknown J's and the energy differences between the rotamers can be obtained (*5*).

We present here a new method of determining these coupling constants by the analysis of the change in the coupling constant with solvent curve. Such changes with solvent have been used quite often, but no detailed analysis of the solvent dependence given. A considerable advantage of this method over the variable-temperature method is that much greater changes in the couplings can be obtained; e.g., in 1-chloro-2-bromoethane, N and L change from 16.55 and -5.43 cps in CCl_4 solution to 13.80 and -1.09 cps in formamide, whilst the corresponding values (in the pure liquid) at -10 and 140°C are 15.12 and -3.15 cps and 14.78 and -2.75 cps, respectively (Figs. 1 and 2). Thus the experimental error of

[1] *Present address:* Laboratori Ricerche di Bollate, Societa Edison, Milan, Italy.

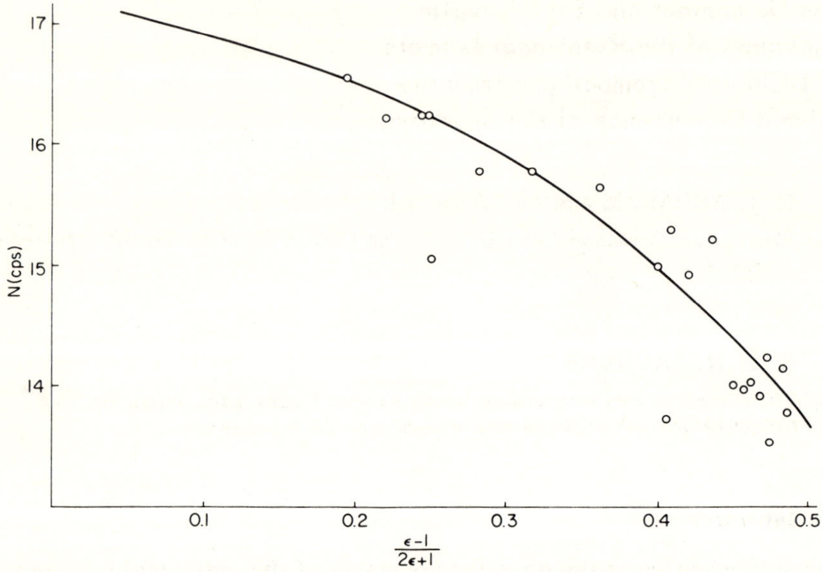

FIG. 1. The value of $N(J + J')$ for 1-chloro-2-bromoethane in various solvents *versus* the dielectric constant of the medium.

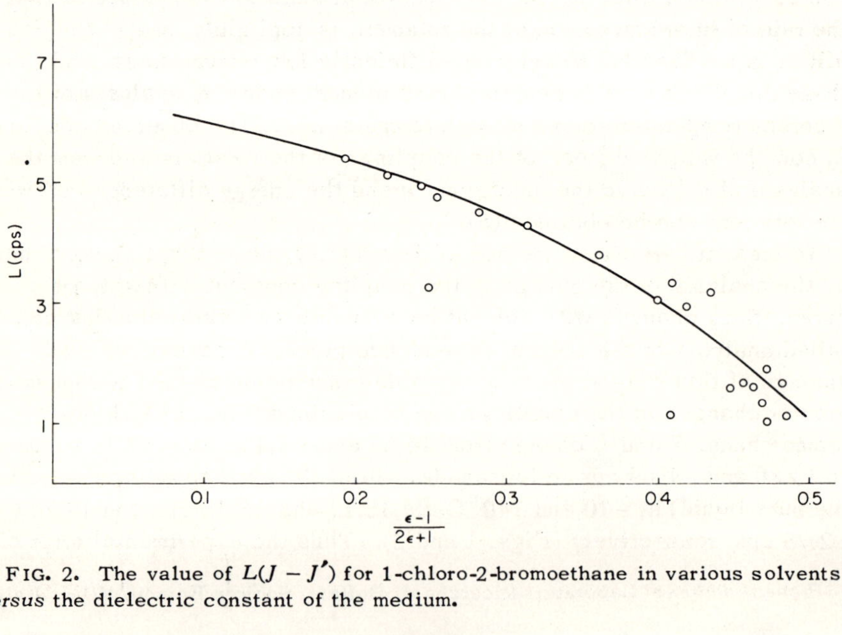

FIG. 2. The value of $L(J - J')$ for 1-chloro-2-bromoethane in various solvents *versus* the dielectric constant of the medium.

measurement (ca. ±0.05 cps) is a much smaller fraction of the total range. This is compensated to some extent by the unavoidable uncertainties in the solvent dependence due to specific solvent-solute interactions which cannot be evaluated in the theory. These result in some scatter of the J versus solvent curve. The other advantage of the solvent method is the very small change in the coupling constants of the rotamers with solvent. This is so small as to be less than the experimental error of measurement for CH · CH, CH · CF, and CF · CF coupling constants. However, there is a change in the coupling constants of the rotamers with temperature. This is very small for CH · CH couplings but is appreciable for CF · CF couplings (*6*). Indeed the results of the analyses of J_{FF} versus temperature curves are now known to be very seriously in error owing to this and other factors (*6, 7*). Thus the solvent method may be of considerable use in this field.

1-Chloro-2-bromoethane was chosen as a "test case" for the use of this method, as most of the relevant physical data are known for this compound and it gives easily obtainable and analyzable A_2B_2-type spectra (*8*) in dilute solution. We present here a brief summary of the relevant theory and some preliminary results which illustrate the agreement with the experimental measurements. The comprehensive theoretical treatment and the detailed fit with experiment (including the results of the variable temperature measurements) will be given later.

Theory

The observed value of $N(J + J')$ in 1-chloro-2-bromoethane in any solvent (s) is the weighted mean of the couplings N_g and N_t of the *gauche* and *trans* rotamers (Fig. 1); i.e.,

$$N = n_g N_g + n_t N_t \tag{1}$$

where n_g and n_t are the ratios of the *gauche* and *trans* isomers in that solvent and are given by

$$n_g + n_t = 1$$

and

$$n_g/n_t = 2\exp(-\Delta E^s/RT) \tag{2}$$

(there are two equivalent *gauche* forms), where $\Delta E^s = E_g^s - E_t^s$ is the excess energy of the *gauche* rotamer in that solvent.

Eliminating n_g and n_t gives

$$N = \{2N_g + N_t \exp(\Delta E^s/RT)\}/\{2 + \exp(\Delta E^s/RT)\}. \tag{3}$$

We need to evaluate ΔE^s. This energy change is due to the influence of the solvent on the electric fields of the solute molecules. The simplest model to consider is that of a nonpolarizable molecule of dipole moment μ in a solvent of dielectric constant ϵ. Then the energy of the dipolar field, which is the difference between the energy of the molecule in the gas (E^g) and in the solvent (E^s), is given by (*9*)

$$E^g - E^s = \frac{\epsilon - 1}{2\epsilon + 1} \cdot \frac{\mu^2}{a^3} \tag{4}$$

where a is the radius of the molecular cavity which is assumed spherical. This gives the well-known rule that the rotamer with the higher dipole moment becomes more stable in media of higher dielectric constant (*10, 11*). For a mixture of rotamers Eq. (4) becomes

$$\Delta E^g - \Delta E^s = \frac{\epsilon - 1}{2\epsilon + 1}\left(\frac{\mu^2 g}{a^3} - \frac{\mu^2 t}{a^3}\right). \tag{5}$$

Equation (5) gives a reasonable quantitative measure of the change in the excess energy of the gauche rotamer on going from gas to liquid (or solution). The difficulty with Eq. (4) [and therefore Eq. (5)] is that the next approximation, that of considering the polarizability of the solute dipole, does not give quantitative agreement with experiment. This gives (*9*)

$$E^g - E^s = \frac{\mu^2}{a^3} \cdot \frac{x}{1 - 2\alpha x/a^3}, \tag{6}$$

where α is the molecular polarizability, and we write x for $(\epsilon - 1)/(2\epsilon + 1)$. Equation (6) will obviously give an increased value of $E^g - E^s$ compared with Eq. (4).

This anomaly has been noted by Wada (*12*), who calculated the energy of the molecular field in the dielectric for 1,2-dichloro- and 1,2-dibromoethane taking into account the spatial positions of the C–X dipoles in the rotamers, and obtained a reasonable agreement with experiment. However, these calculations led to a complex formula which is not suited for general calculations. We present an alternative, simpler method of overcoming this problem.

The experimental observations show that the *trans* form of a $CH_2X \cdot CH_2X$ molecule is more stable in media of high dielectric constant than would be predicted from Eq. (6). The reason for this is, we suggest, that this isomer, having zero dipole moment, is considered to have no interaction with the medium.

However, this isomer has a large quadrupole moment and thus there will be the energy of the quadrupole field to consider. This energy can be calculated in precisely the same manner as the dipolar field (*9*). We obtain, for a nonpolarizable quadrupole q_a of the type (⟵+ +⟶)

$$E^g - E^s = \frac{6q_a^2}{a^5} \cdot \frac{\epsilon - 1}{2 + 3\epsilon} \tag{7}$$

and for a quadrupole q_b of the type (↥ ↧)

$$E^g - E^s = \frac{9q_b^2}{2a^5} \cdot \frac{\epsilon - 1}{2 + 3\epsilon}. \tag{8}$$

The extension of these expressions to include the quadrupole polarizability C (*13*) is simple. Unfortunately no estimate of this quantity for such molecules is known.

Thus as this only introduces an arbitrary constant into an otherwise completely known expression and it is in any case a refinement of a refinement, we will merely use Eqs. (7) and (8) (i.e., we take $C = 0$).

Thus combining the dipole and quadrupole terms we obtain the required expression for ΔE^s, the energy difference between the *gauche* and *trans* isomers in a solvent of dielectric constant ϵ. This is

$$\Delta E^s = \Delta E^0 - \frac{kx}{1 - (2\alpha/a^3 \cdot x} + \frac{3hx}{5 - x}, \tag{9}$$

where $k = \mu^2 g/a^3$ and

$$h = \frac{3}{2a^5}[(4q_a^2 + 3q_b^2)_{\text{trans}} - (4q_a^2 + 3q_b^2)_{\text{gauche}}].$$

The Temperature Dependence of ΔE^s

Equation (9) immediately gives one interesting result. This is that ΔE^s is not, as is usually implicitly implied in such calculations, independent of temperature. The temperature-dependent terms in (9) are x, which is a function of the dielectric constant of the medium, and a^3, the volume of the molecular cavity. If we consider the pure liquid, then

$$\frac{1}{a^3} = 4\pi n\rho/3M = l\rho, \tag{10}$$

where n = Avogadro's number, M the molecular weight, and ρ the density. Also,

$$\frac{\alpha}{a^3} = \frac{n_D^2 - 1}{n_D^2 + 2} = \rho r, \tag{11}$$

where r is the specific refraction which is temperature independent.

We define

$$k' = \frac{k}{\rho} = l\mu_g^2$$

and

$$h' = \frac{h}{\rho^{5/3}} = \frac{3}{2}\cdot I^{5/3}\, f(q_a q_b), \tag{12}$$

where $f(q_a, q_b)$ is the term in parenthesis in Eq. (9). Thus (9) becomes

$$\Delta E^l = \Delta E^g - \frac{k'\rho x}{1 - 2r\rho x} + \frac{3h'\rho^{5/3}x}{5 - x}. \tag{13}$$

It is convenient to write this as

$$\Delta E^l = \Delta E^g - H. \tag{14}$$

The temperature dependence of ΔE^l is now given explicitly in Eq. (13) in terms of ρ and x, and thus can be immediately calculated. It is of more interest to consider how this affects previous measurements of ΔE^l which assume ΔE^l constant. These determine a "constant" ΔE_0^l from the equation

$$\frac{d \ln (n_g/n_t)}{d\,(1/T)} = -\frac{\Delta E_0^l}{R} = -\frac{\Delta E^l}{R} - \frac{T}{R}\cdot\frac{dH}{dT}$$

from (13). Thus

$$\Delta E^l = \Delta E_0^l - T\cdot\frac{dH}{dT}. \tag{15}$$

From Eqs. (13–15) the true value of ΔE^l at any temperature can be calculated.

The Application to 1-Chloro-2-bromoethane

The theory given in the preceding sections can now be used for the compound investigated, 1-chloro-2-bromoethane. We consider for simplicity a model in which μ_{C-X} is 2.03 D [from the dipole moments of the ethyl halides (*14*)] and the C–X distance is 1.86 Å, the mean of the C–Br and C–Cl distances. The dipoles are considered as point dipoles at the midpoint of the C–X bond. The normal tetrahedral angles are used, with the dihedral angle between the two C–X bonds in the *gauche* isomer 70° (*15*). These give $\mu_t = 0$, $\mu_g = 3.12\ D$, and therefore $k = 4.16$ kcal/mole.

The *trans* isomer gives $q_a = 4.83 \times 10^{-26}$ esu, $q_b = 2.95 \times 10^{-26}$ esu, and the corresponding values for the *gauche* isomer are 2.57 and 1.69×10^{-26} esu. Thus h is 5.15 kcal/mole. With n_D 1.4908, Eq. (9) becomes

$$\Delta E^s = \Delta E^g - \frac{4.16\,x}{1 - 0.578\,x} + \frac{15.45\,x}{5 - x}. \tag{16}$$

Thus for the pure liquid (x = 0.402), Eq. (16) gives $\Delta E^g - \Delta E^l = 0.83$ kcal/mole.

The value of $d\rho/dT$ was taken as the mean of these for 1,2-dichloro- and 1,2-dibromoethane (*16*), to give $d\rho/dT = -1.79 \cdot 10^{-13}$ gm/ml/°C. The values of the temperature dependence of the dielectric constant are also known (*17*), and thus Eqs. (13–15) give

$$\Delta E^l = \Delta E_0^l + 0.16.$$

Note that dH/dT is negative as both the density and the dielectric constant decrease with increasing temperature.

Using the value of ΔE_0^l of 0.44 kcal/mole (*11*) gives ΔE^l 0.60 kcal/mole and thus $\Delta E^g = 1.43$ kcal/mole. This is in excellent agreement with the observed value of ΔE^g of 1.43 kcal/mole (*18*); thus the theory predicts quantitatively the change in the energy differences between the isomers in the gaseous and liquid states.

As now ΔE^s is known explicitly as a function of x, the solvent dependence of N and L can be used to obtain the values of these parameters for the individual isomers, from Eq. (3). Furthermore, as there are only two unknown parameters for each curve (i.e., N_g and N_t) the shape of the curve provides a critical test of the theory. It can be seen that the agreement is in general very good, though certain aromatic solvents behave abnormally by interacting preferentially with the *gauche* isomer. These preliminary results give $N_t = 18.4$ and $N_g = 10.3$ cps and $L_t = -8.3$ and $L_g = 4.2$ cps. The significance of these results will be discussed in a subsequent paper.

References

1. N. Sheppard and J. J. Turner, *Proc. Roy. Soc.* (*London*) **A252**, 506 (1959).
2. M. Karplus, *J. Chem. Phys.* **30**, 11 (1959).
3. C. N. Banwell and N. Sheppard, *Discussions Farad. Soc.* **34**, 115 (1962).
4. R. A. Newmark and C. H. Sederholm, *J. Chem. Phys.* **39**, 3131 (1963).
5. H. S. Gutowsky, G. G. Belford, and P. E. McMahon, *J. Chem. Phys.* **36**, 3353 (1962).
6. W. S. Brey and K. C. Ramey, *J. Chem. Phys.* **39**, 844 (1963).
7. R. J. Abraham and L. Cavalli, *Mol. Phys.* **9**, 67 (1965).
8. J. A. Pople, W. G. Schneider, and H. J. Bernstein, "High Resolution Nuclear Magnetic Resonance." McGraw-Hill, New York, 1959.

9. C. J. F. Böttcher, "Theory of Electric Polarisation." Elsevier, Amsterdam, 1952.
10. S. Mizushima, "Structure of Molecules and Internal Rotation." Academic Press, New York, 1954.
11. N. Sheppard, *Advan. Spectr.* **1**, 288 (1959).
12. A. Wada, *J. Chem. Phys.* **22**, 198 (1954).
13. A. Buckingham, *Quart. Rev. (London)* **13**, 193 (1959).
14. "Handbook of Chemistry and Physics," 43rd ed. Chemical Rubber Publishing Co., Cleveland, 1961.
15. J. Ainsworth and J. Kerle, *J. Chem. Phys.* **20**, 425 (1952).
16. "International Critical Tables," Vol. III, p. 28, McGraw-Hill, New York, 1928.
17. C. P. Smyth, R. W. Domten, and E. B. Wilson, Jr., *J. Am. Chem. Soc.* **53**, 4242 (1931).
18. J. Powling and H. J. Bernstein, *J. Am. Chem. Soc.* **73**, 1815 (1951).

Geminal-Substituted Cyclohexanes as Models for Conformational Analysis by NMR

M. ANTEUNIS and D. TAVERNIER

State University of Ghent, Ghent, Belgium

The shifts of the dioxolane–methylene protons derived from cyclohexanone and the 3- or 4-substituted derivatives are almost identical. This shift is unique. Table I illustrates this (differences *versus* the cyclohexanone glycol ketal), and only highly substituted ring compounds, such as that derived from cholestanone, are somewhat different. The methyl

TABLE I
SHIFTS OF SOME GLYCOL KETALS *VERSUS* THE GLYCOL KETAL OF CYCLOHEXANONE[a] ($\delta = 3.86$ AT 56.4 MC) IN 10% CHLOROFORM SOLUTIONS

Product	Shift difference (c/s)
4-Methylcyclohexanone	−0.01 ± 0.05
3-Methylcyclohexanone	+0.10 ± 0.02
2-Methylcyclohexanone	−0.10 ± 0.07
trans-β-Decalone	0.0 ± 0.05
Cholestanone	+0.58 ± 0.08

[a]With the corresponding cycloheptanone or the cyclopentanone derivatives as internal standard.

ketals in contrast give a *doublet* as soon as substituents on the ring prevent the conformational equilibrium from being equipopulated. This is attributed to the fact that different shielding exists for the (rotating) equatorial and axial methoxy group. Let us differentiate the two methoxy groups by I and II, respectively:

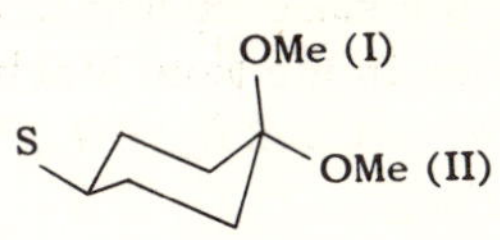

N_e

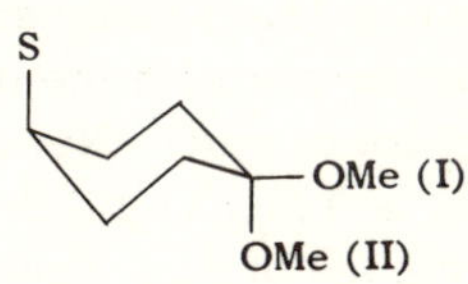

N_a

In the equatorial-substituted conformer (mole fraction N_e) the idealized proton spectrum would be as shown in Fig. 1 (upper part): the two resonance peaks for I and II are separated by some value $\Delta\nu_m$, the axial

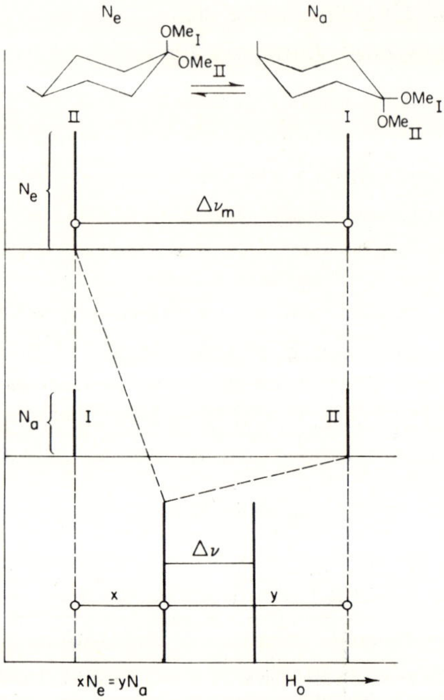

FIG. 1. (See text.)

methoxy group resonating at the higher field value. In the other conformer (mole fraction N_a) an identical shift would be observed if secondary effects (e.g., a field effect caused by the substituent) could be neglected. Here, however, group I stays equatorial and II axial, and peak attribution in this case is to be reversed (Fig. 1, middle part). The peak heights are directly related to the population density in each state (N_e and N_a respectively) so that by rapid interconversion, which is the actual situation, each of two peaks of either methoxy group collapses to a peak position which is the weighted mean. Thus peak I is shifted by an extent x to a lower value and peak II by the same amount to a higher value in comparison with the original position in the more stable conformer (Fig. lower part). Now, when $N_a + N_e = 1$ and $K_a = N_e/N_a$ then $K_{a,e} = (\Delta\nu_m + \Delta\nu)/(\Delta\nu_m - \Delta\nu)$ if $x + y = \Delta\nu_m$.

The knowledge of $\Delta\nu_m$ could give us directly the value of $K_{a,e}$ and this can be achieved in two ways:

(a) Through properly chosen constellations, i.e., frozen ones, in which $\Delta\nu_m$ is the actually observed splitting separation;

(b) Through the study of the splitting separation as a function of the temperature, and especially by observing the spectrum at temperatures below the coalescence temperature.

Frozen Conformers

The shifts of a number of frozen conformers were determined, as illustrated in Table II. The substituents in position 3 are believed to prevent any flipping over so that these derivatives can be used. One is struck by several facts. Variable values for $\Delta\nu_m$ are obtained, and the heavier the substituent in position 3 for instance, the higher the shift is. From a study of Dreiding models it is suspected that the heavier the alkyl substituent is, the greater the magnetic anisotropy could be for the axial methoxy group. These substituents must then cause increased shielding. For some derivatives it is obvious that still another effect is operating. It can be shown that a 4-*tert*-butyl group can be as close as 2.1 Å to the methoxy protons in some of the rotamers, and this overcrowding is more pronounced for the 4-*tert*-butyl than for the 3-*tert*-butyl derivative. We accept that the exceptionally low value for the 4-*tert*-butyl derivative is the result of an altered conformation, and we propose a half-chair form in which the geminal dimethoxy groups receive quasi-equatorial character.

TABLE II
SHIFTS OF THE METHYL PROTONS OF KETALS (FROZEN MODELS) IN C/S FOR 10% CHLOROFORM SOLUTIONS

Compound	Shift
3-Methylcyclohexanone	3.32 ± 0.05
3-Ethylcyclohexanone	3.41 ± 0.09
3-Isopropylcyclohexanone	3.85 ± 0.06
3-*tert*-Butylcyclohexanone	4.09 ± 0.09
4-*tert*-Butylcyclohexanone	2.69 ± 0.05
trans-β-Decalone	2.93 ± 0.05
cis-β-Decalone	2.92 ± 0.07
trans-β-Hydrindanone	3.21 ± 0.06

From all examples given, the methyl substituent seems to cause nearly no secondary effect. This is strengthened by the observation that the midpoint of the doublet for the 4 and the 3 derivative coincides exactly with the shift of the unsubstituted 1, 1-dimethoxy cyclohexane.

In general we can hope to get to a good approximation the value of K_e by taking the value found for the 3 derivative as $\Delta\nu_m$ for the *corresponding* 4 derivative.

To date we have calculated $K_{a,e}$ values for Me and Et at 29 ± 0.5 °C and have found values of respectively 7.2 and 9.2 ($\Delta\nu$ for 10% $CHCl_3$ solutions, respectively 2.54 and 2.74 cps), while the corresponding "A" values are then 1.18 for the 4-methyl and 1.32 for the 4-ethyl derivative.

Temperature Dependence

We have now started with the second part of our program, the study of the temperature dependence. Before interpreting these data, which for instrumental reasons cannot be obtained at low temperatures, it was of interest to look at the effect of higher temperatures on the shift of our frozen models. Table III gives us the values obtained for neat liquids.

TABLE III
TEMPERATURE DEPENDENCE OF SHIFTS
(NEAT LIQUIDS)

Compound (neat)	Temperature (°C)								
	26	29	48	51	75	100	127	130	150
1,1-Dimethoxy-4-Me-cyclohexane	1.97	–	1.77	–	–	1.49	–	1.34	–
1,1-Dimethoxy-4-Et-cyclohexane	2.18	–	2.00	–	–	1.80	–	–	–
1,1-Dimethoxy-4-*tert*-Bu-cyclohexane	–	1.84	–	1.81	1.71	1.68	–	1.67	–
1,1-Dimethoxy-3-Me-cyclohexane	2.84	–	2.70	–	–	2.49	–	2.37	–
1,1-Dimethoxy-3-Et-cyclohexane	–	2.93	–	2.80	–	2.65	–	2.54	–
1,1-Dimethoxy-3-iso-Pr-cyclohexane	3.38	–	3.18	–	3.12	2.90	–	2.71	–
1,1-Dimethoxy-3-*tert*-Bu-cyclohexane	–	3.29	–	3.22	3.21	3.10	–	3.02	–
trans-2,2-Dimethoxy decalane	2.60	–	2.44	–	–	2.28	2.16	–	2.08
trans-5,5-Dimethoxy hydrindane	–	2.84	–	2.83	2.80	2.67	–	2.64	–
cis-2,2-Dimethoxy decalane	2.61	–	2.39	–	–	2.08	2.02	–	1.89
trans- and *cis*-4-*tert*-Bu-cyclohexyl methyl ethers	–	1.44	–	1.36	1.29	1.21	–	1.12	–

As seen, these shift differences do not remain constant, and this immediately illustrates the great danger in employing shift data obtained at another temperature.

We have, however, calculated for each temperature the equilibrium constant once again by using the value at each temperature of the 3 substituent for the corresponding 4 substituent. It is seen in Table IV that this value K_e decreases as the temperature increases, and that the value of free energy change ΔF^0 remains on the contrary merely constant, as

TABLE IV
CONFORMATIONAL EQUILIBRIUM DATA FOR CYCLOHEXANE DERIVATIVES (NEAT)

	$K_{a,e(T)}$	"A Value"	Ref.
A. Methyl substitution			
26	5.5 ± 0.1	0.995 ± 0.02	1.74[a]
48	4.8 ± 0.2	1.000 ± 0.02	1.01[b]
100	4.0 ± 0.2	1.050 ± 0.02	1.8[c]
130	3.6 ± 0.2	1.020 ± 0.02	1.8[d]
B. Ethyl substitution			
29	6.81 ± 0.6	1.145 ± 0.03	1.74[a]
50	6.17 ± 0.6	1.157 ± 0.03	1.86[e]
100	5.24 ± 0.5	1.222 ± 0.03	2.1[d]

[a] A. Lewin and S. Winstein, *J. Am. Chem. Soc.* **84**, 2464 (1962).

[b] J. B. Hendrickson, *J. Am. Chem. Soc.* **84**, 3355 (1962). In fact this is not the A value, but the calculated enthalpy. As entropy differences are in general very small the enthalpy change, however, nearly equals the free energy change.

[c] C. W. Beckett, K. S. Pitzer and R. Spitzer, *J. Am. Chem. Soc.* **69**, 977 (1947).

[d] S. Winstein and N. J. Holness, *J. Am. Chem. Soc.* **77**, 3562 (1955).

[e] N. L. Allinger and S. Hu, *J. Am. Chem. Soc.* **84**, 370 (1962).

it should. It is interesting to compare our values with some found by other procedures (Table IV). It seems that our values in general are lower, with the exception of the calculated data by Hendrickson.

For the conclusion of this paper we point to the fact that although the shift differences used to estimate conformational equilibria are small, the method is still applicable. In order not to be troubled by secondary effects, however (hindered rotation, comparable solvent effects), and to increase appreciably the sensibility of our system on the other hand, we have started with the synthesis of entirely new *gem* substituents, e.g., fluorine. In this case the nuclei directly bonded to the ring should give us differences which could be some powers of ten greater.

We believe that these models are very promising and much more reliable than the NMR methods already described in the literature. We hope to report the results obtained in the near future.

Acknowledgment

One of us (D. T.) is indebted to the N.F.W.O. (Belgian National Fund for Scientific Research) for a scholarship as aspirant.

* * *

Question: Can the 3-substituted derivatives really be considered as frozen models?

Answer: Yes, we are sure that they may. Otherwise the temperature dependence should be much more pronounced as compared with those observed with the really rigid models such as the 4- and 3-*tert*-butylcyclohexanone and the decalone systems. In fact the temperature dependence seems to be somewhat *more* pronounced the larger the 3 substituent actually is.

The Conformation and Flexibility of Unsaturated Seven-Membered Ring Systems[1]

H. FRIEBOLIN and S. KABUSS

Institut für Elektrowerkstoffe Freiburg,
Chemisches Institut der Universität Freiburg, Freiburg, Germany

The boat (a), the twist-boat (b), and the chair (c) are the only conformations of the unsaturated seven-membered ring that can be made with

(a) (b) (c)

models. Unlike the chair, the boat is flexible in the model. It is easy to transform the normal boat into the twist-boat by pseudorotation, i.e., without deformation of valence angles. The pseudorotation demands little energy; for the saturated 7 ring, Hendrickson calculated 2 kcal/mole. The twist-boat is the more stable conformation of these two forms, because the H-H repulsion forces (*Pitzer-Spannung*) here are lower than for the normal boat. The H's are more staggered. So, if we want to discuss the conformations of seven-membered rings, we have to consider mainly the two forms, twist-boat and chair, in which all protons are staggered.

In the model the chair is rigid. There can, however, be an inversion into the other chair conformation, which has equal energy. This inversion takes place via the boat form under strong deformation of valence angles. Therefore we expect a higher free activation enthalpy for the inversion than for the pseudorotation. In the chair the two geminal protons in the 5 position are distinguished by their pseudoaxial and pseudoequatorial positions. Therefore, if the chair is rigid, these two protons give two different signals in NMR spectra, if we ignore spin-spin coupling. However, there is only one signal if it is flexible. In the twist boat the C(5) lies on a twofold symmetry axis. Therefore we expect only one H(5) NMR signal from the frozen-in twist-boat.

We have prepared and investigated with NMR the following substances:

[1]Published in part in Ref. (*1*).

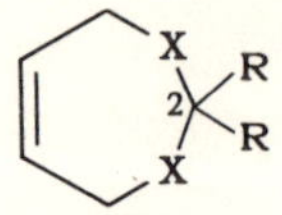 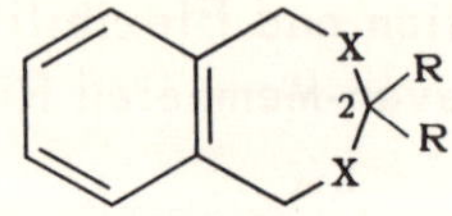

X = CH_2,O,S; R = H,CH_3

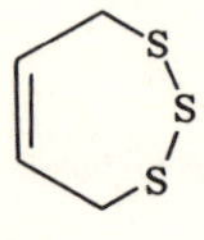 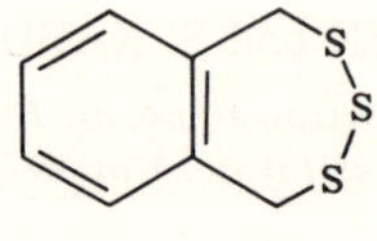

These are unsaturated seven-membered rings with two heteroatoms S, O, or CH_2 in the 1- and 3-positions and benzo-substituted compounds. In the 2-position there are two protons or two CH_3 groups. We also investigated some trisulfides. The spectra of these compounds are not very complex and can be discussed easily. For the interpretation we generally used the signals of the H(2) protons or those of the CH_3 group. By means of the approximate formulas of Gutowsky and Holm (*2*), the rate constant k was estimated at coalescence temperature ($T_a = T_c$). Rate constant k and coalescence temperature T_a gave the free activation enthalpy $\Delta G^{\ddagger}$ via the Eyring equation.

Results

S Compounds (Table I)

T_{min} is the lowest experimental temperature. Because of the poor solubility of some compounds at low temperatures we used mixtures of solvents and so the values of $\Delta G^{\ddagger}$ are not exactly comparable with each other.

(1) All compounds show temperature-dependent proton resonances.

(2) At low temperatures the protons at C(2) and those in the CH_3 groups give rise to quartet and doublet spectra, respectively. Thus, as we have seen above, the chair is the conformation of lowest energy.

(3) The benzo compounds show higher $\Delta G^{\ddagger}$ values than (I) and (II). This might be due to resistance of the rigid benzene ring to deformation of valance angles, which must occur if the chair inverts.

So far, we cannot estimate the influence of the change of energy content of the ground state.

At low temperatures in the spectrum of 2,2-dimethyl-1,3-dithiacycloheptene there appears in addition to the AB quartet of the protons H(4,7) a fifth line and in addition to the doublet of the CH_3 line a third line.

Probably we have an equilibrium of two conformations, chair and twist-boat, which are not very different in energy content. Estimation gives a

TABLE I[a]

		Proton signal observed	τ_{mean} (ppm)	$T_a = T_c$ (°C)	T_{min} (°C)	$\Delta\nu_\infty$ (Hz)	k (sec^{-1})	$\Delta G^\ddagger$ (kcal/mole)	Solvent
(I)	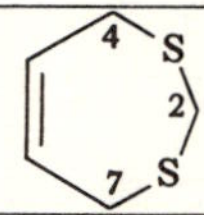	H(2) H(4,7)	6.02 6.57	−100 —	−114	30(Q) 36(Q)	66.7 —	8.5 —	$C_2H_2Cl_2$/ $CHCl_3$
(II)	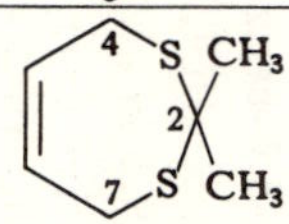	CH_3(2) H(4,7)	8.35 6.73	−110 −105	−119	20(D) 38(Q)	44.4 84	8.2 8.2	CS_2 CS_2
(III)	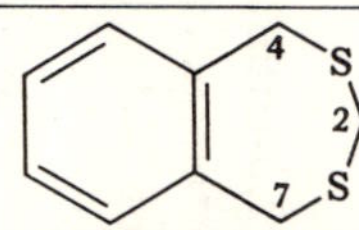	H(2) H(4,7)	6.20 6.13	— −55	−88	— 29(Q)	— 64	— 10.9	$CHCl_3$/ *m*-F-Toluene
(IV)	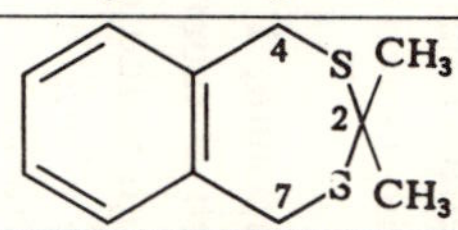	CH_3(2) H(4,7)	8.23 6.06	−33 −26	−72	25.7(D) 39.8(Q)	57 88	12.1 12.2	CH_2Cl_2 CH_2Cl_2

[a] $T_a = T_c$ = coalescence temperature; = lowest temperature of observation; Q = quartet; D = doublet; $\Delta\nu_\infty$ = separation of peaks; k = rate constant; $\Delta G^\ddagger$ = free energy of activation.

difference of 0.2 kcal/mole. In the case of the corresponding benzo compounds there is no indication of such a behavior as we have seen.

O Compounds (Table II)

The resonance lines of the first two compounds do not change with temperature. Here again, the benzo compounds are more suitable for investigation, since they show an effect in the attainable temperature range. All compound (VII) shows, however, is a broadening. Only 2,2-dimethyl-1,3-dioxa-5,6-benzocycloheptene (VIII) could be frozen in. The resonance lines of the H(4,7) protons form an AB quartet below $-76°C$, whereas the CH_3 line does not split. Thus we cannot exclude the twist form for this compound, since the geminal protons H(4) and H(7) also can be distinguished in the twist-boat conformation.

Benzocycloheptene Compounds (Table III)

At low temperatures the three benzo compounds give the proton resonance spectra of the rigid, probably chair, forms. T_a is in the same range for the two methyl compounds, which means that either the CH_3 groups do not influence the inversion mechanism or various effects might compensate each other.

In the case of benzocyloheptene we prepared the compound tetradeuterated in the 4,6 positions, in order to get T_a and $\Delta G^‡$ values for this system, too (*4*). Naturally, the proton resonance spectrum of the normal benzocycloheptene is very complex and not suitable for our investigations.

Among all the substances we investigated, the di- and trisulfides have the highest $\Delta G^‡$ values (Table IV).

This stiffening of the ring might be due to the S–S bonds. We have found similar effects in six-ring disulfides (*3*). $\Delta G^‡$ of (XIV) is much lower than $\Delta G^‡$ of the other compounds, so another inversion mechanism must be taken into acount, probably a mechanism which is not possible in the case of the benzo compound because of the stiffness of the benzene ring.

Only one compound, 1,2,3-trithia-5,6-benzocycloheptene is rigid for proton resonance at room temperature. H(4,7) give an AB quartet at $+25°C$. The high value of $\Delta G^‡ = 19.3$ kcal/mole is near the potential of 23 kcal/mole where optically active compounds can be separated chemically. This separation of optically active disulfides was the start of our investigations.

TABLE II[a]

		Proton signal observed	τ_{mean} (ppm)	$T_a = T_c$ (°C)	T_{min} (°C)	$\Delta\nu_\infty$ (Hz)	k (sec^{-1})	$\Delta G^\ddagger$ (kcal/mole)	Solvent
(V)	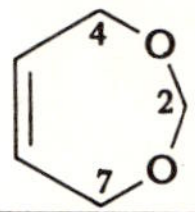	H(2) H(4,7)	5.33 5.93	<−120 —	−120	Spectrum unchanged			CS_2/ Toluene
(VI)	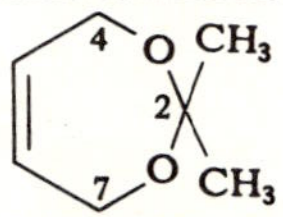	CH_3(2) H(4,7)	8.7 5.98	<−128 —	−128	Spectrum unchanged			CS_2/ CH_2Cl_2/ Toluene
(VII)	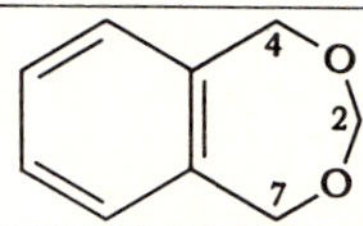	H(2) H(4,7)	4.98 5.14	<−139 —	−139	Signal broadened			Vinyl chloride
(VIII)	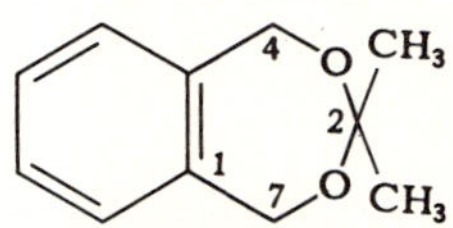	H(4,7) CH_3(2)	5.21 8.55	−76 <−127	−112 −127	35(Q)	7.8 CH_3 Signal broadened	9.7	$CHCl_3$/ *m*-F-Toluene/ CH_2Cl_2/ Toluene

[a]See Table I, footnote *a*.

TABLE III[a]

		Proton signal observed	τ_{mean} (ppm)	$T_a = T_c$ (°C)	T_{min} (°C)	$\Delta\nu_\infty$ (Hz)	k (sec^{-1})	$\Delta G^\ddagger$ (kcal/mole)	Solvent
(IX)	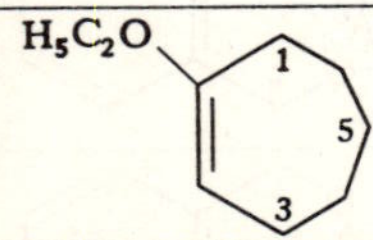	H(2)	5.2	<−99	−99	–	–	<8	CS_2
(X)	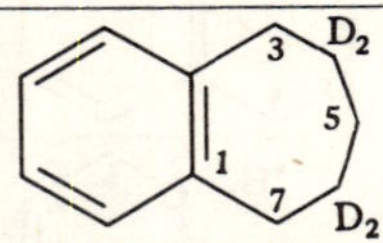	H(5)	8.25	−56	−103	26.0	58	10.9	CS_2
(XI)	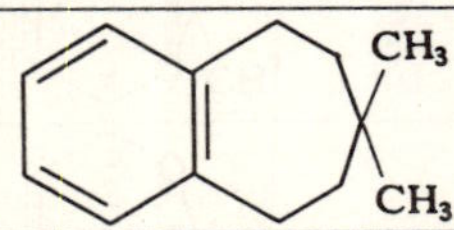	CH_3	9.03	−45	−79	12.5	28	11.8	CS_2
(XII)	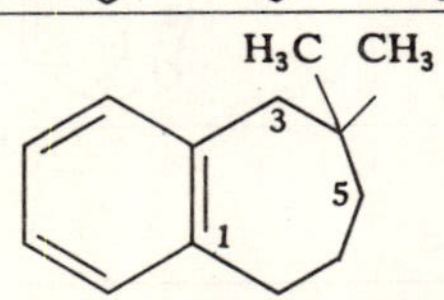	CH_3 CH_2(3)	9.18 7.45	−45 ~−40	−101	28 39	62 87	11.4 11.5	CS_2

[a]See Table I, footnote *a*.

TABLE IV[a]

		Proton signal observed	τ_{mean} (ppm)	$T_a = T_c$ (°C)	T_{min} (°C)	$\Delta\nu_\infty$ (Hz)	k (sec^{-1})	$\Delta G^\ddagger$ (kcal/mole)	Solvent
(XIII)	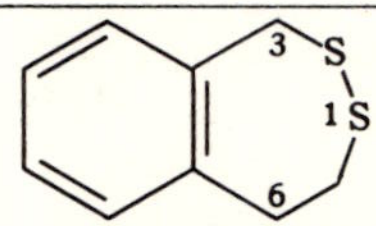	H(3)	5.95	−3	−46	25.9	58	13.6	CS_2
(XIV)	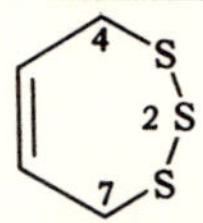	H(4,7)	6.13	−95	−124	54	120	8.6	CS_2
(XV)	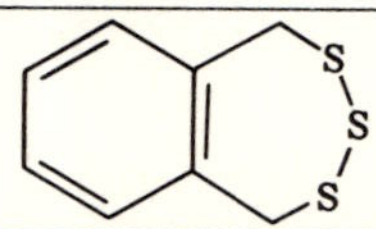	H(4,7)	5.7	+110 ± 5	+25	41.7	–	19.3	CS_2

[a]See Table I, footnote *a*.

References

1. H. Friebolin, R. Mecke, S. Kabuss, and A. Lüttringhaus, *Tetrahedron Letters* **29**, 1929 (1964).
2. H. S. Gutowsky and C. H. Holm, *J. Chem. Phys.* **25**, 1228 (1956).
3. A. Lüttringhaus, S. Kabuss, W. Maier, and H. Friebolin, *Z. Naturforsch.* **16b**, 761 (1961).
4. S. Kabuss, H. Friebolin, and H. Schmid, *Tetrahedron Letters* p. 469 (1965).

Spectres de résonance magnétique nucléaire (RMN) de quelques couples de diastéréoisomères simples aliphatiques

PIETRO BUCCI

Istituto di Chimica-Fisica dell' Università di Pisa, Pisa, Italy

et

RENZO ROSSI

Istituto di Chimica Organica Industriale dell' Università di Pisa, Pisa, Italy

La détermination de la configuration relative des atomes de carbone asymétriques contenus dans des molécules diastéréoisomériques, et aussi de la pureté stérique, est très intéressante pour l'étude du mécanisme de certaines réactions de chimie organique et dans la chimie des hauts polymères.

Parmi les méthodes physiques traditionelles, quelques-unes (infrarouge et ultraviolet) sont, dans la plupart des cas, inefficaces, tandis que des autres (rayons X), qui en principe pourraient fournir de très bons résultats, sont gênées par des difficultés notables dues à la complexité de l'élaboration analytique des données de l'expérience.

La méthode de la RMN à haute résolution a été appliquée pour l'étude de diastéréoisomères de haut poids moléculaire (*1*), bien que dans ce cas là, la complexité extrême des spectres n'ait permis qu'une interprétation grossière. Il nous a paru intéressant, alors, d'étudier avec cette technique-là des composés diastéréoisomériques particulièrement simples, pour lesquels fûtil possible d'obtenir des spectres bien définis et capables d'être interprétés clairement.

Nous avons donc examiné les couples suivantes de diastéréoisomères:

(I) 2,3-diméthyl-succinate-diméthylique *méso* (Ia) et *d, 1* (Ib)

(II) 3,4-diméthyl-adipate-diméthylique *méso* (IIa) et *d, 1* (IIb)

(III) 2,4-diméthyl-glutarate-diméthylique *méso* (IIIa) et *d, 1* (IIIb)

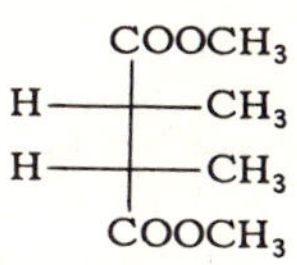

(Ia)

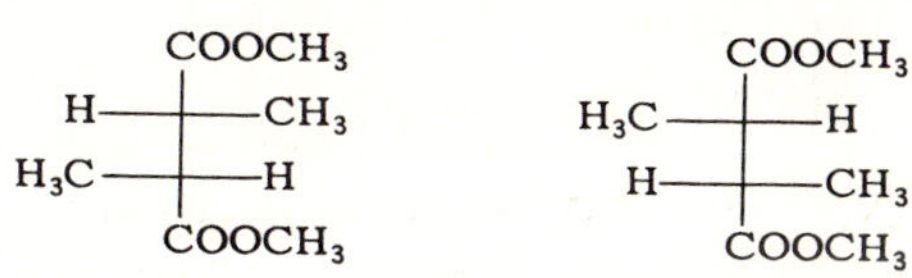

(Ib)

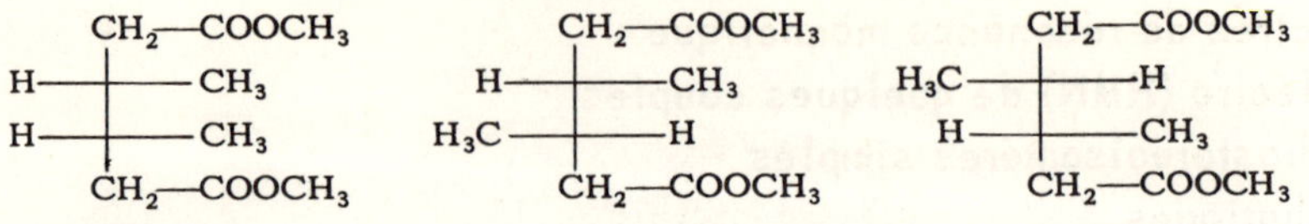

(IIa) (IIb)

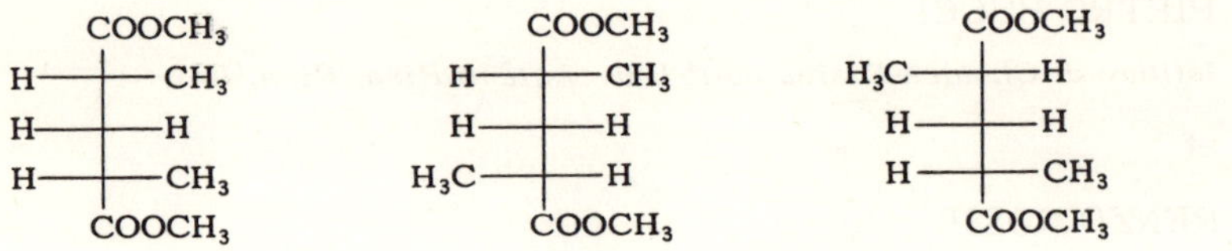

(IIIa) (IIIb)

Les spectres obtenus pour les couples diverses de diastéréoisomères ont étés reportés dans les Fig. 1, 2 et 3.

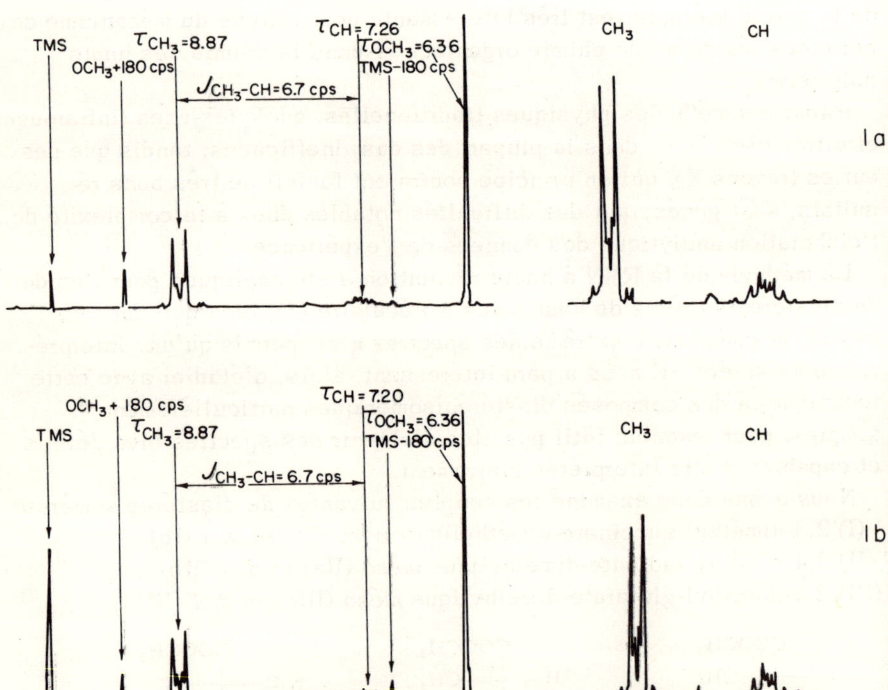

FIG. 1. (a) Spectre RMN, en $CHCl_3$, du 2,3-diméthyl-succinate-diméthylique *méso*; (b) Spectre RMN, en $CHCl_3$, du 2,3-diméthyl-succinate-dimétilique *d, l*.

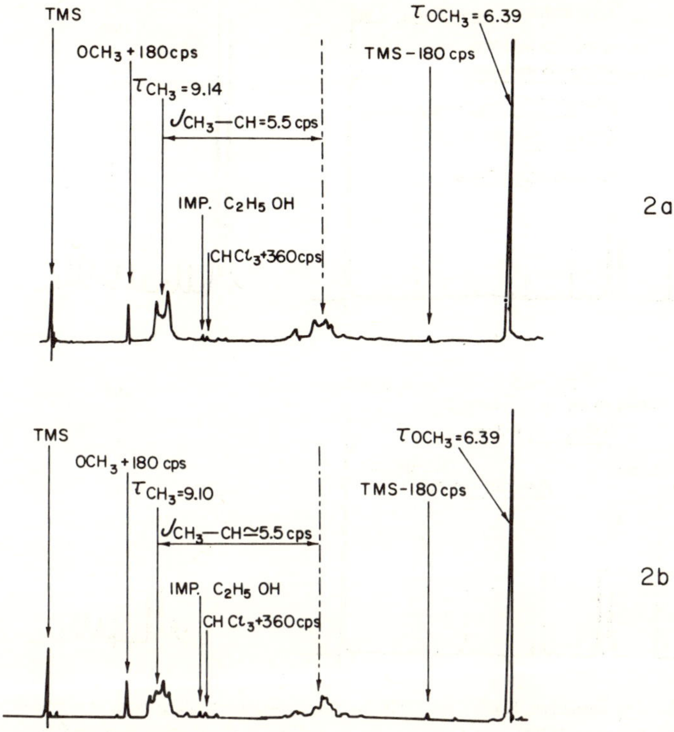

FIG. 2. (a) Spectre RMN, en $CHCl_3$, du 3,4-diméthyl-adipate-diméthylique *méso*. (b) Spectre RMN, en $CHCl_3$, d'un mélange des deux esters 3,4-diméthyl-adipiques diastéréoisomères.

(I) 2,3-Diméthyl-succinate-diméthylique (Ia) et (Ib)

On peut observer que la seule différence entre les spectres de (Ia) et (Ib) est le déplacement du groupement de raies des méthynes, qui, dans le diastéréoisomère *d, l* (Ib) sont déplacées vers les champs faibles. L'aspect du groupement de raies correspondant aux méthyles des deux diastéréoisomères est celui de la partie "A" d'un spectre A_2X_2 dans lequel on a:

$$J_{AA'} \simeq 0; \quad J_{XX'} \simeq 7 \text{ cps}; \quad J_{AX} = J_{A'X'} = 6.7 \text{ cps}; \quad J_{A'X} = J_{AX'} \simeq 0$$

Cet aspect est raisonnable, puisqu'au premier ordre la partie A d'un système $A_nA_{n'}X_2$ ou $J_{AA'} = 0$ est égale à la partie A d'un système A_2X_2 sous la même condition. Le spectre correspondant aux méthynes est, naturellement, plus compliqué.

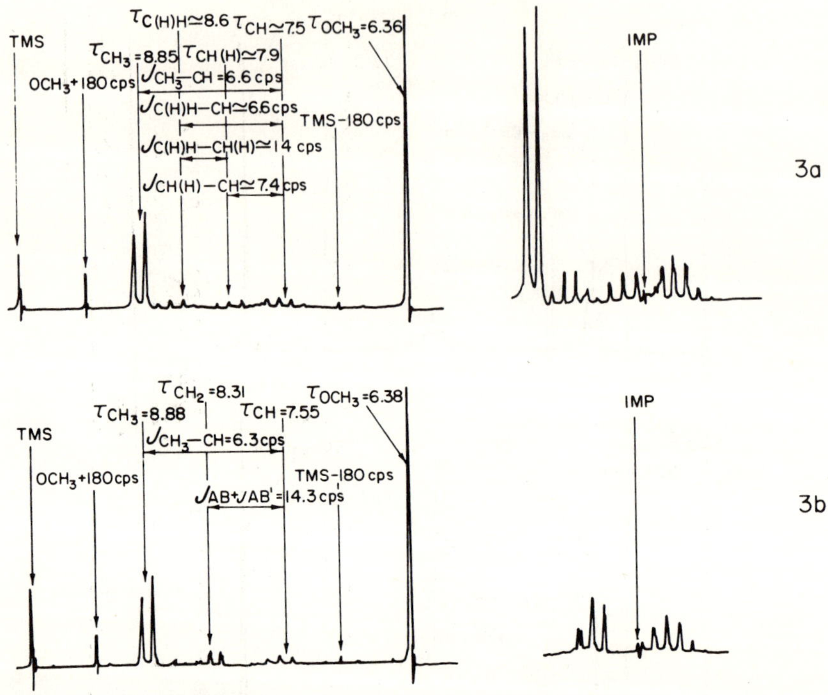

FIG. 3. (a) Spectre RMN, en $CHCl_3$, du 2,4-diméthyl-glutarate-diméthylique *méso*. (b) Spectre RMN, en $CHCl_3$, du 2,4-diméthyl-glutarate-diméthylique *d, l*.

On peut interpréter le déplacement vers les champs faibles du groupement de raies du diastéréoisomère *d, l* si l'on considère le trois rotamères "staggered" de chaque diastéréoisomère.[1]

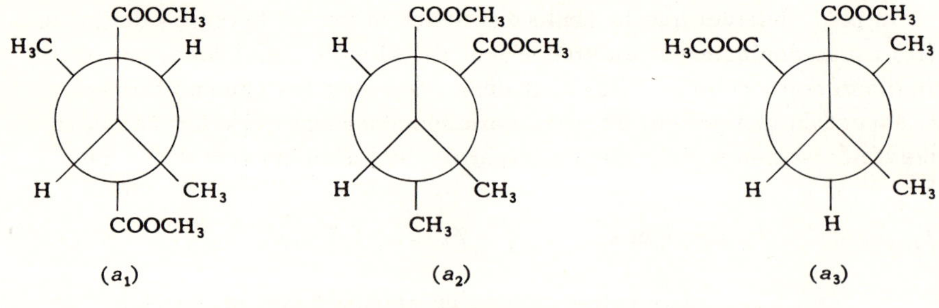

(a_1) (a_2) (a_3)

(Ia)

[1]Les considérations suivantes représentent une extension du travail de H. S. Gutowsky, *J. Chem. Phys.* **37**, 2196 (1962).

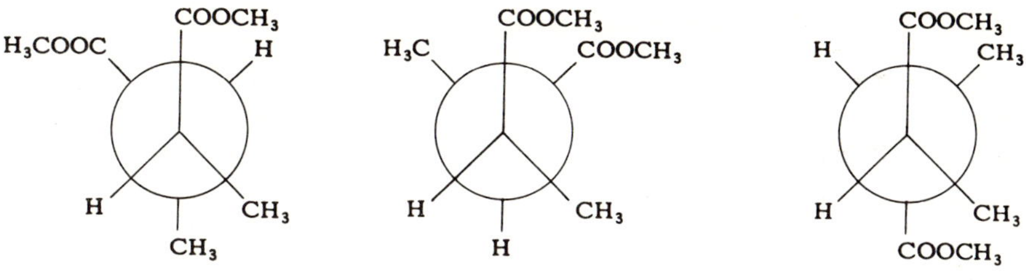

(b_1) (b_2) (b_3)

(Ib)

Dans le diastéréoisomère *méso* (Ia) les protons méthyniques sont tous les deux en position "gauche" par rapport au groupement —CO— dans le conformère a_1, tandis qu'ils sont l'un *gauche* et l'autre *trans* dans les conformères a_2, a_3, qui ont les même poids. Indiquant avec $\Delta\tau_g$ la contribution au déplacement chimique du —CH— fourni par la —CO— *gauche*, et avec $\Delta\tau_t$ celui donné par le groupment —CO— *trans*, on a:

$$\Delta\tau_{méso} = P_{a_1}\cdot\Delta\tau_g + (P_{a_2} + P_{a_3})\left(\frac{\Delta\tau_g + \Delta\tau_t}{2}\right) = \Delta\tau_g(P_{a_1} + P_{a_2}) + \Delta\tau_t\cdot P_{a_2} = \Delta\tau_g + (\Delta\tau_t - \Delta\tau_g)\cdot P_{a_2}$$

où P_{a_1}, P_{a_2} et P_{a_3} sont les poids des trois rotamères a_1, a_2 et a_3. On rappelle que $P_{a_2} = P_{a_3}$ et $P_{a_1} + P_{a_2} + P_{a_3} = 1$.

Avec un symbolisme analogue on a, pour le diastéréoisomère *d, l*:

$$\Delta\tau_{d,l} = \Delta\tau_g + (\Delta\tau_t - \Delta\tau_g)\cdot P_{b_2}$$

La différence du déplacement chimique entre le deux diastéréoisomères, causée par le groupement —CO—, est donnée par:

$$\Delta\tau = \Delta\tau_{méso} - \Delta\tau_{d,l} = (\Delta\tau_t - \Delta\tau_g)(P_{a_2} - P_{b_2})$$

Comme l'on peut observer (Fig. 1), le groupement de raies correspondant au —CH— du diastéréoisomère *meso* est déplacé vers les champs forts et on a donc $\Delta\tau = 0.06$ ppm.

Cela va indiquer que les protons méthyniques du diastéréoisomère *d, l* sont plus sujet à l'action du groupement —CO— qui tend à diminuer l'écran du proton, c'est-à-dire à déplacer la raie de résonance vers les champs faibles.

Le conformère b_2 aurait donc, dans l'équilibre conformationel du diastéréoisomère *d, l*, un poids plus petit que celui qui correspond au conformère a_2 (ou a_3) dans l'équilibre conformationel du diastéréoisomère *meso*.

(II) 3,4-Diméthyl-adipate-diméthylique (IIa et IIb)

Tandis que la synthèse de l'acide 3,4-diméthyl-adipique *meso* et donc de son ester (IIa) est bien simple (*2*), le diastéréoisomère *d, l* épimériquement pur n'est pas obtenible par cristallisation fractionnée (*2*).

Toutefois, puisqu'il est possible d'évaluer, même si d'une façon approximative, les paramètres du 3,4-diméthyl-adipate-diméthylique *d, l* (IIb) par comparaison du spectre du *méso* avec le spectre d'un mélange des deux diastéréoisomères, nous n'avons pas retenu nécessaire, pour le moment, de synthétiser l'acide *d, l* épimeriquement pur, selon Ställberg-Stenhagen (*2*).

De l'examen comparatif des deux spectres de (IIa) et du mélange (IIa + IIb) (Fig. 2), on peut voir que les deux diastéréoisomères présentent des bandes complexes correspondant aux méthyles, qui sont semblables à celles des esters succiniques, mais qui sont déplacées entre elles d'environ 0.04 ppm. Le couplage $J_{CH_3—CH}$ est de 5.5 cps dans le *méso* et presque le même dans le *d, l.* La partie du spectre des protons méthyniques et méthyléniques est trop compliquée d'être facilement interprétée. Il s'agit en effect d'un spectre du type $A_2B_4X_6$, les protons méthoxyliques donnant lieu à deux raies dont la position est presque identique dans les deux spectres (τ = 6.39 ppm).[2]

(III) 2,4-Diméthyl-glutarate-diméthylique (IIIa) et (IIIb)

Les spectres des deux diastéréoisomères présentent les raies correspondant aux méthyles, de structure identique mais déplacées entre elles de 0.03 ppm (Fig. 3).

On observe aussi une différence remarquable dans les spectres qui correspondent aux deux atomes d'hydrogène du groupement méthylénique. Cela peut être expliqué si l'on va examiner les conformères possibles en faisant tourner les deux groupements asymétriques autour des liaisons du —CH_2—.

De la symétrie du système on peut déduire l'équivalence des deux protons méthyléniques du diastéréoisomère *d, l* (IIIb), indépendamment de l'équilibre conformationel; ce sont toutefois liés à ce dernier les constants de couplage spin-spin entre l'un des protons méthyléniques et les protons méthyniques vicinaux, qui sont en général différentes entre elles, et qui devraient devenir égales à température infinie, c'est-à-dire si la rotation était libre du tout autour des deux liaisons considérées.

Le spectre, si l'on peut négliger, pour les protons méthyléniques, l'effet du couplage entre les protons méthyliques et méthyniques, se

[2]La raie du méthoxyle dans le mélange est un doublet séparé par ~ 0.25 cps.

présente comme la partie A d'un spectre de type A_2B_2 :

$$J_{AA'} \simeq 12 \div 14 \text{ cps}; \qquad J_{BB'} \simeq 0; \qquad J_{AB} = J_{A'B'} \neq J_{AB'} = J_{A'B}$$

Malheureusement il n'est pas possible, sans calculs compliqués, d'obtenir par les données expérimentales, toutes les constants de couplage. On peut toutefois obtenir le deplacement chimique et la somme $J_{AB} + J_{AB'}$. Les protons méthyniques se conduisent comme 1 proton couplé à 5 protons à peu près équivalents, c'est-à-dire qu'ils donnent lieu à un spectre formé par 6 bandes symétriques, en position, autour de la raie du proton non perturbé.

Pour le diastéréoisomère *méso*, au contraire, les deux protons méthyléniques ne sont pas équivalents et, à l'égard du couplage spin-spin, on a (si l'on appelle A un des protons méthyléniques, B l'autre, C et C′ les protons méthyniques):

$$J_{AC} = J_{AC'} \neq J_{BC} = J_{BC'}$$

Il n'est pas facile d'obtenir exactement la valeur des paramètres, le spectre étant du type ABC_2, donc très compliqué.

On peut toutefois interpréter trés grossièrement le spectre, si l'on considère qu'il donne trois groupements de raies qui peuvent être assignés respectivement aux protons A, B et C_2, dont la structure est semblable à celle d'un spectre du type AMZ_2, où l'on pose:

$$J_{AX} \simeq 2\, J_{AZ} \simeq 2\, J_{XZ} \simeq 14 \text{ cps}$$

Les résultats obtenus démontrent que, dans quelques cas, il est possible, regardant attentivement les spectres RMN, de déterminer la configuration relative des atomes de carbone asymétriques qui se trouvent dans des molécules diastéréoisomères. Le cas des 2,4-diméthyl-glutarates-diméthyliques est un exemple très clair.

En plus, à cause des différences notables que l'on observe dans les spectres de différentes couples de diastéréoisomères, la RMN de haute résolution est une technique très bonne et efficace pour la détermination de la pureté épimérique.

On doit noter que les différences entre les spectres ne sont pas seulement en relation avec les groupements des raies qui correspondent aux hydrogènes directment liés aux atomes de carbone asymétriques, mais aussi avec ceux qui correspondent aux hydrogènes liés à des atomes de carbone vicinaux aux carbones asymétriques.

En particulier, ce sont les spectres des méthyles qui sont très importants, car ils donnent des raies très fortes et car ils interagissent directement, au maximum, avec un hydrogène.

L'importance de ces considérations est déjà évidente dans les spectres des esters 3,4-diméthyl-adipiques, et elle paraît encore plus évidente dans la détermination que nous avons fait, sur la base de l'intensité relative des raies des protons méthyliques, de la composition épimérique des mélanges des deux diastéréoisomères *érythro* (IVa) et *thréo* (IVb) du 3-méthyl-2-pentanole.

Les spectres de la Fig. 4 sont ceux que l'on obtient d'un premier mélange (A) enrichi du diastéréoisomère *érythro* (IVa), d'un deuxième (B) plus riche en (IVb) (*3*) et d'un troisième (C) obtenu en mélangeant (A) et (B) en rapport 1 : 1.

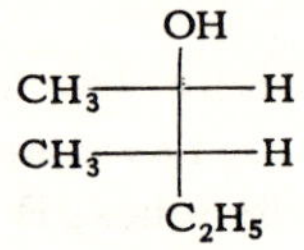

(IVa)

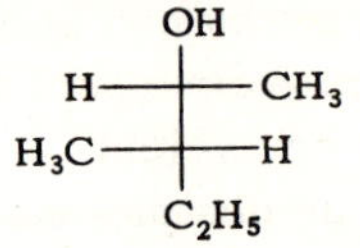

(IVb)

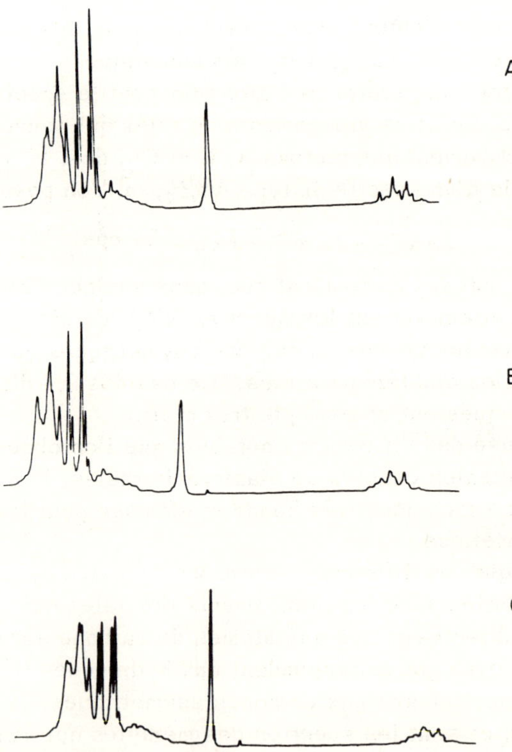

FIG. 4. (a) Spectre RMN, en $CHCl_3$, d'un mélange des deux 3-méthyl-2-pentanoles, enrichi dans le diastéréoisomère *érythro*. (b) Spectre RMN, en $CHCl_3$, d'un mélange des deux 3-méthyl-2-pentanoles, enrichi dans le diastéréoisomère *thréo*. (c) Mélange, en rapport 1 : 1, de (A) et (B).

Partie Expérimentale

Instrument

Les spectres RMN ont étés obtenus par un spectrographe Varian DP-60, avec les composés en solution de $CHCl_3$ et employant le tétraméthyl-silane comme référent interne.

On a employé la technique des "bandes latérales" pour la détermination quantitative des valeurs des déplacements chimiques. La température était de 25° ± 1 °C.

Composés

Un melange de diastéréoisomères de l'acide 2,3-diméthyl-succinique a été obtenu par réaction du α-bromo-propionate éthylique avec méthyl-malonate-diéthylique, selon Cason et Schmitz (*4*).

Par cristallisation fractionnée de la solution aqueuse d'un tel mélange, on a obtenu un échantillon d'acide *méso*-2,3-diméthyl-succinique qui fondait à 194 °C (*5, 6*).

Par estérification avec diazométhane on a obtenu l'ester méthylique correspondant, avec: Eb/23 mm Hg: 100 °C et n_D^{25} 1.4207 (*5*).

De la phase liquide, après trois recristallisations successives, on a obtenu l'acide *d, l*-2,3-diméthyl-succinique, trois fois plus soluble du *méso*, qui fondait à 128 °–129 °C (*5, 6*).

L'ester méthylique correspondant, encore obtenu par réaction avec CH_2N_2, avait: Eb/23 mm Hg: 100 °C et n_D^{25} 1.4220 (*5*).

L'acide 3,4-diméthyl-adipique *méso* (IIa) ayant $F = 131$ °C (*2*) a été préparé selon Ställberg-Stenhagen (*2*).

L'ester méthylique, obtenu par estérification de (IIa) avec méthanol et traces de H_2SO_4, avait: Eb/25 mm Hg: 135 °C et n_D^{25} 1.4345 (*2*).

Un mélange des acides 3,4-diméthyl-adipiques diastéréoisomères, ayant $F =$ 107 °C, enrichi seulement partiellement dans le diastéréoisomère *d, l* (IIb), par cristallisation fractionée de éther éthylique et après de l'eau, a donné, par estérification avec méthanol et H_2SO_4, un mélange d'esters méthyliques ayant Eb/25 mm Hg: 135 °C. L'acide 2,4-dimethyl-glutarique *d, l* (IIIb) ayant $F =$ 140 °C (*7*) a été préparé par cristallisation fractionnée des sels de calcium d'un mélange d'acides 2,4-diméthyl-glutariques diastéréoisomères, selon Auwers et Thorpe (*7*). L'ester méthylique a été obtenu par estérification avec CH_2N_2 et avait: Eb/25 mm Hg: 109 °–110 °C.

L'acide 2,4-diméthyl-glutarique *méso* (IIIa) ayant $F = 127$ °–128 °C (*7*) a été obtenu par hydrolyse de l'anhydride correspondant préparé d'un mélange d'acides 2,4-diméthyl-glutariques diastéréoisomères chauffé au reflux avec anhydride acétique.

Par estérification avec CH_2N_2 on a obtenu l'ester méthylique correspondant, avec : Eb/25 mm Hg: 109 °–110 °C.

Un mélange de (–) (3S)-3-méthyl-2-pentanoles (IVa) et (IVb), diastéréoisomères, ayant Eb/135 °C (*8*) a été obtenu par réaction entre (+) (S)-2-méthyl-butanale et bromure de méthyl-magnesium.

Des essais préliminaires, pour avoir des indications sur la composition des fractions enrichies par rectification de ces alcools, ont étés effectués par chromatographie en phase gazeuse, en utilisant des colonnes de 4 mètres de longueur et ayant comme phase stationnaire poly-(éthylène-glycol) sur Celite. La température était 75 °C, le débit 160 cm^3/mn, la pression 1.7 kg/cm^3.

Remercients

Les auteurs remercient les Professeurs P. Pino et E. Scrocco de l'intérêt pris à ce travail, et le C.N.R. d'avoir partiellement financé ces recherches.

Bibliographie

1. F. A. Bovey, *J. Polymer Sci.* **46**, 59 (1960) ; J. C. Woodbrey, *J. Polymer Sci.* **B2**, 315 (1964) ; F. C. Stehling, *J. Polymer Sci.* **A2**, 1815 (1964) ; S. Brownstein, S. Bywater et D. J. Worsfold, *J. Am. Chem. Soc.* (*Div. Polymer Chem., Preprints*) **3** (2), 134 (1962).
2. S. Ställberg-Stenhagen, *Arkiv Kemi* **3**, 249 (1951).
3. C. L. Arcus, J. M. J. Page et J. A. Reid [*J. Chem. Soc.* p. 1213 (1963)] ont observé que dans la chromatographie en phase gazeuse le (2RS, 3RS)-3-méthyl-2-pentanole a un temps de rétention plus bas du diastéréoisomère correspondant (2RS, 3SR) quand on utilise comme phase stationnaire le poly (éthylène-glycol)-400 sur Celite.
Pour celà nous croyons que le mélange (A) plus riche en diastéréoisomère ayant un temps de retention plus bas, soit constitué d'un excès du diastéréoisomère *érythro*.
4. J. Cason et F. Schmitz, *J. Org. Chem.* **28**, 555 (1963).
5. E. Berner et R. Leonardsen, *Ann. Chem.* **538**, 1 (1939).
6. W. A. Bone et C. H. G. Sprankling, *J. Chem. Soc.* **75**, 852 (1899).
7. K. Auwers et J. F. Thorpe, *Ann. Chem.* **285**, 315 (1895).
8. F. H. Norton et H. B. Hass, *J. Am. Chem. Soc.* **58**, 2147 (1936).

Long-Range $^{13}C—H$ Coupling Constants

H. DREESKAMP and E. SACKMANN

Laboratorium für physikalische Chemie,
Technische Hochschule, Stuttgart, Germany

This paper reports on some of our recent work in the investigation of the indirect nuclear spin-spin coupling between proton and carbon-13 separated by more than one chemical bond (e.g., a two-bond coupling denoted by J_{C-C-H}). Typically these constants are a few cycles per second and can have either sign. Owing to their small magnitude and the low natural abundance of the only carbon isotope with spin 1/2 they will be observed only at perfect resolution of the spectrometer (*1, 2*), by using substances enriched in carbon-13 (*3*) or else by a special H-{C} double irradiation technique (*4*).

Experimentally the signs of these coupling constants are measured relative to those of the one-bond ^{13}C-H coupling constants J_{C-H} (*5*), so it is essential to be certain about the absolute signs of these, from which the absolute signs of long-range coupling constants follow. The carbon–proton coupling over one bond has been found to be proportional to the probability density of the binding electron at the carbon nucleus (*6*) and may therefore be interpreted in terms of the hybridization at the carbon atom (*7, 8*). In particular for CH_3X compounds, where X is Li, $BeCH_3$, or $B(CH_3)_2$, the J_{C-H} have been measured and are plotted together with the other compounds where X is any of the first-row elements as a function of the product of the electronegativity of X (*9*) and the distance of C–X (see Fig. 1) (*10*).[1]

This good correlation further supports a maximum overlap calculation given earlier (*1*) and in particular justifies the assumption made in the following that all one-bond C–H couplings are positive as follows from theory (*6*) and from some experiments on partially oriented molecules (*11*). Essentially three different methods have been applied to investigate signs and magnitudes of long-range C–H couplings:

(1) Exact analysis of the ^{13}C satellites, which may become possible by a judicious choice of the effective chemical shift through selection of the static magnetic field or selective dilution shift as outlined in previous work (*5, 12*). The data of 2, 3, and 4 of Fig. 4 were obtained in this way.

(2) H—{H} decoupling of the ^{13}C satellites while observing the long-range ^{13}C satellites. By this method the data of 5, 6, 7, and 8 of Fig. 4

[1]For Li, the covalent radius was taken.

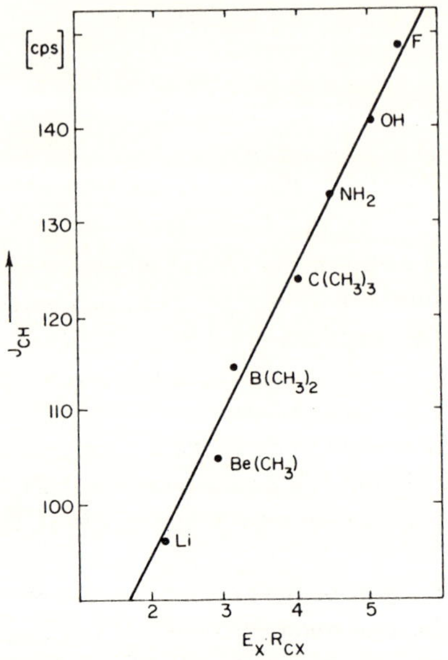

FIG. 1. Values of carbon–proton coupling constants in CH_3X compounds plotted against the product of electronegativity times distance of *X*. Neat liquids, except $LiCH_3$ and $Be(CH_3)_2$ as 30% solution in ether.

were obtained. As an example the case of acetaldehyde is presented in Fig. 2. In the upper half of Fig. 2 the spectrum of the aldehyde proton is presented with satellites due to a ^{13}C at the methyl group. In the left-hand part the low-field methyl-^{13}C satellite (B_1') is irradiated, while in the right-hand part the high-field methyl-^{13}C satellite (B_1) is irradiated. In the lower half of Fig. 2 the spectra of the methyl protons of acetaldehyde are similarly presented. The left-hand part presents the spectrum obtained while irradiating the low-field aldehyde proton satellite (A_1'), the middle part without perturbing field, and the right-hand part while irradiating the high-field aldehyde proton satellite (A_1). These results, together with the sign of the H–H coupling constant from another experiment not covered in this work, are summarized in Fig. 3.

(3) Even when the long-range satellites are completely obscured by the resonance of protons not coupled to ^{13}C, the sign and magnitude of a long-range coupling was observed by a "tickling experiment" (*13*). Here the effect of a second small rf field close to the resonance of the long-range satellite is observed on the rather strong and well-separated one-bond satellite.

From all the results obtained thus far it is apparent that the two-bond C–C–H coupling of an ethane-like fragment is determined by essentially

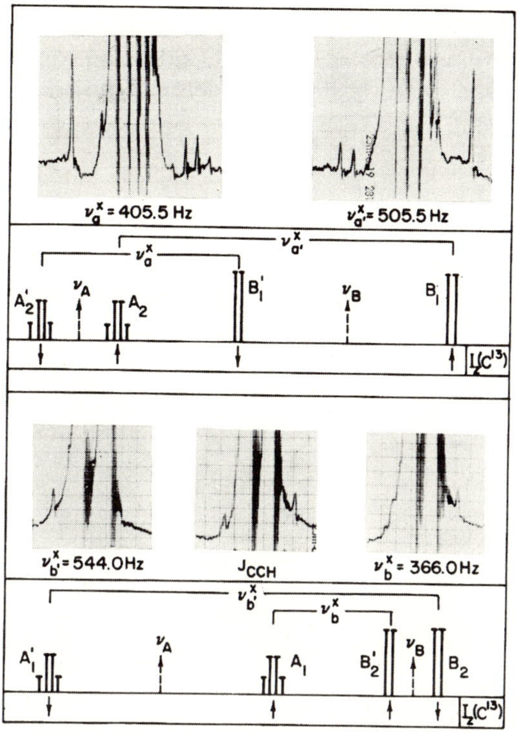

FIG. 2. Double resonance spectra of acetaldehyde. See text for explanation.

two factors: The s character of the coupling carbon in the C–C bond and the hybridization of the intermediate carbon atom. Such a behavior had been expected from the similar case of the geminal H–H coupling in substituted methanes (*14*) and the heteronuclear coupling of tetramethyl compounds of group IV elements (*15*).

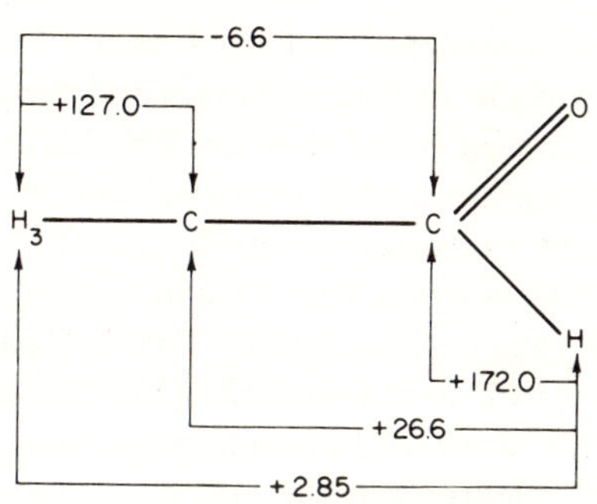

FIG. 3. Spin coupling constants in acetaldehyde.

In Fig. 4 all geminal $^{13}C_2$–C_1–H coupling constants whose magnitude and sign are known are plotted as a function of the one-bond J_{C_1-H} after correction has been made for the *s* character of the coupling carbon in the C–C bond, i.e.,

$$J^* = J \frac{\rho_{C_2C_1}(sp^3)}{\rho_{C_2C_1}(sp^n)},$$

where J is the measured coupling constant and $\rho_{C_2C_1}(sp^n)$ the *s* character in the C–C bond estimated from C–H coupling constants (*16*).

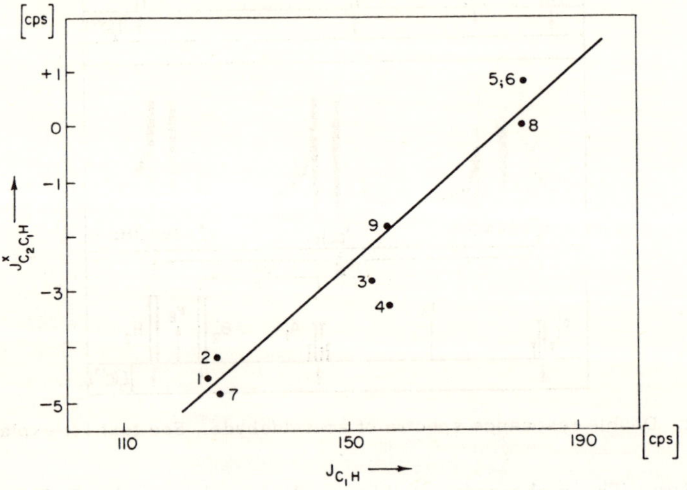

FIG. 4. Two-bond carbon–proton coupling constants adjusted for 25% *s* character of the coupling carbon in the C–C bond plotted against J_{C_1-H}. (1) Ethane (*18*); (2) ethyliodide (*5*); (3) 1,2-dichloroethane; (4) 1,2-dibromoethane; (5) 1,1,2,2-tetrachloroethane; (6) 1,1,2,2-tetrabromoethane (*19*); (7) acetaldehyde; (8) dichloroacetaldehyde; (9) ethylene (*18*).

Since the correlation presented in Fig. 4 seems to be of quite general nature an attempt was made to find a theoretical formulation which reproduces these results. It was found that the VB theory of spin-spin coupling, using the Dirac vector model as formulated by Alexander (*17*) and semiempirical values of exchange integrals, gives $J_{C-C-H} = -5$ cps in ethane (exp $= -4.5$ cps), roughly zero in ethylene (exp $= -2.4$ cps), and $+20$ cps in acetylene (exp $= +49.4$ cps). The quoted experimental values are those from Lynden-Bell and Sheppard (*18*).

A full account of this work will be published later. The authors wish to thank Prof. Dr. Th. Förster and the Deutsche Forschungsgemeinschaft for support of this work.

References

1. H. Dreeskamp and E. Sackmann, *Z. Physik. Chem. (Frankfurt)* **34**, 273 (1962).
2. N. Muller, *J. Chem. Phys.* **36**, 359 (1962).
3. G. J. Karabatsos, J. D. Graham, and F. M. Vane, *J. Am. Chem. Soc.* **84**, 37 (1962).
4. W. A. Anderson, *J. Chem. Phys.* **37**, 1373 (1962). See also the paper by W. A. Anderson and R. Freeman, in this volume.
5. H. Dreeskamp and E. Sackmann, *Z. Physik. Chem. (Frankfurt)* **34**, 261 (1962).
6. M. Karplus and D. M. Grant, *Proc. Nat. Acad. Sci. US* **45**, 1269 (1959).
7. N. Muller and D. E. Pritchard, *J. Chem. Phys.* **31**, 768, 1471 (1959).
8. J. N. Shoolery, *J. Chem. Phys.* **31**, 1427 (1959).
9. M. L. Huggins, *J. Am. Chem. Soc.* **75**, 4123 (1953).
10. "Tables of Interatomic Distances." Chem. Soc., London, 1958.
11. A. D. Buckingham and E. G. Lovering, *Proc. Chem. Soc.*, p. 144 (1963).
12. H. Dreeskamp, E. Sackmann, and G. Stegmeier, *Ber. Bunsenges. Physik. Chem.* **67**, 860 (1963).
13. R. Freeman and W. A. Anderson, *J. Chem. Phys.* **37**, 85 (1962).
14. H. J. Bernstein and N. Sheppard, *J. Chem. Phys.* **38**, 3012 (1963).
15. H. Dreeskamp, *Z. Naturforsch.* **19c**, 139 (1964).
16. K. Frei and H. J. Bernstein, *J. Chem. Phys.* **38**, 1216 (1963).
17. S. Alexander, *J. Chem. Phys.* **34**, 106 (1961).
18. R. M. Lynden-Bell and N. Sheppard, *Proc. Roy. Soc. (London)* **A269**, 385 (1962).
19. R. Freeman and W. A. Anderson, *J. Chem. Phys.* **39**, 806 (1963).

Stereospecificity of Long-Range Coupling Constants in Fluorinated Alkenes

N. BODEN, J. W. EMSLEY,[1] J. FEENEY,[2] and L. H. SUTCLIFFE

The Donnan Laboratories, The University of Liverpool, Liverpool, England

Introduction

There is a good deal of interest at present in the effects contributing to fluorine–fluorine coupling constants. A study of vicinal coupling constants in substituted fluoroalkanes has shown that the averaged coupling constant J_{FF} tends to zero when the sum of the electronegativities of the substituents is at its highest value (*1, 2*). It is assumed that vicinal coupling takes place only by a through-bond mechanism but that all other fluorine–fluorine nuclear interactions may have a contribution from a through-space mechanism if the fluorine nuclei in question are separated by less than about 2.7 Å (*3*). A difficulty attendant on resolving the relative importance of the various coupling mechanisms is that of determining the exact conformation of a particular molecule. We have selected for study *cis*- and *trans*-iso-$C_3F_7CH{=}CHF$ and $CF_2{=}CFSF_5$. The geometric isomers lend themselves to characterization of the individual rotamers by studying the temperature variation of the coupling constants, while the sulfur hexafluoride derivative gives rise to no conformational problems.

Experimental

^1H and ^{19}F resonance spectra of the compounds were obtained using a Varian V-4300B spectrometer operating at 56.4 and 40.0 Mc/s. The samples were examined both as neat liquids and as 20% v/v solutions in $CFCl_3$. A thermostated 40 Mc/s probe was used in the temperature studies: the sample temperature was held constant to within ±1°K over the range 170° to 380°K. Corrections were made to allow for temperature differences between the sample and the sensing thermocouple in the probe (*4*). In these experiments extra care was taken, a minimum of 40 spectra being recorded at each temperature.

About 10 recordings of each spectrum were made for the usual room temperature measurements. The *cis*- and *trans*-iso-$C_3F_7CH{=}CHF$ samples were kindly provided by Professor W. K. R. Musgrave of Durham University and the sample of $CF_2{=}CFSF_5$ was generously supplied by Dr. H. L. Roberts of the Alkali Division of I.C.I. Ltd.

[1] *Present address*: Chemistry Department, The University of Durham, South Road, Durham City, England.

[2] *Present address*: Varian Research Laboratory, Walton-on-Thames, Surrey, England.

Results

cis- and *trans-*iso-$C_3F_7CH{=}CHF$

The ^1H and ^{19}F spectra of both these compounds contain many bands since all but one of the possible coupling constants manifest themselves. All the nuclei, with the exception of H_1 and H_2 in the *trans* isomer, are weakly coupled and the first-order spectra can be analyzed simply (*5*).

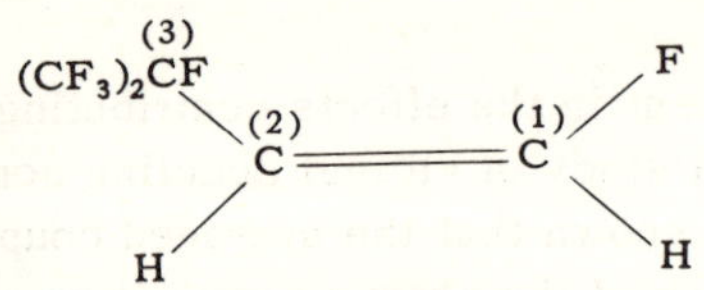

(A)

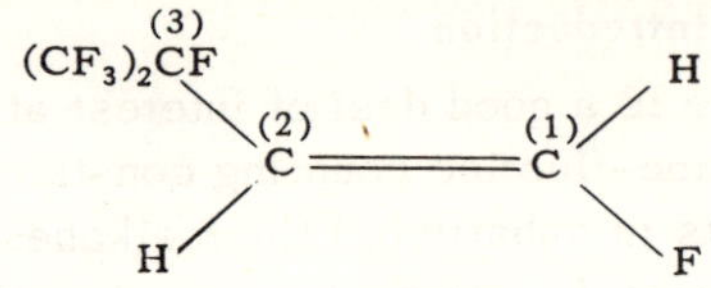

(B)

The room temperature spectral parameters are listed in Table I and II.

Information on the conformational behavior of the two isomers was obtained mainly from temperature variation of the $\langle J_{H_2F_3}\rangle$ coupling constants as determined from the ^1H spectra. These coupling constants are particularly suitable for this purpose because they change by 2 to 3 c/s over the temperature range available.

In both compounds, separate spectra of the individual rotamers could not be frozen out at 170°K. Hence the experimentally observed coupling constants and chemical shifts are time-weighted averages of the spectral parameters of the contributing rotamers. The relevant experimental data are shown in Tables III and IV; similar data (see Table V) were obtained for $\langle J_{H_1F_3}\rangle$, $\langle J_{F_1F_3}\rangle$, and $\langle J_{F_1(CF_3)}\rangle$ of the *cis* isomer and for $\langle J_{F_1F_3}\rangle$ of the *trans* isomer.

Microwave spectra of fluoropropenes (*6, 7*) have been interpreted on the basis that one of the methyl group substituents lies in the plane of the

TABLE I

ROOM TEMPERATURE COUPLING CONSTANTS (C/S)[a]

	(A)	(B)
$J_{H_1H_2}$	5.9 ± 0.1	11.2 ± 0.2
$J_{H_1F_3}$	2.10 ± 0.05	~0
$J_{H_1F_1}$	77.3 ± 0.1	77.7 ± 0.3
$J_{H_1(CF_3)}$	~0	~0
$J_{H_2(CF_3)}$	0.7 ± 0.1	0.68 ± 0.05
$J_{H_2F_3}$	22.8 ± 0.1	21.1 ± 0.1
$J_{H_2F_1}$	36.8 ± 0.2	16.5 ± 0.1
$J_{F_3(CF_3)}$	7.6 ± 0.3	7.6 ± 0.3
$J_{F_1F_3}$	30.8 ± 0.2	5.3 ± 0.3
$J_{F_1(CF_3)}$	5.8 ± 0.1	0.56 ± 0.07

[a]See text for (A) and (B).

TABLE II

^{1}H AND ^{19}F CHEMICAL SHIFTS IN *cis* AND *trans*-ISO-$C_3F_7CH{=}CHF$ AND ISO-$C_3F_7CH{=}CF_2$ AT 25°C[a,b]

	iso-$C_3F_7CH{=}CHF$		
	cis	*trans*	iso-$C_3F_7CH{=}CF_2$
τ (H_2)	3.40 ± 0.01	3.04 ± 0.01	–
τ (H_2)	5.10 ± 0.01	4.41 ± 0.01	5.51 ± 0.01
δ CF(1)	108.27 ± 0.05	115.98 ± 0.05	71.64 ± 0.04[c]
δ CF(3)	184.64 ± 0.05	186.73 ± 0.05	185.93 ± 0.5
δ CF_3(3)	78.55 ± 0.05	78.66 ± 0.05	78.98 ± 0.05

[a]Conversion of these shifts to CF_3COOH external reference may be made by using the relationship: $\delta^{ext.}_{CF_3COOH} = \delta^{int.}_{CCl_3F} - 76.5$, in parts per million.

[b]The ^{19}F chemical shifts are given in parts per million relative to CCl_3F as internal standard; measurements were made on 20% v/v solutions of the compound in CCl_3F: positive shifts are to high field of the reference signal.

[c]Position of the $=CF_2$ resonance.

TABLE III

THE TEMPERATURE DEPENDENCE OF $\langle J_{H_2F_3} \rangle$ OBSERVED IN *cis*-ISO-$C_3F_7CH{=}CHF$[a]

Temperature (°K)	$\langle J_{H_2F_3} \rangle$ (c/s)	Temperature (°K)	$\langle J_{H_2F_3} \rangle$ (c/s)
176	23.73 ± 0.20	311	22.66 ± 0.15
190	23.65 ± 0.17	317	22.59 ± 0.12
204	23.58 ± 0.20	324	22.49 ± 0.16
219	23.48 ± 0.18	332	22.45 ± 0.17
233	23.40 ± 0.17	337	22.40 ± 0.13
250	23.25 ± 0.10	350	22.24 ± 0.12
269	23.08 ± 0.16	363	22.15 ± 0.10
294	22.86 ± 0.15	373	22.09 ± 0.12
307	22.73 ± 0.20	379	21.98 ± 0.10

[a]These measurements were made on the neat compound using a resonance frequency of 40 Mc/s. The values of the coupling constants are averages of about 40 determinations and the errors shown are standard deviations. The temperatures are considered accurate to within ±1°C.

vinyl group. We have assumed that only two of the six possible rotamers (5), labeled (I) and (II) in Fig. 1, are significantly populated: in these the large substituents are staggered with respect to the plane of the vinyl group. Systems of this type have two potential energy minima and the observed averaged coupling constant can be expressed by

TABLE IV
THE TEMPERATURE DEPENDENCE OF $\langle J_{H_2F_3} \rangle$ OBSERVED IN *trans*-ISO-$C_3F_7CH{=}CHF$[a]

Temperature (°K)	$\langle J_{H_2F_3} \rangle$ (c/s)	Temperature (°K)	$\langle J_{H_2F_3} \rangle$ (c/s)
183	23.32 ± 0.20	273	21.66 ± 0.20
193	23.11 ± 0.18	283	21.37 ± 0.21
203	23.06 ± 0.21	293	21.23 ± 0.18
213	22.84 ± 0.14	303	21.04 ± 0.14
223	22.60 ± 0.24	313	20.96 ± 0.17
233	22.55 ± 0.25	323	20.80 ± 0.20
243	22.29 ± 0.24	333	20.65 ± 0.17
253	21.98 ± 0.23	343	20.57 ± 0.23
263	21.76 ± 0.15	353	20.47 ± 0.17

[a]These observations were made on a 20% v/v solution of the compound in CCl_3F and at a resonance frequency of 40 Mc/s. The values of the coupling constants are averages of about 40 determinations and the errors shown are standard deviations. The temperatures are considered accurate to with- within ±1°C.

TABLE V
EFFECT OF TEMPERATURE ON COUPLING CONSTANTS INFLUENCED BY INTERNAL ROTATION

	Temperature (°K)	Coupling constants (c/s)	
		Observed[a]	Calculated[b]
cis-iso-$C_3F_7CH{=}CHF$			
$\langle J_{H_1F_3} \rangle$	172	2.18 ± 0.05[c]	(2.18)
	295	2.10 ± 0.05[c]	2.08
	372	1.98 ± 0.05[c]	(1.98)
$\langle J_{F_1(CF_3)} \rangle$	203	4.40 ± 0.12	(4.40)
	297	5.81 ± 0.10	5.84
	353	6.54 ± 0.13	(6.54)
$\langle J_{F_1F_3} \rangle$	203	32.05 ± 0.23	(32.05)
	297	30.81 ± 0.16	30.80
	353	29.98 ± 0.14	(29.98)
trans-iso-$C_3F_7CH{=}CHF$			
$\langle J_{F_1F_3} \rangle$	203	7.32 ± 0.16	(7.32)
	297	5.34 ± 0.30	(5.34)
	393	4.41 ± 0.20	4.40

[a]The values of the coupling constants are averages of about 40 determinations and the errors shown are standard deviations.
[b]The values quoted in parentheses were used to calculate the third value.
[c]These values were taken from a curve drawn through values obtained at 17 different temperatures.

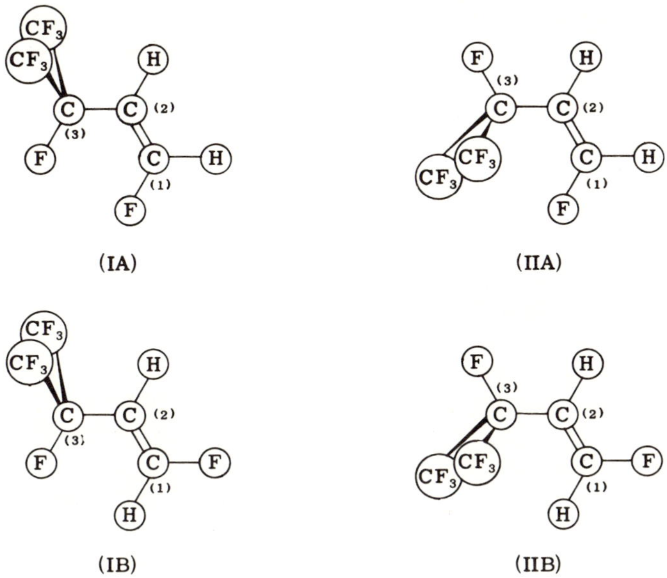

FIG. 1.

$$\langle J \rangle = P_{\mathrm{I}} J_{\mathrm{I}} + P_{\mathrm{II}} J_{\mathrm{II}}\,, \tag{1}$$

where J_{I} and J_{II} are the coupling constants associated with rotamers (I) and (II), and P_{I} and P_{II} are the corresponding fractional populations. The latter are given by

$$P_{\mathrm{I}} = [\exp(-\Delta H/RT)]/[1 + \exp(-\Delta H/RT)] \tag{2}$$

and

$$P_{\mathrm{II}} = [1 + \exp(-\Delta H/RT)]^{-1}\,, \tag{3}$$

where the enthalpy difference, ΔH, between the rotamers may be written $\Delta H = H_{\mathrm{I}} - H_{\mathrm{II}}$. Substituting expressions (3) and (2) into Eq. (1) leads to an equation containing the three unknowns J_{I}, J_{II}, and ΔH:

$$\langle J \rangle = [J_{\mathrm{I}} \exp(-\Delta H/RT) + J_{\mathrm{II}}]/[1 + \exp(-\Delta H/RT)]. \tag{4}$$

It has been assumed that the partition functions for each pair of rotamers are equal.

The whole procedure has yet to be checked by deriving component coupling constants by "freezing-out" each rotamer at low temperatures and then checking these values by increasing the temperature until both rotamers have equal populations, thus giving the mean J values. The results should show whether one is justified in assuming that coupling constants in the rotamers are temperature invariant. Unfortunately, this procedure could not be adopted here because the available temperature range (173° to 373°K) was too small. To date, no one has succeeded in making a complete investigation of a temperature range leading to limiting values of coupling constants at each end[3] of this range. We extracted values for the individual rotamer coupling constants by a least-squares fitting of the observed coupling constants at different temperatures to coupling constants derived from Eq. (4). A computer program was written to enable values of $\langle J_{H_2F_3} \rangle$ to be generated for a wide range of ΔH values using values of J_I and J_{II} coupling constants appropriate to each value of ΔH. The derived enthalpy differences are -1700 ± 100 cal/mole and -900 ± 200 cal/mole for the *cis* and *trans* isomers, respectively. The coupling constants of the individual rotamers are presented in Table VI together with the corresponding coupling constants calculated from $\langle J_{H_1F_3} \rangle$, $\langle J_{F_1F_3} \rangle$, and $\langle J_{F_1(CF_3)} \rangle$ using these values of ΔH. A striking feature of Table VI is the marked similarity between the individual coupling constants calculated from $\langle J_{H_2F_3} \rangle$ for the rotamers of both the *cis* and *trans*

TABLE VI

COUPLING CONSTANTS (C/S) IN ROTAMERS OF *cis*- AND *trans*-ISO-$C_3F_7CH{=}CHF$[a]

	cis		*trans*	
	(IA)	(IIA)	(IB)	(IIB)
$J_{H_2F_3}$	23.89 ± 0.06	4.1 ± 2	25.0 ± 0.7	3.9 ± 1
$J_{F_1F_3}$	32.5 ± 0.1	2.6 ± 4	±9.8 ± 1	∓14.9 ± 2
$J_{H_1F_3}$	±2.20 ± 0.01	∓0.1 ± 0.3	<0>	
$J_{F_1(CF_3)}$	3.9 ± 4	39 ± 1	<0.56 ± 0.7>	

[a]See Fig. 1.

isomers. This agreement justifies to some extent the assumption that only two rotameric forms need to be taken into account. Furthermore, the excellent agreement between the observed and calculated values (see Table V) of four different coupling constants of the two isomers gives added confidence in the potential function chosen for the barrier to internal rotation.

$CF_2{=}CFSF_5$

All the possible F–F spin-spin interactions are featured in the ^{19}F spectrum of the compound. Some of the nuclei are strongly coupled and a detailed analysis would require the molecule to be treated as an AB_4PQX spin system. There are four main regions of the spectrum, two belonging

[3]Newmark and Sederholm (*8*) attempted to make a complete investigation but the temperature range they used was too small (*1*).

to the A and B_4 parts of the spectrum characteristic of the SF_5 group, one arising from the strongly coupled PQ spin system of the CF_2 group, and the remaining one to the CF group that behaves like an X-type nucleus. A complete analysis of an AB_4PQX spin system is a formidable undertaking: fortunately our spectrum lends itself to an approximate method of analysis. Details of the analysis of the ^{19}F spectrum of $CF_2{=}CFSF_5$ are given elsewhere (*9*). The spectral parameters of the compound are given in Table VII.

TABLE VII

^{19}F CHEMICAL SHIFTS AND COUPLING CONSTANTS OF

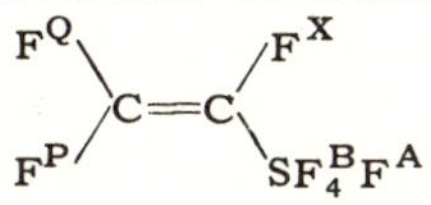

Coupling constants (c/s)[a]		Chemical shifts (ppm)[b]	
J_{PQ}	52.5	δ_A	−69.3
{ J_{QX}	±57.2	δ_B	−59.2
J_{PX} }	∓117.6	δ_P	+99.9
J_{AB}	151.0	δ_Q	+100.4
J_{BX}	6.1	δ_X	+163.2
{ J_{BP}	±15.3		
J_{BQ} }	±11.1		
J_{AX}	3.0		
{ J_{AP}	±1.8		
J_{AQ} }	±0.88		

[a] The relative signs of the coupling constants apply only to pairs in braces. The accuracy of J_{AB} is ±0.5 c/s while the remainder of the coupling constants are accurate to within ±0.2 c/s.

[b] The chemical shifts were measured relative to the internal reference CCl_3F and are accurate to within ±0.2 ppm: positive values are to high fields of the reference.

Discussion

cis- and *trans-iso-*$C_3F_7CH{=}CHF$

Because some of the coupling constants collected in Table VI are of opposite sign, it is clear that a knowledge of the relative signs of all the coupling constants would be of considerable help in understanding the stereospecificity of coupling mechanisms.

The magnitudes of the long-range coupling constants can be seen to be sensitive to the geometry of the molecules. A comparison of the H–F and

F–F coupling constants indicates that a different mechanism operates in each case.

The small value of the long-range coupling constant $J_{H_1F_3}$ observed in rotamer (IB) (Fig. 1) indicates that H–F coupling is not controlled by spatial interactions. Furthermore, this coupling in rotamer (IA) is numerically larger than that in (IIA) where the nuclei are *trans-trans* oriented. In H–H coupling, the *trans-trans* nuclei usually have the largest coupling (*10, 11*).

The really interesting couplings are those between fluorine nuclei which allow the through-space postulate of Petrakis and Sederholm (*3*) to be tested. Undoubtedly, there are some large coupling constants for fluorine nuclei in close spatial proximity but separated by five bonds ($J_{F_1(CF_3)}$ = 39 c/s) in (IIA). Nevertheless, there are some appreciable spin-spin interactions for fluorine nuclei well-separated spatially and interacting through four bonds [$J_{F_1F_3} = \mp 14.9$ c/s in (IIB)] and five bonds ($J_{F_1(CF_3)}$ = 3.9 c/s). It is also interesting that $J_{F_1F_3}$ is large (= 32.5 c/s) in isomer (IA), where the two interacting nuclei are separated by only 2.4 Å. The corresponding coupling constant in a frozen-out rotamer of the structurally related compound $CF_2{=}CFCOF$ is also large (= 84.5 c/s) (*12*). The opposite signs found in this work emphasize the danger of simply comparing magnitudes of coupling constants, particularly when attempting to establish the presence of a spatial contribution to the coupling since contributions from such a mechanism would be expected to have the same sign.

$CF_2{=}CFSF_5$

This compound presents no ambiguities as regards the possible conformations. In the stable form, the basal fluorines of the SF_5 group are staggered with respect to the plane of the vinyl group. In looking at the four-bond coupling between the basal SF_5 fluorines and the fluorine in β *cis* position (J_{BP} = 15.3 c/s), we estimate the corresponding internuclear spatial separation to be 2.59 Å. According to the Petrakis and Sederholm relationship (*3*) the contribution from the spatial coupling mechanism should be 20 c/s (= 40/2 c/s) in good agreement with the observed value. However, the similar four-bond coupling constant J_{BQ} is as great as 11.1 c/s even though the spatial separation of the B and Q nuclei is too large for a through-space mechanism to be operative. Furthermore, the corresponding coupling constants of $CF_2{=}CFCF_3$ are similar in magnitude and probably in sign (*13*) (J_{CF_3CF} *cis* = 22.0 c/s and J_{CF_3CF} *trans* = 8.7 c/s) despite the fact that in this molecule all the nuclei involved are more than the critical 2.73 Å apart. There are several inadequacies in the correlation of internuclear separations with J_{FF} values and a criticism of this correlation has been made elsewhere (*14*). The importance of including the signs of coupling constants in discussions of this kind must be emphasized once again. For example, in F_5SOF J_{OF-SF_4} (= 18.3 c/s) is of opposite sign to J_{OF-SF} (= 0.6 c/s) (*15*). We hope to establish the sign relationship in our compound in the near future.

References

1. R. J. Abraham and L. Cavalli, *Mol. Phys.*, **9**, 67 (1965).
2. S. Ng and C. H. Sederholm, *J. Chem. Phys.* **40**, 2090 (1964).
3. L. Petrakis and C. H. Sederholm, *J. Chem. Phys.* **35**, 1243 (1961).
4. H. S. Gutowsky, G. G. Belford, and P. E. McMahon, *J. Chem. Phys.* **36**, 3353 (1962).
5. N. Boden, J. W. Emsley, J. Feeney, and L. H. Sutcliffe, *Proc. Roy. Soc. (London)* **A282**, 559 (1964).
6. S. Siegel, *J. Chem. Phys.* **27**, 989 (1957).
7. R. A. Beaudet and E. B. Wilson, *J. Chem. Phys.* **37**, 1133 (1962).
8. R. A. Newmark and C. H. Sederholm, *J. Chem. Phys.* **39**, 3131 (1963).
9. N. Boden, J. Feeney, and L. H. Sutcliffe, *Spectrochimica Acta.* **21**, 627 (1965).
10. A. A. Bothner-By and H. Gunther, *Discussions Farad. Soc.* **34**, 127 (1962).
11. C. N. Banwell and N. Sheppard, *Discussions Farad. Soc.* **34**, 115 (1962).
12. W. S. Brey and K. C. Ramey, *J. Chem. Phys.* **39**, 844 (1963).
13. J. D. Swalen and C. A. Reilly, *J. Chem. Phys.* **34**, 2122 (1961).
14. N. Boden, J. Feeney, and L. H. Sutcliffe, *J. Chem. Soc.* in press (1965).
15. R. K. Harris and K. J. Packer, *J. Chem. Soc.*, p. 4736 (1961); p. 3077 (1962).

Side Chain Long-Range Couplings in the NMR Spectra of Pyridinecarboxaldehydes

VALDEMAR J. KOWALEWSKI and DORA G. DE KOWALEWSKI

Department of Physics, Faculty of Exact Sciences, University of Buenos Aires, Buenos Aires, Argentina

In a previous paper (*1*) it was shown that in pyridine-3-carboxaldehyde the aldehydic proton was coupled with the ring proton in position 5 across five bonds, without coupling to the nearest proton in position 4. This situation is similar to the one found in benzaldehydes (*2*) and other aromatic ring systems (*3, 4, 5*). This aroused our interest to know if a similar situation holds in other pyridinecarboxaldehydes, but up to now the spectra of these substances have defied all attempts at interpretation. Experience gained by the authors with other pyridines (*1, 6*), improved technique by judicious use of solvent effects (*7*), and an improved computer program (*6*) at last allowed these difficulties to be overcome.

Experimental Details

The spectra were taken with a Varian DP-60 instrument operating at 60 Mc/sec and, eventually, at 19.25 Mc/sec. The spectra were usually field swept with the field increasing from left to right. Some of them were frequency swept with a device similar to the one of Freeman and Anderson (*8*) based on Primas' (*9*) method. In the present case the field was locked either on the solvent line (acetone) or the signal of hexamethyldisiloxane, a small amount of which was added for reference, instead of tetramethylsilane, due to its higher boiling point (99 °C). The locking signal was obtained from the lower sideband of 2000-cps modulation obtained from the crystal standard oscillator of a Muirhead-Wigan high precision decade oscillator. Another sideband obtained from a modified Hewlett-Packard 202-A low frequency function generator was used for sweeping the spectra. The function generator, which is basically a frequency linear instrument, was improved by the following means: improving the power supply stability; the use of a 15 turn potentiometer to increase frequency resolution; extension of the frequency range up to and beyond 2000 cps; and careful adjustment of linearity. Several partial ranges were provided and the span of each one was made adjustable so the scale on the recorder could be made exactly 1, 2, 4, or 8 mm/cps, according to the speed of the synchronous motor drive of the potentiometer. Some difficulties were encountered due to short circuits between turns in the latter.

Pyridine-2-carboxaldehyde

Figure 1 shows the experimental spectrum of this substance in acetone solution and at room temperature.

Several solutions of this substance at different concentrations in acetone were tried until one was found with the lines sufficiently spread out to give some hope of interpretation (*7*). As this is a five-spin strongly coupled system, many second-order effects are present which make the initial assignment rather difficult. There are several combination lines (marked Co in the drawings). Some of the high-field group (*A*) of lines are split and some are not. The signal of the aldehyde is a triplet with a central line that we now know is characteristic of a strongly coupled spectrum. Tentative assignments were made using previous experience

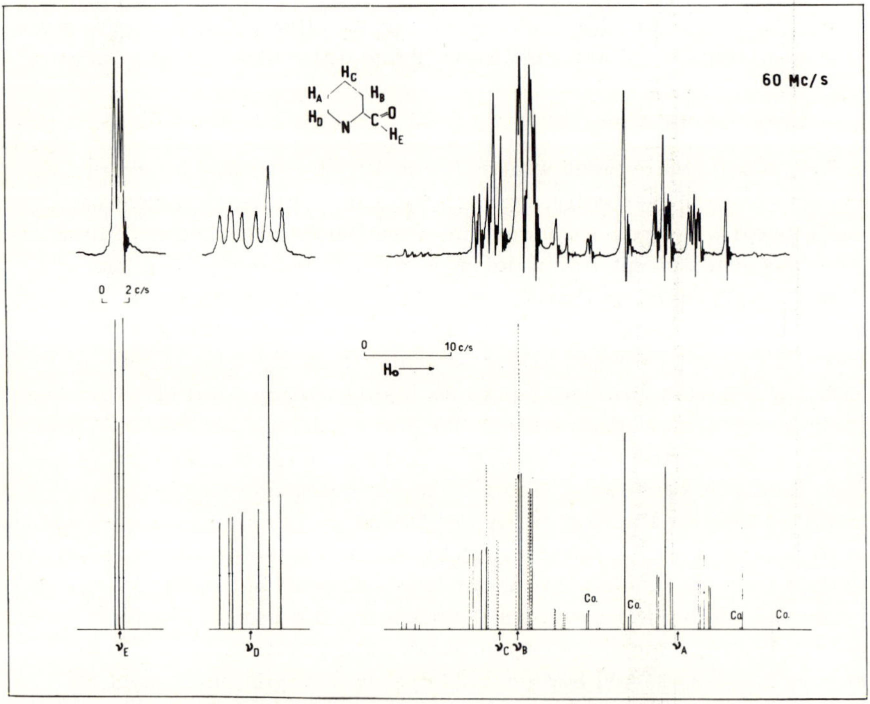

FIG. 1. Sixty Mc/s spectrum of 2-pyridinecarboxaldehyde together with the one calculated. It shows important second order effects, like the aldehyde "triplet," the splitting of the A lines even in the absence of a spin-spin coupling, and the combination doublets marked Co.

with the 2-substituted pyridines (*6*), assuming that coupling of the aldehydic proton was nonexistent and using the iterative ABCD program. The aldehyde coupling was then introduced with the noniterative ABCDE program and the theoretical spectrum was slightly improved by another iteration. Since there is no uncertainty as to the assignment of the *D* lines to proton 6 due to its proximity to nitrogen, the possible ambiguity of a four-spin system with two *ortho* couplings each is overcome. Now the splittings of the *A* lines can be accounted for. In a five-spin system, especially if it is a strongly coupled one, it is not necessary for a *J* coupling to exist in order to observe the splittings. With respect to the combination lines it is interesting to note that the computer shows the existence of four "combination doublets" symmetrically distributed among the *A* lines (three of these are visible in the experimental spectrum). An analysis of these doublets shows (*8*), quite similarly as for benzaldehydes, that they are functions not only of J_{CE} but also of J_{BE} and J_{AE}. It was verified in this way that the aldehydic proton has no other L.R.C. (long-range coupling), or at least none greater than 0.1 cps, except with proton 4 (J_{CE} = 0.69 cps). Figure 2 shows the splittings S_i of the four doublets for different values of the eventual long-range couplings.

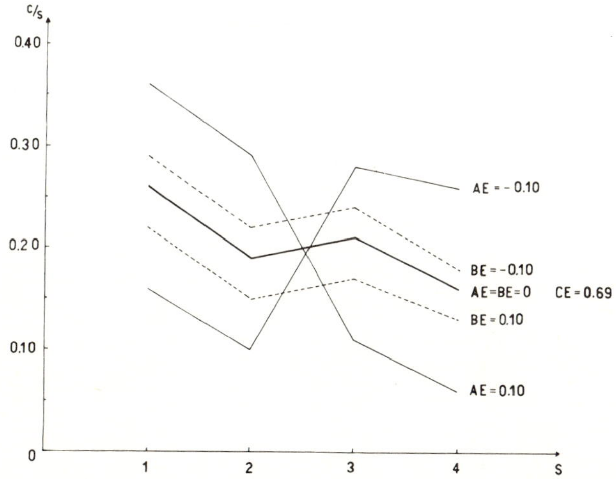

FIG. 2. Diagram showing the values taken by the splittings S_i of the combination doublets Co for several sets of values of the possible long range couplings. J_{CE} was obtained from the theoretical ABCDE spectrum. A second long range coupling J_{AE} or J_{BE} of even 0.1 c/s would introduce a marked change in the values of the splittings S_i of the combination doublets. This shows that J_{CE} = 0.69 c/s is the only long range coupling present in this case.

6-Methylpyridine-2-carboxaldehyde

The ring protons of this substance show a strongly coupled spectrum of rather wide lines which could not be interpreted until a dilute solution in acetone was prepared (Fig. 3). Owing to the narrowing of the lines and the solvent effect, the spectrum showed up in its full complexity. This spectrum was frequency swept. The aldehyde signal is a doublet of fine lines with strong wiggles. The methyl signal is a triplet with an unresolved fine structure. Definitely all ring protons are involved in several L.R.C., one probably with the aldehyde and two with the methyl group.

The analysis of the spectrum began with the study of the ABC spectrum which results when we discount the L.R.C., averaging out and counting as single lines the several doublets and quartets. The resulting spectrum was analyzed with the ABC iterative program assuming

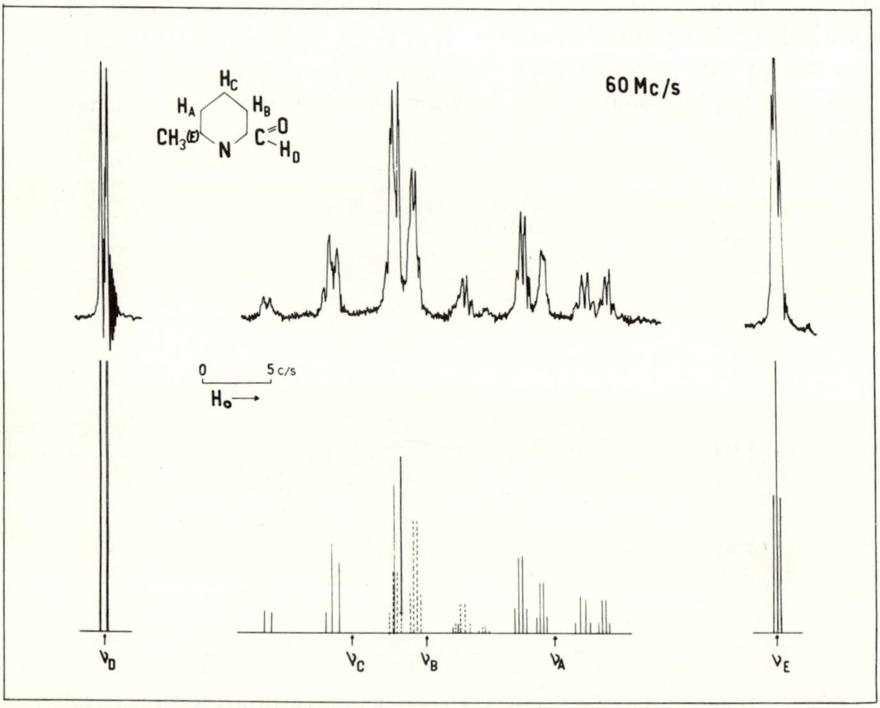

FIG. 3. Sixty Mc/s spectrum of 6-methyl-2-pyridinecarboxaldehyde showing: (1) a long range coupling between the aldehyde and the ring proton in position four $J_{CD} = 0.91$ c/s; (2) a long range coupling between the methyl proton E and the ring protons A and B in positions three and five with $J_{AE} = 0.46$ and $J_{BE} = 0.48$ c/s.

that the first four lines on the high-field side were due to one proton, and the four on the low-field side to another proton, which had two couplings *ortho*. Once this assignment for the ABC part had been verified, the effect of the aldehydic proton was introduced and the spectrum calculated with the iterative ABCD program. The effect of the methyl proton was now introduced with the ABCDE program in the way described before. It is evident that the aldehydic proton is coupled as usual to the proton in position 4 across five bonds (J_{CD} = 0.90), while the methyl group is coupled to both protons 3 and 5 (0.48 and 0.46 cps) but not to proton 4. To decide whether this rather unusual behavior of the methyl group was due to the presence of the aldehyde group, the spectra of some other 2-substituted 6-methylpyridines were studied (6-methyl-2-amino-, 6-methyl-2-acetyl-, and 6-methyl-2-cyanopyridine). Only the last one showed a nonclustered spectrum. Except for a slight difference in chemical shifts, it is similar to the previous one and shows that the L.R.C. of the methyl group occurs with its neighboring 5 proton (0.41 cps) and with proton 3 (0.54 cps). The aldehyde group itself thus does not seem responsible for this particular behavior.

References

1. V. J. Kowalewski and D. G. de Kowalewski, *J. Chem. Phys.* **36**, 266 (1962).
2. D. G. de Kowalewski and V. J. Kowalewski, *J. Chem. Phys.* **37**, 1009 (1962).
3. S. Gronowitz and R. A. Hoffman, *Acta Chem. Scand.* **13**, 1683 (1959).
4. R. A. Hoffman, B. Gestblom, and S. Gronowitz, *J. Mol. Spectr.* **11**, 454 (1963).
5. S. Forsen and B. Akemark, *Acta Chem. Scand.* **17**, 1712 (1963).
6. V. J. Kowalewski and D. G. de Kowalewski, *J. Chem. Phys.* **37**, 2603 (1962).
7. S. Clough, *Mol. Phys.* **2**, 349 (1959).
8. R. Freeman and W. A. Anderson, *J. Chem. Phys.* **37**, 2053 (1962).
9. H. Primas, *Proc. 5th European Congr. on Mol. Spectr., Amsterdam* (1961).
10. D. G. de Kowalewski and V. J. Kowalewski, *Mol. Phys.* (in press).

Couplages à longue distance dans des hétérocycles oxygénés et détermination des signes

D. GAGNAIRE, E. PAYO-SUBIZA et A. ROUSSEAU

Laboratoire de Chimie Organique Physique et de Résonance Magnétique du Centre d'Etudes Nucléaires, Grenoble, France

Introduction

Les couplages "à longue distance" sont ceux qui se manifestent à travers plus de trois liaisons, et pour lesquels est adoptée la notation classique ^4J ou ^5J, s'ils ont lieu à travers quatre ou cinq liaisons.

On peut noter tout de suite qu'il en existe de deux types principaux:

— ceux pour lesquels les liaisons entre noyaux couplés sont toutes des liaisons σ. Dans ce cas, pour les couplages ^4J, l'hypothèse de Meinwald (*1*) s'est révélée très satisfaisante : les conditions géométriques des couplages et une disposition en "M" favorisant un recouvrement de "petits lobes d'orbitales opposés à la liaison σ";

— ceux pour lesquels une ou plusieurs liaisons intermédiaires est une double liaison: les couplages allyliques ^4J (H—C=C—C—H) et "homoallyliques" ^5J (*2*) (H—C—C=C—C—H) sont alors prévus théoriquement (*3*) et observés dans de nombreux cas.

Le but de cette étude est la mise en évidence de couplages de ces différents types dans des hétérocycles oxygénés, tout en déterminant dans les cas où cela est possible les signes de ces couplages.

En effet, ces couplages sont en général faibles, et leurs signes très importants dans l'interprétation théorique que l'on peut en faire.

Mise en évidence des couplages

Les hétérocycles oxygénés, monocycliques ou bicycliques, se prêtent relativement bien à l'analyse des spectres. Les déplacements chimiques sont influencés par l'oxygène, ce qui permet le plus souvent des analyses totales des spectres et met à l'abri d'erreurs d'interprétations comme dans le cas où des "couplages virtuels" (*4*) apparaissent.

Les spectres dont il sera question s'analysent soit au premier ordre, soit rentrant dans des catégories classiques facilement analysables: A_2X_2, ABX, ABXY, ...

Le détail des spectres et de leur analyse figure dans le texte de la thèse de E. Payo-Subiza (*5*), et ne sera pas, en général, discuté ici.

Couplages ^5J homoallyliques dans les systèmes dihydrofuranniques

Ces couplages sont observés dans plusieurs cas (*6*), et l'oxygène hétérocyclique, par extension du système conjugué, semble y jouer un rôle important.

L'exemple le plus simple est l'α-angélica lactone (I), pour laquelle l'analyse du spectre AK_2X_3 donne:

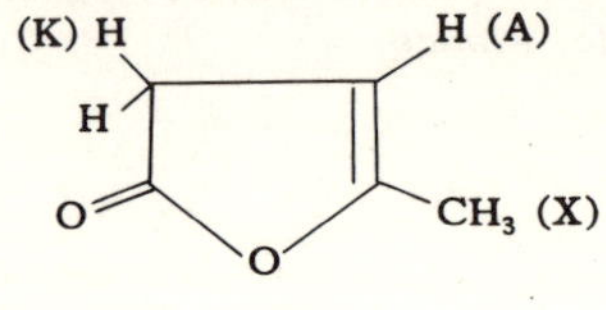

$J_{AK} = 2.5$ cps
$J_{AX} = 1.6$ cps
(allylique ^4J)
$J_{KX} = 2.7$ cps
(homoallylique ^5J)

(I)

C'est cet exemple qui a été choisi pour la détermination des signes (voir plus loin).

Couplages ^4J à travers l'oxygène du système oxa-7-bicyclo (2-2-1) heptanique

Les deux hydrogènes en tête de pont sont couplés dans tous les exemples étudiés (7),

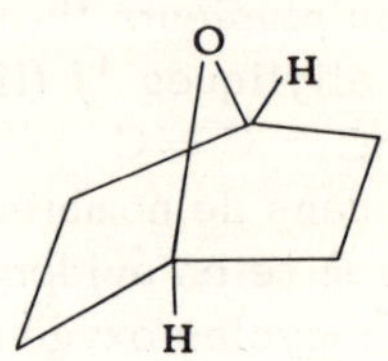

les valeurs des couplages variant de 1 à 1.4 cps.

On a cherché à voir si la présence *d'une double liaison supplémentaire* dans le cycle modifiait la valeur de ce couplage: il garde en effet une géométrie semblable, mais les deux protons sont *également en position homoallylique.* L'exemple choisi est le produit (II) de la synthèse diénique entre le furanne et l'acide maléique:

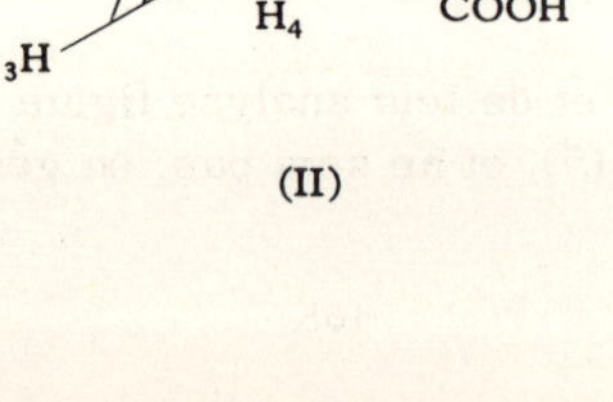

(II)

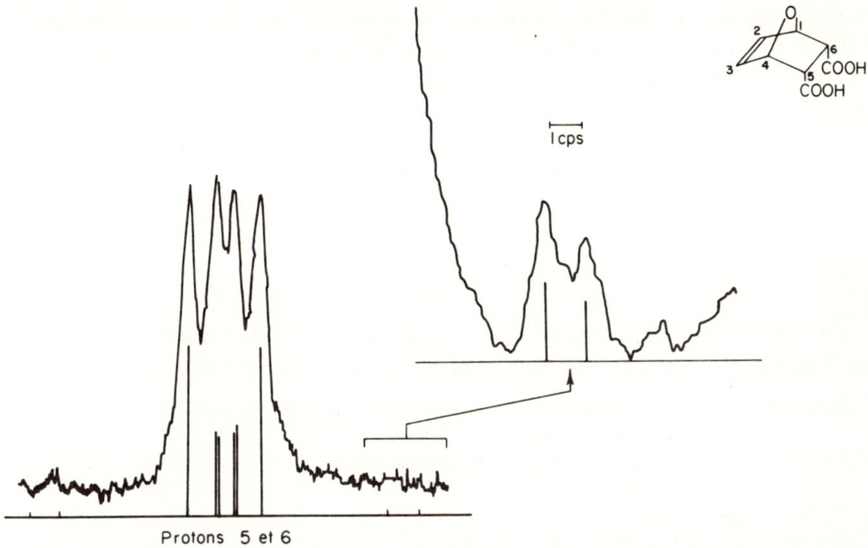

FIG. 1. Spectre des protons 5 et 6 du dérivé (II), et développement de la partie résonant à champ fort par un analyseur multicanaux.

Dans cet exemple, les protons 5 et 6 ne sont couplés qu'aux protons 4 et 1, et leur spectre est donc constitué par la partie X d'un spectre AA′XX′ (δ = 3.31 ppm, spectre dans D_2O).

L'analyse de ce spectre demande la détermination précise des raies "extérieures," noyées,dans le bruit de fond. L'utilisation d'un *analyseur multicanaux* permet cependant de les repérer avec précision (Fig. 1), ce qui conduit au résultat suivant de l'analyse:

J_{56} = 10 cps; $J_{16} = J_{45}$ = 4.3 cps; $J_{15} \neq 0$; J_{14} = 1 cps.

Ce dernier couplage ^{4}J est donc comparable à ceux obtenus en l'absence de la double liaison.

Couplage ^{4}J à travers trois atomes de carbone dans le système oxa-7-bicyclo (2-2-1) heptanique

L'hypothèse de Meinwald précédemment rappelée permet d'envisager dans les composés de cette série un couplage ^4J de type différent, entre un proton *exo* du cycle et un proton du pont, soit:

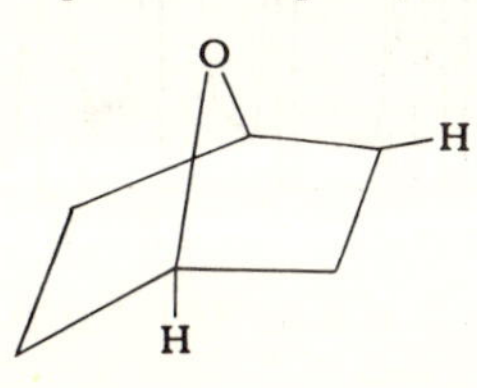

Quoique de valeurs très faibles, à la limite de résolution de l'appareil utilisé (Varian A.60), un tel couplage a été observé dans deux cas voisins, l'anhydride (III) et le diacide correspondant

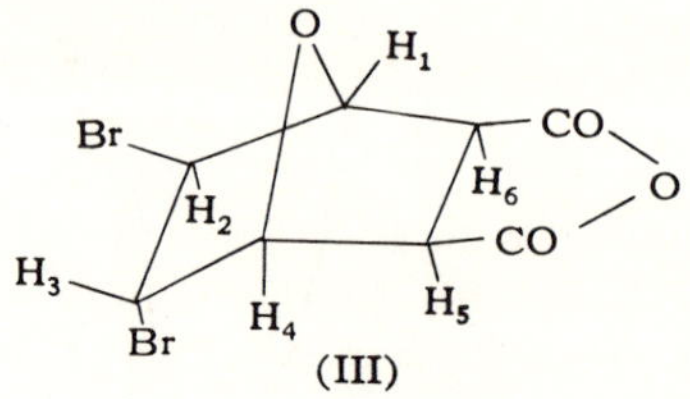

(III)

Le spectre de (III), réalisé dans l'acétone deutériée, conduit en effet au résultat suivant des couplages et déplacements chimiques:

Protons 5 et 6: spectre AB sans couplage avec les autres protons; 4.05 et 4.36 ppm, $J = 7.5$ cps

Proton 1: $\delta = 5.21$ ppm	$J_{13} = 0.6$ cps (couplage ^4J H—C—C—C—H)
Proton 2: $\delta = 4.58$ ppm	$J_{14} = 1.1$ cps
Proton 3: $\delta = 4.71$ ppm	$J_{23} = 3.4$ cps
Proton 4: $\delta = 5.41$ ppm	$J_{34} = 5$ cps
	(autres valeurs des couplages trop faibles pour être détectées)

La Fig. 2 montre les spectres théoriques et expérimentaux des protons 2 et 3.

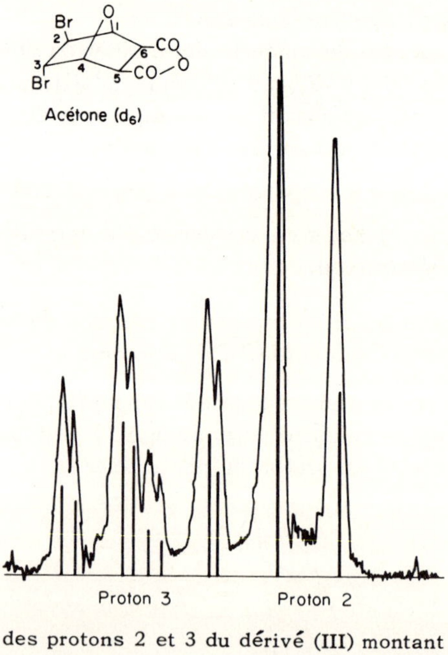

FIG. 2. Spectre des protons 2 et 3 du dérivé (III) montant le couplage J_{13}.

Couplage ^4J *de type* H—C—O—C—H *dans un système bicyclique oxygéné non ponté*

Le produit de photocondensation de la benzophénone et du furanne donne le dérivé (IV) (*8*) dont le spectre s'analyse au premier ordre sans ambiguïté (*6*) pour les quatre protons non aromatiques.

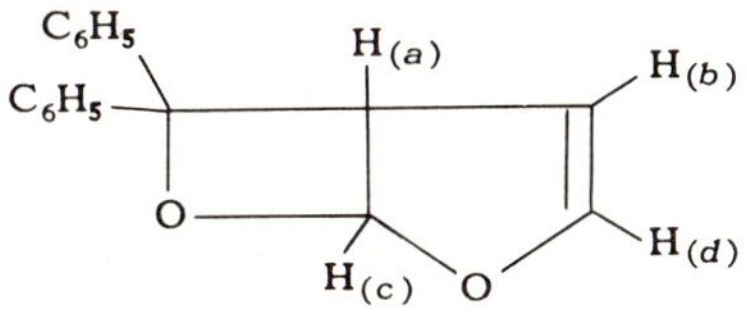

(IV)

$J_{ab} = 3.1$ cps	$\delta_a = 4.30$ ppm
$J_{ac} = 4.5$ cps	$\delta_b = 4.79$ ppm
$J_{ad} = 1.3$ cps	$\delta_c = 6.40$ ppm
$J_{bc} = 0$	$\delta_d = 6.31$ ppm
$J_{bd} = 2.9$ cps	
$J_{cd} = 0.8$ cps	

Un couplage $J_{cd} = 0.8$ cps existe donc dans ce dérivé, mais *disparaît lors de l'hydrogénation de la double liaison* (le spectre du proton $H_{(c)}$ devient un doublet correspondant au couplage avec $H_{(a)}$).

Détermination des signes

Résultats expérimentaux

Le principe est celui de la double irradiation sélective des raies correspondant aux mêmes états de spins de certains noyaux considérés (*9, 10*).

La technique expérimentale est celle des bandes latérales (*11*) et son application aux deux exemples considérés ici sera décrite par ailleurs (*12*). Les résultats sont les suivants:

Angélica lactone (composé 1): J_{AK} et J_{KX} sont de même signe, et J_{AX} de signe contraire. L'hypothèse la plus vraisemblable est donc J_{AK} (couplage vicinal ^3J) positif, J_{AX} (couplage allylique ^4J) négatif, et J_{KX} (couplage homoallylique ^5J) positif.

Produit de photocondensation (IV) : Tous les couplages non nuls sont de même signe sauf J_{ad}. L'hypothèse la plus vraisemblable est donc que seul ce couplage est négatif (couplage allylique ^4J), ce qui donne J_{ac}, J_{ab} et J_{bd} positifs (couplages ^3J) *ainsi que* J_{cd} (couplage ^4J, H—C—O—C—H).

Discussion des résultats

Les couplages allyliques sont prévus théoriquement comme étant négatifs (*3*), ce que confirme l'expérience dans les deux cas précédents, et d'autres résultats expérimentaux (*13*).

Les couplages homoallyliques sont prévus théoriquement comme étant positifs (*7*), ce que l'expérience confirme donc dans l'exemple de l'α-angelica lactone (I).

Quant aux couplages à travers quatre liaisons σ, la théorie et quelques exemples (H—C—C—C—H) donnent des couplages *négatifs* (*14*), contrairement au résultat trouvé ici dans un cas H—C—O—C—H.

On peut rapprocher de ce résultat le fait observé expérimentalement de l'absence de couplage observable ^4J dans l'éther diméthylique (observation des satellites C_{13}) :

$$CH_3—O—C_{13}H_3$$

contrairement à ce qui se passe dans l'acétone (*15*) et ceci bien que des mesures par micro-ondes (*16*) aient montré une conformation privilégiée où existe la disposition en "M".

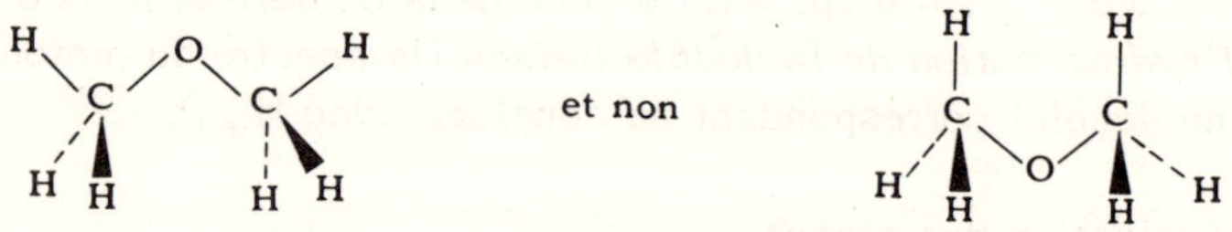

On peut donc envisager une moyenne nulle par l'existence de couplages *positifs ou négatifs* selon la conformation, cette particularité étant due à l'oxygène.

Conclusion

Les couplages à longue distance sont d'autant mieux interpretables théoriquement que l'on en connaît les signes.

La technique de double irradiation sélective utilisée ici, demande cependant l'existence minimum de trois protons *sans couplages nuls entre deux d'entre eux.* Cette condition du "triangle" limite les expériences réalisables (bien que des possibilités expérimentales plus complexes existent dans les autres cas, *17*), puisqu'elle nécessite souvent *deux* couplages à longue distance (comme c'est le cas dans les exemples considérés).

Les résultats expérimentaux sont encore beaucoup trop limités pour permettre une discussion générale des signes obtenus.

Bibliographie

1. J. Meinwald et A. Lewis, *J. Am. Chem. Soc.* **83**, 2769 (1961).
2. J. T. Pinhey et S. Sternhell, *Tetrahedron Letters* **4**, 275 (1963).
3. M. Karplus, *J. Chem. Phys.* **33**, 1842 (1960).
4. J. I. Musher et E. J. Corey, *Tetrahedron* **18**, 791 (1962).
5. E. Payo-Subiza, Thèse d'université, Grenoble, France (1964).
6. D. Gagnaire et E. Payo-Subiza, *Bull. Soc. Chim. France* **1963**, 2623.
7. D. Gagnaire et E. Payo-Subiza, *Bull. Soc. Chim. France* **1963**, 2627.
8. G. O. Schenck, W. Hartmann et R. Steinmetz, *Chem. Ber.* **96**, 498 (1963).
9. J. P. Maher et D. F. Evans, *Proc. Chem. Soc.* **1961**, 208.
10. R. Freeman et D. H. Whiffen, *Mol. Phys.* **4**, 321 (1961).
11. R. Freeman et D. H. Whiffen, *Proc. Phys. Soc. (London)* **79**, 794 (1962).
12. D. Gagnaire, A. Rousseau et P. Servoz-Gavin, *J. Chim. Phys.* **61**, 1207 (1964); D. Gagnaire, E. Payo-Subiza et A. Rousseau, *J. Chim. Phys.* **62**, 42 (1965).
13. D. D. Elleman et S. L. Manatt, *J. Chem. Phys.* **36**, 2346 (1962).
14. S. L. Manatt et D. D. Elleman, *J. Am. Chem. Soc.* **84**, 1579 (1962).
15. J. R. Holmes et D. Kivelson, *J. Am. Chem. Soc.* **83**, 2959 (1961).
16. U. Blukis, P. H. Kasal et R. J. Myers, *J. Chem. Phys.* **38**, 2753 (1963).
17. A. D. Cohen, R. Freeman, K. A. McLauchlan et D. H. Whiffen, *Mol. Phys.* **7**, 45 (1963).

NMR Spectra of Some *N*-Heterocyclic Compounds

G. GIACOMETTI and G. RIGATTI

Istituto di Chimica Fisica, Università di Padova, Padova, Italy

Introduction

Recently Gil and Murrell (*1*) interpreted in a semiquantitative way the contributions to the proton chemical shifts in the pyridine molecule arising from the excess electron densities of the atoms participating in the conjugated system, from the diamagnetic anisotropy of the nitrogen atom, and from the polarization of the C–H bond due to the electric field of the lone pair electrons.

In the present paper the effects of the azo group on the pyridine ring protons are studied. This group has been chosen because (i) the electronic structure of its nitrogen atoms is not too different from that of pyridine nitrogen, both having sp^2 hybridization; (ii) there are two nitrogen atoms contributing to the proton chemical shifts, compensating for the fact that the magnetic anisotropy and bond polarization effects vanish rapidly with distance; (iii) it is possible to examine in this case whether the diamagnetic susceptibility tensor values calculated by Gil and Murrell have an absolute meaning, or just a relative one with respect to the C–H fragment (*1*).

Therefore we have run the spectra of pyridine and of the three symmetric azopyridines, we have calculated and assigned the chemical shifts, and finally we have estimated the magnitude of the above mentioned effects.

Experimental

The proton NMR Spectra were run in ethanol solution with TMS as internal standard, with a Varian A-60 spectrometer operating at 60 Mc/sec. The chemical shifts were measured from the TMS by the sideband technique using a Hewlett-Packard frequency meter, and are assumed to be positive to low field. The three isomeric symmetrical azopyridines were kindly supplied by Prof. M. Colonna of the University of Bologna. We must acknowledge that his generous gift made possible this research and sincerely thank him.

The other products are commercially available.

The chemical shifts and the coupling constants measured from the spectra were refined by trial and error, calculating the spectra with a program for the Olivetti Elea 6001 computer of the University of Padua. The spectrum of the γ, γ' isomer is of the A_2B_2 type and was not completely analyzed because it is not sufficiently resolved.

The assignment was made on the basis that the protons near the nitrogen atom resonate at lower field and that the coupling constants between these protons and those in the *ortho* position are low (about 5 cps) and that the coupling constants between *meta* protons are greater than those between *para* protons (*2*, *3*).

The results are reported in Tables I and II.

TABLE I
CHEMICAL SHIFTS OF THE AZOPYRIDINES (PPM FROM TMS)

Compound	Proton[a] 1	2	3	4	5
α, α'-Azopyridine	—[b]	8.78	7.67	8.13	7.97
β, β'-Azopyridine	9.21	—[b]	8.79	7.73	8.37
γ, γ'-Azopyridine	7.85	8.82	—[b]	8.82	7.86
Pyridine	—[b]	8.64	7.43	7.84	—

[a]The atoms are numbered round the ring, with the azo substituent in position 6.
[b]The nitrogen atom is in this position.

TABLE II
CHEMICAL SHIFT DIFFERENCES OF THE AZOPYRIDINES FROM THE CORRESPONDING ONES OF THE PYRIDINE

Compound	Proton[a] 1	2	3	4	5
α, α'-Azopyridine	—[b]	0.14	0.24	0.32	0.54
β, β'-Azopyridine	0.57	—[b]	0.15	0.30	0.56
γ, γ'-Azopyridine	0.43	0.18	—[b]	0.18	0.43
Average value	0.50	0.16	0.19	0.26	0.51

[a]The atoms are numbered round the ring, with the azo substituent in position 6.
[b]The nitrogen atom is in this position.

Electron Density Contribution

We have calculated the electron densities for all the molecules we are considering. The molecular orbitals were approximated by a linear combination of atomic orbitals and the variational problem was solved following the SCF treatment of Pariser and Parr and of Pople (*4*) in the zero differential overlap approximation, the numerical calculation being performed with the electronic computer Olivetti Elea 6001 of the University of Padua. The β core integrals between C atomic orbitals were calculated with an exponential formula, whereas those involving N atoms were obtained with the aid of the overlap integral tables (*7*) using the "magic formula" of Mulliken (*5*). The atomic energies data necessary to calculate the monocenter Coulomb repulsion integrals were taken from the tables of Skinner and Pritchard (*7a*). The uniformly charged spheres approximation (*6*) was used to calculate the two center integrals.

The electron densities obtained are reported in Table III. The densities depend mainly on the relative position with respect to the heterocyclic nitrogen atom: the values found are 0.95 for the *ortho* carbon atom, 1.00 for the *meta*, and 0.97 for the *para*. The inefficiency of the substituent azo group in modifying such values is consistent with its electronic and molecular structure: the N–N double bond is almost localized, as is evident from the high bond order (0.95) and the small bond distance (*8*).

TABLE III

SCF π-ELECTRON DENSITIES OF THE SYMMETRIC AZOPYRIDINES

	Atom[a]						
Compound	1	2	3	4	5	6	7
α, α'-Azopyridine	1.1202[b]	0.9467	1.0029	0.9784	1.0002	0.9499	1.0067
β, β'-Azopyridine	0.9438	1.1228[b]	0.9402	1.0046	0.9734	1.0029	1.0127
γ, γ'-Azopyridine	1.0000	0.9458	1.1202[b]	0.9465	1.0007	0.9761	1.0097

[a]The atoms are numbered round the ring, with the azo substituent in position 6. The molecular dimensions are taken for the azogroup from Eichorn (*8*) and for the pyridine ring from Bak *et al.* (*9*).

[b]The nitrogen atom is in this position.

Assuming the usual value for the proportionality constant of the relation (*10, 11*)

$$\Delta\sigma = k\Delta\rho,$$

the contribution to the chemical shift differences ($\Delta\sigma$) by the differences in electron densities ($\Delta\rho$) for the protons of the pyridine molecule and the corresponding ones of the azopyridines are only a few parts per hundred million.

Diamagnetic Anisotropy Contribution

The screening constant of an atom A depends on the diamagnetic anisotropy of the neighbor atom B on a dipole basis according to the relation (*11, 12, 13*)

$$\sigma_{AB} = \frac{1}{3\,\mathbf{R}_{AB}^3 N} \sum_{i=1}^{3} \chi_\beta^i \,(3\cos^2\gamma_i - 1),$$

where $\mathbf{R}_{AB}$ is the vector joining A to B, χ_β^i is a principal component of the diamagnetic susceptibility tensor, and γ_i is the angle between **R** and the directions of χ_β^i. Tentatively we assumed for the azo group the same susceptibility values as for the heterocyclic nitrogen atom (*1*). We calculated the contribution to the screening constants for two molecular models: first, supposing a rigid molecule with all atoms in the same plane, and second, a molecule in which the pyridine ring is free to rotate about the C–N bond.

The integration over the rotation angle was performed analytically for the contribution of the *N* atom lying on the axis of rotation, and graphically for the other atom. The values obtained for the two models are reported in Table IV. The very large calculated difference between

TABLE IV

MAGNETIC ANISOTROPY AND BOND POLARIZATION CONTRIBUTIONS OF THE AZO GROUP TO THE CHEMICAL SHIFTS

	Proton				
	1	2	3	4	5
Fixed rings					
Anisotropy effect	0.169	0.039	0.028	0.048	0.317
Field effect	0.256	0.041	0.027	0.038	0.889
Total effect	0.425	0.080	0.055	0.086	1.406
Rotating rings					
Anisotropy effect	0.036	0.031	0.028	0.031	0.036
Field effect	0.449	0.039	0.027	0.039	0.449
Total effect	0.485	0.070	0.055	0.070	0.485

positions 1 and 5, which are both *ortho* to the azo group, rules out the fixed structure if other effects are not present, because the spectrum of the γ, γ' isomer is of the A_2B_2 type.

Electric-Field Contribution

The effect of an electric field, such as that arising from the dipole moment due to the lone pair electrons of an sp^2 nitrogen atom, has been considered by Buckingham (*14*). He derived the relation

$$\Delta\sigma = A \cdot E_{C-H} + B \cdot E^2,$$

where A and B are proportionality constants and E_{C-H} is the component of the field strength E along the C–H bond. Putting (*15*) $A = 2.9 \cdot 10^{-12}$ and $B = 0.74 \cdot 10^{-18}$ we calculate the contribution to the chemical shifts for the two models as listed in Table IV. It is not necessary in a crude calculation like this to consider the total field arising from the various atoms because the second term in the Buckingham formula is contributing only to a few percent. The dipole approximation is not a good one at small distances; nevertheless we hoped to get at least the order of magnitude of the chemical shifts.

Conclusions

The chemical shifts of the protons of the azopyridines measured from the corresponding ones of the pyridine molecule itself can be attributed only to the effects of the azo group. They are consistent with each other, apart the *meta* protons of α, α'-azopyridine which differ by 0.15 ppm. These chemical shifts, which are all downfield, indicate that the azo group shifts the *ortho* protons by about 0.5 ppm, the *meta* protons by 0.3 ppm, and the *para* protons by 0.2 ppm. The agreement with the calculated contributions is extremely satisfactory for the *ortho* protons if we assume the rotating ring model, but the calculated effect is underestimated for the other positions; nevertheless it is of the right order.

This agreement is surprising considering that the C–H bond polarization is the main contribution, and that usually this effect does not follow the Buckingham relation when it is too great, especially when the dipole approximation is applied to short distances.

References

1. V. M. S. Gil and J. N. Murrell, *Trans. Farad. Soc.* **60**, 249 (1964).

2. W. G. Schneider, H. J. Bernstein, and J. A. Pople, *Can. J. Chem.* **35**, 1487 (1957); *Ann. N.Y. Acad. Sci.* **70**, 806 (1958).

3. S. Matsuura and T. Goto, *J. Chem. Soc.*, p. 1773, (1963).

4. R. Pariser and R. G. Parr, *J. Chem. Phys.* **21**, 466, 767 (1953); J. A. Pople, *Trans. Farad. Soc.* **49**, 1375 (1953).

5. R. S. Mulliken, *J. Chim. Phys.* **46**, 500, 521 (1949).

6. R. G. Parr, *J. Chem. Phys.* **20**, 1499 (1952).

7. R. S. Mulliken, C. A. Rieke, and H. Orloff, *J. Chem. Phys.* **17**, 1248 (1949).

7a. H. Pritchard and H. Skinner, *Chem. Rev.* **55**, 745 (1955).

8. E. L. Eichorn, *Acta Cryst.* **12**, 746 (1959).

9. B. Bak, L. Hansen, and J. Rastrup-Anderson, *J. Chem. Phys.* **22**, 2013 (1954).

10. B. P. Dailey, A. Gawer, and W. C. Neickam, *Discussions Farad. Soc.* **34**, 18 (1962).

11. J. A. Pople, *J. Chem. Phys.* **37**, 53 (1962); *Discussions Farad. Soc.* **34**, 147 (1962).

12. N. F. Ramsay, *Phys. Rev.* **78**, 699 (1950); **86**, 243 (1952).
13. H. M. McConnell, *J. Chem. Phys.* **27**, 226 (1957).
14. A. D. Buckingham, *Can. J. Chem.* **38**, 300 (1960).
15. J. I. Musher, *J. Chem. Phys.* **37**, 34 (1962).

Proton Magnetic Resonance of Aliphatic Sulfides, Sulfoxides, and Sulfones

F. TADDEI and C. ZAULI

Istituto di Chimica Fisica ed Elettrochimica, Università di Bologna, Bologna, Italy

In the literature many papers have appeared dealing with the perturbation of aliphatic protons by substituents and with attempts to correlate the chemical shifts with inductive (and anisotropic) effects of the substituents (*1–5*).

Since the situation does not yet seem to be completely understood, we have undertaken a systematic investigation of aliphatic derivatives in order to accumulate empirical knowledge which may throw some light on the effects determining the chemical shifts of aliphatic groups. Here we shall refer mainly to the results obtained for aliphatic sulfides, sulfoxides, and sulfones (Table I).

The first conclusion that can be drawn from the data collected in Table I is that a change in structure of the aliphatic chain does not significantly affect the chemical shifts of methyl, methylene, and methyne groups. The corresponding resonances fall in fact in narrow ranges (Table II).

Empirically we may express the change in chemical shift due to substitution by a $-\mathrm{C}\!<$ group of a proton bonded to the *i*th carbon atom ($i = \alpha, \beta, \gamma, \ldots$) with respect to the absorbing protons as follows:

$$\delta\tau_i = A(\mathrm{C}_i\!-\!\mathrm{H}) - A(\mathrm{C}_i\!-\!\mathrm{C}\!<), \qquad (1)$$

where $A(\mathrm{R})$ is comprehensive of all the perturbations induced by the R group.

Thus the empirical observation that a change in structure of the aliphatic chain does not perturb significantly the τ value for the various aliphatic groups implies that $\delta\tau_i$ falls rapidly to zero even when $i = \beta$, that is, when the substitution $-\mathrm{H} \longrightarrow -\mathrm{C}\!<$ does not concern the carbon bonded to the absorbing protons.

As expected, the perturbation induced by substituents decreases with distance: Table II shows clearly this behavior for CH_3 and CH_2, the

TABLE I

CHEMICAL SHIFTS IN τ UNITS OF ALIPHATIC SULFIDES, SULFOXIDES, AND SULFONES R′—X—R″, MEASURED IN CCl_4 AT 56.4 MC

R′	R″	Group	X = S	X = SO	X = SO_2
CH_3	CH_3	CH_3	7.94 (7.91[a]; 7.97[b])	7.49	7.15
	C_2H_5	CH_3 (α)	7.97 (7.90[c])	7.59	7.20
		CH_2 (α)	7.56 (7.47[c])	7.40	7.04
		CH_3 (β)	8.76 (8.73[c])	8.71	8.62
	n-C_3H_7	CH_3 (α)	8.00	7.54	7.18
		CH_2 (α)	7.61	7.42	7.04
		CH_2 (β)	8.49	8.22	8.17
		CH_3 (γ)	9.02	8.92	8.91
	iso-C_3H_7	CH_3 (α)	7.98	7.65	7.25
		CH (α)	7.32	7.47	7.02
		CH_3 (β)	8.76	8.80	8.64
	n-C_4H_9	CH_3 (α)	7.99	7.56	7.16
		CH_2 (α)	7.58	7.42	7.03
		CH_2 (β)	8.5	8.4	8.3
		CH_2 (γ)	8.5	8.4	8.4
		CH_3 (δ)	9.08	9.02	9.00
	tert-C_4H_9	CH_3 (α)	8.04	7.76	7.26
		CH_3 (β)	8.73	8.82	8.61
	n-C_5H_{11}	CH_3 (α)	— (7.92[c])	—	—
		CH_2 (α)	— (7.50[c])	—	—
		CH_2 (β)	—	—	—
		CH_2 (γ)	—	—	—
		CH_2 (δ)	—	—	—
		CH_3 (ε)	— (9.07[c])	—	—
C_2H_5	C_2H_5	CH_2	7.53 (7.51[ab])	7.43	7.14
		CH_3	8.77 (8.76[a]; 8.77[b])	8.73	8.64
n-C_3H_7	n-C_3H_7	CH_2 (α)	7.59 (7.57[a])	7.50	7.12
		CH_2 (β)	8.43 (8.41[a])	8.15	8.18
		CH_3	9.03 (9.02[a])	8.92	8.92
iso-C_3H_7	iso-C_3H_7	CH	7.14 (7.07[a])	7.41	6.92
		CH_3	8.76 (8.76[a])	8.79	8.70
n-C_4H_9	n-C_4H_9	CH_2 (α)	7.58	7.46	7.10
		CH_2 (β)	8.5	8.3	8.4
		CH_2 (γ)	8.6	8.4	8.4
		CH_3	9.08	9.02	9.01
iso-C_4H_9	iso-C_4H_9	CH_2	7.68	7.7	7.25
		CH	8.26	7.7	7.74
		CH_3	9.03	8.91	8.89
tert-C_4H_9	*tert*-C_4H_9	CH_3	8.61	8.73	8.56
n-C_5H_{11}	n-C_5H_{11}	CH_2 (α)	7.57	7.49	7.15
		CH_2 (β)	8.6	8.4–8.6	8.5
		CH_2 (γ)	8.6	8.4–8.6	8.5
		CH_2 (δ)	8.6	8.4–8.6	8.5
		CH_3	9.09	9.06	9.06

(continued)

TABLE I (continued)

R′	R″	Group	X = S	X = SO	X = SO_2
iso-C_5H_{11}	iso-C_5H_{11}	CH_2 (α)	7.57	7.46	7.19
		CH_2 (β)	8.55	8.3–8.4	8.36
		CH	8.4–8.5	8.3–8.4	8.2–8.4
		CH_3	9.09	9.09	9.03

[a]Ref. (*2*) [b]Ref. (*3*) [c]Ref. (*6*).

TABLE II

AVERAGE CHEMICAL SHIFTS OF ALIPHATIC GROUPS AT VARIOUS DISTANCES FROM THE SUBSTITUENTS

Group	Subst.	α	β	γ	δ	ϵ
CH_3	S	7.98 ± 0.04	8.76 ± 0.09	9.03 ± 0.01	9.08 ± 0.01	9.08 ± 0.01[a]
	SO	7.59 ± 0.15	8.76 ± 0.06	8.92 ± 0.01	9.05 ± 0.04	9.06[b]
	SO_2	7.20 ± 0.04	8.63 ± 0.07	8.91 ± 0.02	9.02 ± 0.02	9.06[b]
CH_2	S	7.57 ± 0.04[c]	8.49 ± 0.06[d]	8.56 ± 0.02[d]	8.6[e]	—
	SO	7.47 ± 0.04[c]	8.28 ± 0.12[d]	8.43 ± 0.01[d]	8.6[e]	—
	SO_2	7.10 ± 0.06[c]	8.28 ± 0.10[d]	8.40 ± 0.01[d]	8.6[e]	—
CH	S	7.23 ± 0.09	8.26[b]	8.4–8.5[b]	—	—
	SO	7.44 ± 0.03	7.7[b]	8.3–8.4[b]	—	—
	SO_2	6.97 ± 0.05	7.74[b]	8.2–8.4[b]	—	—

[a]Average between the value relative to di-*n*-amyl and that derived from the literature for methyl-*n*-amyl sulfide (*6*).

[b]Single values.

[c]With exception of iso-butyl derivatives, $\tau = 7.68$ (X = S); $\tau = 7.7$ (X = SO); $\tau = 7.25$ (X = SO_2).

[d]The *n*-amyl derivatives are not included because of experimental uncertainties.

[e]Indicative values obtained from the *n*-amyl derivative.

limiting values being very close to those observed for hydrocarbons ($\tau_{CH_3} = 9.11$, $\tau_{CH_2} = 8.74$).

The perturbation of the substituent thus seems to be transmitted mainly through the bonds, or if direct effects are operating, then the statistical weight of conformations where the chain folds back toward the substituent is low.

Generally the perturbation of the substituent X, as measured by the chemical shift of aliphatic groups, follows the series S < SO < SO_2 at all distances: this seems connected, at least qualitatively, with the intensity and behavior of the potentials (and thus of charge densities) of these sulfurated groups at various distances. In fact the perturbation of chemical shifts due to first-row elements (—O—, NH, and $-\mathrm{C}\!\lessgtr$; for the latter group see the above discussion) falls rapidly to zero, while that due to groups containing second- or higher-row elements such as $\rangle S$, $\rangle SO$, $\rangle SO_2$, —Cl, —Br, and —I is not negligible even at distances greater than a C–C bond length from the absorbing protons (Figs. 1 and 2). $\delta\tau_\alpha$, i.e., the contribution due to the substitution of —H by $-\overset{|}{\underset{|}{C}}-$ at the α carbon, is practically constant at various distances from X (Table III), as first noticed by Cavanaugh and Dailey (2) for the

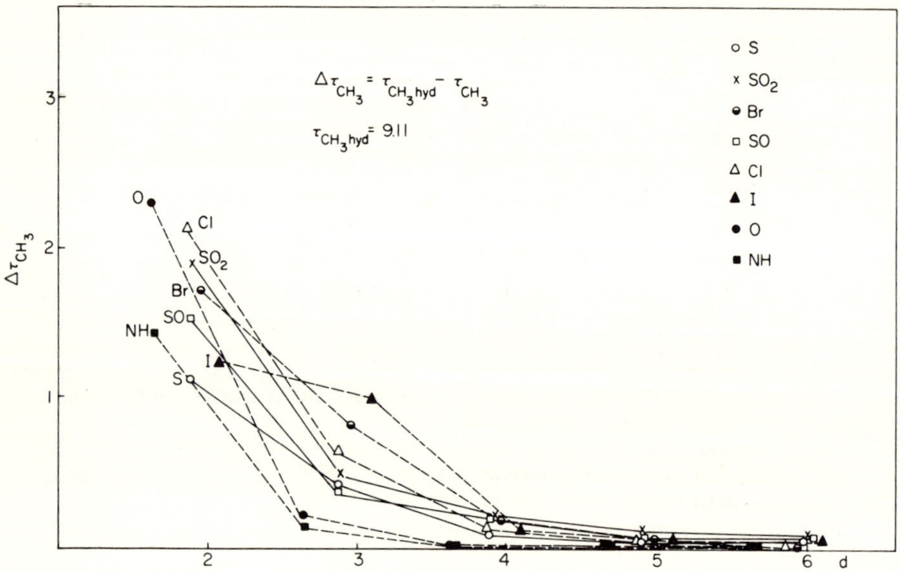

FIG. 1.

TABLE III
SUBSTITUTION OF A C–H BY A C–C BOND

Δ	X	Position[a] α	β	γ	δ	ω
$\tau_{CH_3}-\tau_{CH_2}$	S	0.41 (0.39)	0.27 (0.29)	0.5	0.5	0.37 (0.41)
	SO	0.12 (0.09)	0.48 (0.52)	0.5	0.4–0.6	—
	SO_2	0.10 (0.11)	0.35 (0.41)	0.5	0.5	—
$\tau_{CH_3}-\tau_{CH}$	S	0.75 (0.78)	0.50 (0.58)	0.5–0.6	—	0.85 (0.82)
	SO	0.15 (0.18)	1.09 (1.04)	0.5–0.6	—	—
	SO_2	0.23 (0.22)	0.89 (.082)	0.5–0.7	—	—

[a]Referred to hydrocarbons. Values in parentheses are obtained by taking the effect due to the substitution H ⟶ CH_3 as additive.

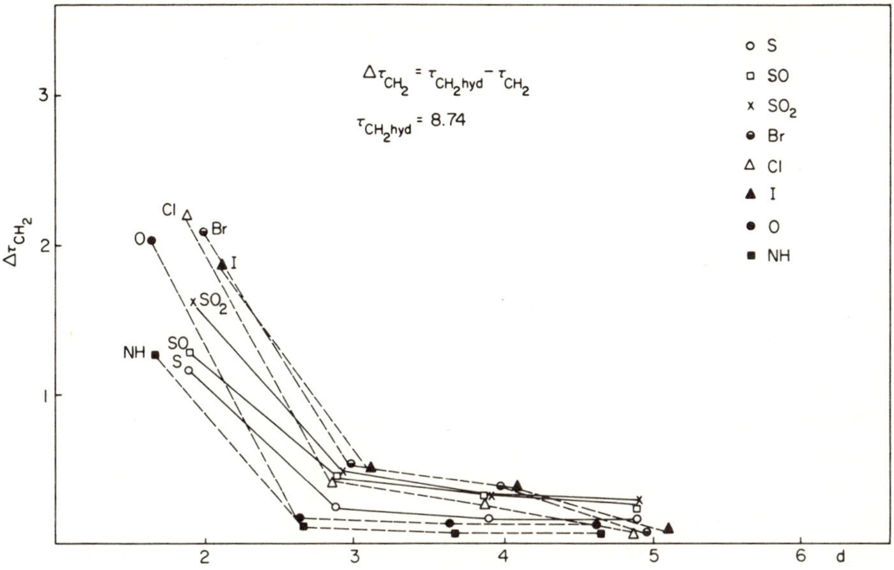

FIG. 2.

carbon atom directly bonded to the substituent. Moreover, $\delta\tau_\alpha$ is seen to be affected by the distance of the proton from the substituent and by the nature of the latter. The above authors interpret $\delta\tau_\alpha$ as due only to the second term in Eq. (1), i.e., to "C–C as affected by the various substituents." Apart from other considerations, this does not seem likely since $\delta\tau_\alpha$ does not decrease monotonically with distance; moreover a linear relation exists between $\delta\tau_\alpha(1)$ and $\delta\tau_\alpha(2)$, where (1) and (2) indicate the neighboring and next-neighboring carbon atom from X, respectively.

The relation is

$$\delta\tau_\alpha(1) = 0.82 - 1.59\,\delta\tau_\alpha(2) \qquad (\Delta = 0.09),$$
$$X = \text{—O—, Hal, —CH}_3, \ \rangle S,\ \rangle SO,\ \rangle SO_2, \tag{2}$$

which can be used together with that reported by Dailey and Shoolery (*1*),

$$\tau_{CH_3(\alpha)} = 9.45 - 1.28\,\delta_{ij} \qquad (\Delta = 0.15), \tag{3}$$

where δ_{ij} is the "internal shift" $\tau_{CH_3(\beta)} - \tau_{CH_2(\alpha)}$, to obtain relation (4) between $\tau_{CH_2(\beta)}$, $\tau_{CH_3(\beta)}$, and $\tau_{CH_3(\alpha)}$:

$$\tau_{CH_2(\beta)} - 0.37\,\tau_{CH_3(\beta)} = 4.13 + 0.14\,\tau_{CH_3(\alpha)}. \tag{4}$$

The above authors have also related the electronegativity of the first atom (after Huggins, *7*) with the "internal shift" δ_{ij}. In this scale we obtain for $>S$, $>SO$, $>SO_2$ (and Cl— for comparison) the values given in the tabulation.

Group	χ
S	2.57
SO	2.69
SO_2	2.89
Cl	3.22

These electronegativities do not seem very reasonable from the point of view of the chemical evidence, since SO and SO_2 are strongly polarizing substituents. Since δ_{ij} is an approximately linear function of $\tau_{CH_3(\alpha)}$, the electronegativity of X can be derived more simply from the latter value. By using Huggins' or Pritchard and Skinner's (*8*) electronegativity scales, the sets of values tabulated here are obtained.

Group	χ_H	χ_{PS}
S	2.54	7.60
SO	2.83	8.42
SO_2	3.11	9.14
Cl	3.27	9.55
CH_3	1.70	6.51

These data seem to fit the chemical evidence better. The use of Pritchard and Skinner's electronegativity scale seems more convenient because it yields good results also for X = —CH_3 ($\chi_{calc.}$ = 5.61, identical with the value given by Pritchard and Skinner).

References

1. B. P. Dailey and J. N. Shoolery, *J. Am. Chem. Soc.* **77**, 3977 (1955).
2. J. R. Cavanaugh and B. P. Dailey, *J. Chem. Phys.* **34**, 1099 (1961).
3. H. Spiesecke and W. G. Schneider, *J. Chem. Phys.* **35**, 722 (1961).
4. P. Biscarini, F. Taddei and C. Zauli, *Boll. Sci. Fac. Chim. Ind. Bologna* **21**, 169 (1963).
5. A. G. Moritz and N. Sheppard, *Mol. Phys.* **5**, 361 (1962).
6. N. S. Bhacca, L. F. Johnson, and J. N. Shoolery, N.M.R. Spectra Catalog, Varian Assoc., 1962.
7. M. L. Huggins, *J. Am. Chem. Soc.* **75**, 4123 (1953).
8. H. O. Pritchard and H. A. Skinner, *Chem. Rev.* **55**, 745 (1955).

Some Correlations for the Chemical Shifts and Coupling Constants of Organo-element Hydrides of Group IVb Elements

H. SCHMIDBAUR

Institut für Anorganische Chemie der Universität Marburg, Marburg, Germany

The "Fifteenth Year of NMR in Chemistry" in many respects is characterized by prolonged efforts dedicated to finding various possible correlations between chemical shifts and also coupling constants with fundamental atomic and molecular parameters, among which the atomic number, electronegativity, bond length and angle, and hybridization are the most attractive. While for a goodly number of organic compounds these efforts have already been successful, our knowledge concerning analogous correlations for compounds with heavier elements is still at its very beginning.

In the course of our present investigations on PMR spectra of group IVb element hydrides and organyls (*1–3*), we have determined the chemical shifts and coupling constants of the methylsilyl and methylgermanyl hydrides having the general formulas $(CH_3)_nXH_{4-n}$, where $n = 1, 2, 3, 4$ and $X = Si, Ge$. The results obtained and others reported in the literature for analogous alkanes, stannanes, and plumbanes were subjected to a comparative study. The δ and J values of the hydrides were measured in dilute solution in anhydrous carbon tetrachloride at 35 °C and 60 Mc, with tetramethylsilane as internal standard.

A simple correlation is easily detected between δ and n of the two homologous series, for both hydridic and methyl protons, as shown in Fig. 1. The shielding and deshielding effects are of different magnitude and direction at both types of protons within each series. Evidently in both cases an additivity rule for the substituent effect of a methyl group is obeyed, similar to that reported for various types of organic compounds. An interpretation of the trends of the chemical shifts δ/n in terms of electronegativity and Si–C bond shifts is possible, but not convincing.

The coupling constants J_{vic}(HCXH) show a similar relation to the substitution number n, as proved by Fig. 2. Again an additivity of the substituent effect of methyl groups for the J values could be observed. The magnitude of the substituent effect is very similar for X = Si, Ge, as is

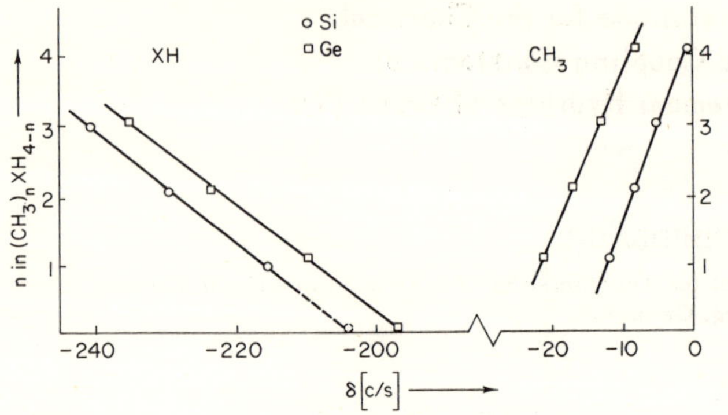

FIG. 1. Variations of chemical shifts of hydridic and methyl protons with substitution number n in $(CH_3)_nXH_{4-n}$, where X = Si, Ge.

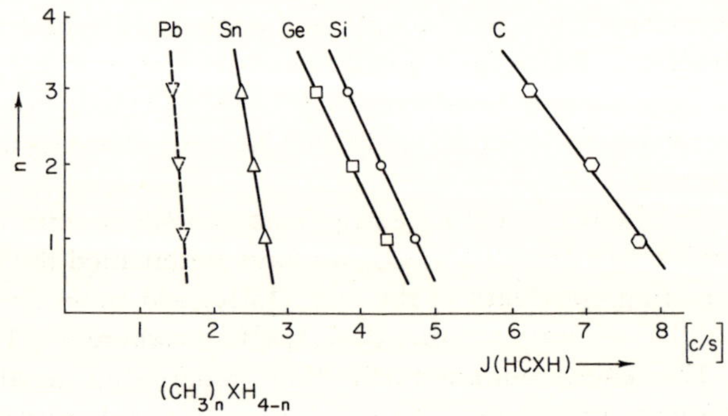

FIG. 2. Variation of coupling constants J(HCXH) with the substitution number n in $(CH_3)_nXH_{4-n}$, where X = C, Si, Ge, Sn, Pb.

seen from the parallelism of the relative lines in Fig. 2. Analogous effects for the alkanes are significantly greater; for stannanes, smaller. For plumbanes values other than that for Me_3PbH are not yet known and must be extrapolated.

A combination of these data with the molecular dimensions of the compounds leads to the observation of a strict proportionality between J(HCXH) and the bond lengths d(HCXH) = d(HC) + d(CX) + d(XH) (angstroms) (Fig. 3). This means, however, that in analogous compounds having identical hybridization at all atoms which "lie on the way" of the spin-spin interaction the magnitude of spin-spin coupling is primarily determined by the over-all distance d(HCXH) of the coupling nuclei, at least when the atoms involved (X) belong to the same group of the periodic table. Investigations of a series of compounds with other magneti-

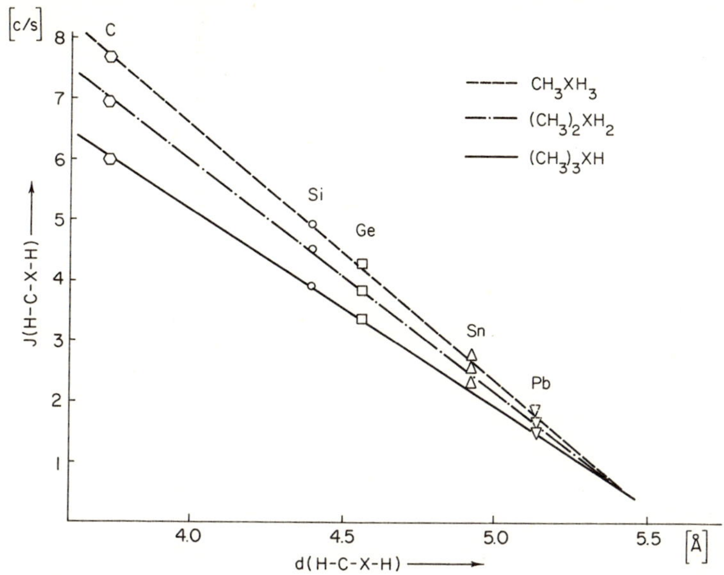

FIG. 3. Correlation of bond lengths d(HCXH) with coupling constants J(HCXH) for methyl element hydrides of group IVb elements.

cally active nuclei (^{207}Pb, ^{119}Sn) interacting, for the determination of J(HXCPb) and J(HXCSn), respectively, are not yet completed, but show similar trends.

References

1. H. Schmidbaur, *Chem. Ber.* **97**, 270, 1639, 3381 (1964); *J. Am. Chem. Soc.* **85**, 2336 (1963).
2. H. Schmidbaur and I. Ruidisch, *Inorg. Chem.* **3**, 599 (1964).
3. H. Schmidbaur and H. Hussek, *J. Organometal. Chem.* **1**, 235 (1964).

NMR and Ion Pair Association of Ions Containing Elements in Groups III and V

R. J. CHUCK, A. G. MASSEY, E. W. RANDALL, and D. SHAW
Queen Mary College, London, England

Introduction

Most of the elements of the main groups III and V of the Periodic Table have nuclear electric quadrupole moments. In order to determine their spin coupling constants with protons in their alkyl derivatives, it is necessary to study symmetrical tetrahedral ions so that quadrupolar relaxation effects are minimized. This was tried previously for the ions BEt_4^-, NEt_4^+, $AsEt_4^+$, and $AsMe_4^+$ (*1*).

The internal proton chemical shift, Δ, between the methyl and methylene groups in these ions depends upon the nature and the electric charge of the central nucleus, X, and was used to compare the electronegativities of the central atoms. These studies have been extended to the cases X = Sb, Sb^+, Al, and Al^-, and the effects of varying solvent and anion on Δ have now been investigated. The Δ values for two different ethyl groups in the same ion have been studied for the case of the 1,4-diethylpyridinium ion.

In the case of the arsonium cations the above variables were previously found to produce changes in the line shapes of the proton spectra. The changes were interpreted as due to variations in electric quadrupole effects of the central nucleus. This behavior is shown to be typical of antimony also.

Previous investigations of linewidths and relaxation rates of quadrupolar nuclei in ionic species have been confined to the monatomic ions ^{127}I (*2*), ^{23}Na, ^{133}Cs, ^{79}Br, and ^{81}Br (*3*). Our experiments involved polyatomic ions, $[XR_n]^{\pm}$, in which the resonance spectra investigated were not those of the quadrupolar nucleus X, but arise from nuclei in the group R which are spin coupled to X. A similar approach has since been used for the case AsF_6^- (*4*).

Experimental and Results

The NMR spectra were obtained at 60 Mc/s using a Varian Associates A-60 spectrometer at 35 °C. They were calibrated by the standard audio sideband technique (*5*).

In the decoupling experiments a second spectrometer at approximately 9400 gauss was employed. It will be described elsewhere (*6*).

Standard preparations were used for the tetraalkyl and trialkyl compounds (*7*) and also for the ethyl pyridinium compounds (*8*).

The proton spectra of the trialkyl compounds were recorded for the neat liquids. The quarternary salts were used as approximately 10% solutions by weight in water and chloroform unless otherwise stated. The internal chemical shifts for the ethyl compounds are given in Table I.

TABLE I

VALUES OF Δ FOR XEt_3 AND $XEt_4^{\pm}$

	(CH_3-CH_2), ppm[a]		
Element, X	XEt_3	$XEt_4^{\pm}$ in $CHCl_3$	$XEt_4^{\pm}$ in H_2O
N	1.42	2.10[b]	2.00
P	0.292 (*8*)	1.22[c]	1.00
As	0.30	1.35[c]	1.05
Sb	0.1	1.18[c]	0.97
Al	−0.71 (*24*)	−1.3[d]	—

[a] Δ is positive if the CH_3 resonance is at higher field than the CH_2 resonance.
[b] As bromide. [c] As iodide. [d] In benzene.

The shifts for the tetraethylstibonium ion relative to the water solvent are +3.32 ppm for the methyl group and +2.35 ppm for the methylene group. The linewidths in $SbEt_4^+$ resemble the $AsEt_4^+$ case (*1*). The methyl resonance is 19 c/s wide at half-height in water solution, the proton-proton splitting of 8 c/s being unresolved. $J_{(H-H)}$ is resolved on the methylene group, however, and each component of the spin multiplet is 7 c/s wide. In chloroform solution the proton-proton splittings are resolved in each group and the linewidths are reduced to 3 c/s in the methyl and 2 c/s in the methylene group.

The chemical shifts of the tetramethyl quarternary ions in ppm relative to water solvent are for NMe_4I, +1.40, and for $SbMe_4I$, +2.95. A splitting $J_{(^{14}N-H)}$ of 0.55 ±0.05 c/s was found for the nitrogen compound. The symmetrical stibonium cation had no resolved Sb-H coupling but the linewidth was large (11 c/s). The asymmetrical compound $SbMe_3Br_2$ had a much smaller linewidth of 1 c/s and a shift of −2.61 ppm relative to tetramethylsilane in carbon tetrachloride.

Various anionic compounds of the tetraethylarsonium ion were prepared and were studied in solutions about 10% by weight in both chloroform and water. The internal chemical shifts obtained are given in Table II.

TABLE II
VALUES OF Δ FOR THE TETRAETHYLARSONIUM SALTS, $AsEt_4Z$

$As\overset{+}{Et_4}\overset{-}{Z}$	(CH_3-CH_2) ppm ± 0.05	
	in H_2O	in $CHCl_3$
F^-	1.05	1.0[a]
Cl^-	1.05	1.25
Br^-	1.05	1.33
I^-	1.05	1.35
OH^-	1.05	1.20
CH_3COO^-	1.05	1.25
NO_3^-	1.05	1.15
ClO_4^-	1.05	Insoluble

[a]Not very soluble.

The effect of changing the composition of a mixed solvent on the spectrum of the tetraethylarsonium ion was studied using an arsenic trichloride-chloroform system (Table III and Fig. 1). Arsenic trichloride was chosen because it has a high dielectric constant, ε, of 12.8, is protonless, and is completely miscible with chloroform.

The effect of a very large increase in concentration (from 10% to 137% by weight) of $AsEt_4^+I^-$ in water was to decrease the linewidth of the methylene resonance from 6.0 c/s to 3.5 c/s. The methyl signal was similarly narrowed, sufficiently so to allow resolution of the proton-proton couplings. The internal chemical shift increased from 1.07 ppm to 1.12 ppm.

Temperature effects were observed both on a 10% and on a 137% aqueous solution. The temperature range studied was small, viz., 35°

TABLE III
VARIATION OF LINEWIDTHS IN THE SPECTRUM OF TETRAETHYLARSONIUM IODIDE WITH THE $AsCl_3$-$CHCl_3$ SOLVNET COMPOSITION

% $AsCl_3$ (by vol.)	Linewidths, c/s		$\Delta(CH_3-CH_2)$, ppm
	CH_3	CH_2	
100	Unresolved	3.5 ± 0.2	1.05 ± 0.02
66.6	8.5 ± 0.3	3.0	1.09
50.0	6.3	3.0	1.11
40.0	5.0	3.0	1.12
28.5	4.0	3.0	1.13
20.0	3.0	3.0	1.16
12.5	3.0	3.0	1.17
0	3.0	2.5	1.35

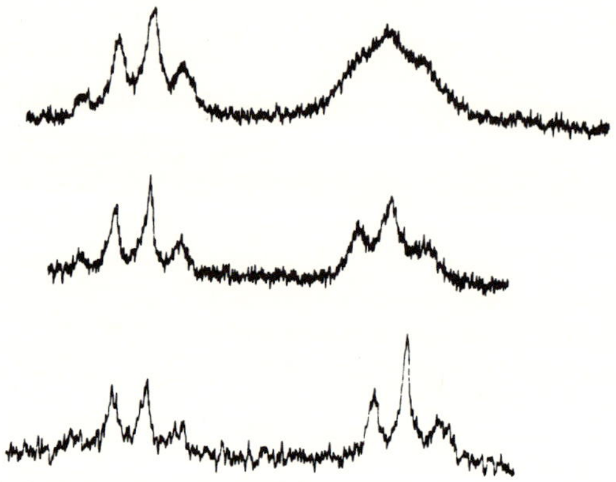

FIG. 1. The proton magnetic resonance spectrum of the tetraethylarsonium ion at 60 Mc/s in (a) arsenic trichloride, (b) 1:1 arsenic trichloride–chloroform (by volume), and (c) 1:8 arsenic trichloride–chloroform (by volume).

to 90 °C. For the dilute solution the broad lines were increased slightly in linewidth. The couplings $J_{(As-H)}$ were not resolved, however. For the more concentrated solution a pronounced increase in linewidth was found; the proton-proton couplings observable at the low temperature could not be resolved at the high temperature.

A double resonance experiment, designated H{As}, involving iradiation of a 10% aqueous sample of tetraethylarsonium iodide at a frequency of approximately 6.85 Mc/s in a magnetic field of approximately 9400 gauss, caused narrowing of the proton signals to less than 1 c/s.

Anion and solvent effects on the two different ethyl groups in 1,4-diethylpyridinium ion are reported in Table IV.

TABLE IV
ANION EFFECTS ON Δ VALUES[a] FOR THE TWO ETHYL GROUPS IN 1,4-DIETHYLPYRIDINIUM ION

	Water solution[b]		Chloroform solution (1.00 *M*)	
Anion	4-Et	1-Et	4-Et	1-Et
F^-	1.67	2.96	1.60	3.18
Cl^-	—	—	1.61	3.25
Br^-	1.65	2.95	1.62	3.28
I^-	1.67	2.96	1.64	3.25

[a]Parts per million ± 0.01.
[b]Shifts independent of concentration (0.1 *M* to 1.0 *M*)

Discussion

Line Shapes

It has previously been suggested that the broadening of the proton resonance lines of tetraethylarsonium bromide in aqueous solution is due to spin-spin interaction between the protons and the arsenic atom for which the quadrupole relaxation occurs at such a rate in the symmetrical ions that these spin-spin effects are not completely averaged out (*1*). This hypothesis was consistent with the fact that the unsymmetrical triethylmethylarsonium ion gives sharp lines in water solution (*1*). Another test is the H{As} decoupling experiment: irradiation at the arsenic resonance frequency sharpens the proton lines of tetraethylarsonium bromide in water. These results exclude the possibility that broadening is due to exchange effects. This conclusion is supported by the temperature results: raising the temperature of the aqueous solutions would be expected to sharpen the signals on the exchange idea.

The line shapes in the NMR spectra of nuclei of spin 1/2 coupled to a nucleus, X, of spin I have been discussed for a wide range of quadrupole relaxation rates of X (*9*). The case where X has a spin of 3/2 has been investigated (*10*) and the results are shown in Fig. 2. The variable χ is

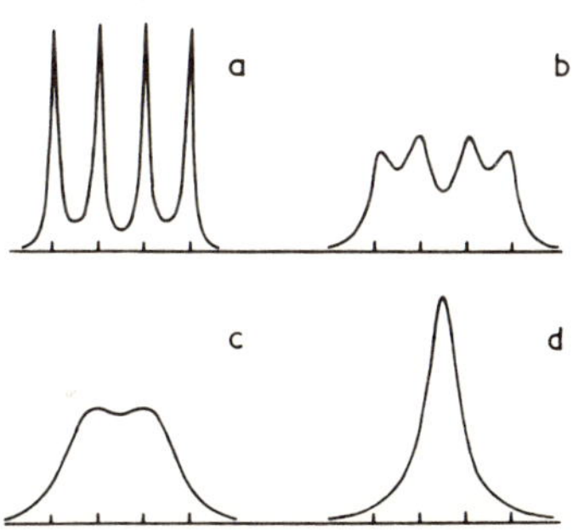

FIG. 2. Theoretical line shapes for the spectra of nuclei of spin 1/2 coupled to a nucleus of spin 3/2: (a) $\chi^2 = 100$, (b) $\chi^2 = 10$, (c) $\chi^2 = 1$, (d) $\chi^2 = 0.1$. The vertical lines indicate the positions of the quartet in the absence of quadrupole relaxation (*10*).

given by $\chi = 2\pi J\tau$, where J is the coupling constant in cycles per second and τ, the lifetime of each spin state, is given in the extreme narrowing case by

$$\frac{1}{\tau} \propto \left(\frac{e^2qQ}{\hbar}\right)^2 \tau_q, \tag{1}$$

where e = the charge on the electron,
eq = the electric field gradient at the nucleus X,
eQ = the quadrupole moment of X,
τ_q = the correlation time for molecular reorientation.

A variation in the linewidth of a species XR_4^+ may arise therefore by a change in eq, Q, τ (or τ_q), or J.

Changes in τ_q are associated with changes of the viscosity of the solution (η). Valiev (*11*), using an ion-solvent model in which the fluctuating electric gradients at the quadrupole nuclei are produced by solvent molecules deduced that

$$\frac{1}{\tau} \propto (eqQ)^2 \frac{\eta}{T}, \tag{2}$$

where T is the temperature of the solution.

Linear variations of linewidths with viscosities at a fixed temperature have been found for aqueous solutions of sodium bromide, hydrogen bromide, and calcium bromide (^{79}Br and ^{81}Br resonances) (*3*), and for sodium iodide and potassium iodide (^{127}I resonances) (*2*). The viscosity changes were produced by varying the concentration of the electrolyte. Thus in these solutions it can be inferred from the ion-solvent model that the quadrupole coupling constant does not vary appreciably with concentration. Moreover, the slopes of the plots, in accordance with Eq. (2), are a measure of eqQ.

The viscosity relation is not adequate for all cases. Richards and Yorke illustrate this for the linewidths of bromine resonances in aqueous of cesium bromide. Deviations from linearity of plots of linewidth against viscosity were found and were attributed to changes in eqQ. Ion-ion interactions may be important in this case (*3*).

Variations of X. In the case of the ions we have studied, an additional variable is J. The effect of changes in J upon the line shapes of the ions XEt_4 (where X = B^-, Al^-, N^+, As^+, Sb^+) does not, however, appear to be as important as the other factors involved. Thus the highest values of χ occur for tetraethylboronate and tetraethylammonium ions for which the three bond couplings of $J_{(^{11}B\text{-}H)}$, 3.0 c/s, and $J_{(^{14}N\text{-}H)}$, 1.7 c/s, are smaller than the values expected for the arsenic and antimony cases (*11a*).

Variations in τ associated with viscosity changes are also unlikely to be large since the viscosity differences in the dilute solutions used are, by comparison with other cases (*3*), expected to amount to less than a factor of 2.

Larger changes occur in both the nuclear quadrupole moments, Q, and the antishielding factors, which are comparative measures of eq for two different ions in the same environment (*12*). Thus $Q = 2 \times 10^{-26}$ cm^2 and 3.5×10^{-26} cm^2 for ^{14}N and ^{11}B, respectively, and -1.0×10^{-24} cm^2 for ^{123}Sb. Similar variations by factors of 10 or more occur in the antishielding factors (*12*).

The largest variations in χ as X is changed appear therefore to be caused by the quadrupole coupling factor.

Solvent Effects upon Line Shape. We have found no solvent effects upon χ for NEt_4^+ or NMe_4^+ but pronounced effects for $AsEt_4^+$, $AsMe_4^+$, $SbEt_4^+$, and $SbMe_4^+$.

No solvent variation of $J_{(X-H)}$, nor of $J_{(H-H)}$, was found in the cases BMe_4^-, NMe_4^+, and $[PMe_nEt_{4-n}]^+$ where $0<n<3$ (*7*). Since the proton-proton couplings are invariant for both $AsEt_4^+$ and $SbEt_4^+$, we infer that $J_{(X-H)}$ is similarly constant in the latter cases. If so, the linewidth variations are not caused by the spin coupling constant factor.

Furthermore, the viscosity factor alone is not sufficient to account for the change in linewidth. For example, the proton line shape of $AsMe_4^+$ in water ($\eta = 0.7$ centipoise) corresponds approximately to Fig. 2c. A decrease in the viscosity of the solution by virtue of changing the solvent to chloroform ($\eta = 0.5$ centipoise) should produce spectra corresponding to greater χ (Figs. 2a and 2b). Instead, however, the change is toward smaller χ (Fig. 2d). Moreover, the spectra derived from ions dissolved in liquid sulfur dioxide, which has an even smaller viscosity than chloroform, resemble those obtained from the aqueous solutions.

The inference is that the most important factor is the quadrupole coupling constant which changes from solvent to solvent because the mean square electric field gradient, q^2, varies. q is then held to be larger in chloroform than in water, liquid sulfur dioxide, or arsenic trichloride solution.

We have previously suggested (*1*) that ion pair formation is important in determining the quadrupolar coupling in solutions of low dielectric constant like chloroform ($\varepsilon = 4$ at 20°). For more polar solvents of higher dielectric constant (water, $\varepsilon = 80.4$; liquid sulfur dioxide, $\varepsilon = 14$; arsenic trichloride, $\varepsilon = 12.3$; all at 20°C) ion pair formation is expected to be less extensive and the quadrupole relaxation rate to be smaller. The importance of ion-ion interactions even in aqueous solution has been either assumed or discussed by other authors (*13–14*).

Variation of Concentration. In the case of a dissociating solvent the possibility of increasing ion-ion effects by raising the concentration was tested for the system $AsEt_4^+I^-$ in water (137% solution). As expected, the effect was to decrease the proton linewidths.

On consideration of viscosity changes alone, one would predict line broadening.

We cannot, of course, exclude the possibility that ion-solvent interactions contribute to the linewidth even in the nondissociating solvents. It is also important to recognize that the two forms of interaction, ion-ion and ion-solvent, are not necessarily independent. The interaction between a given ion and a given solvent is presumably different when that ion is "free," or solvated, from when it is associated with a second ion. Moreover, the formation of ion pairs may increase the relaxation rate not only because of an ion-ion interaction but also because of the change in the ion-solvent interaction. Thus in water and chloroform solutions of the tetraethylarsonium salts, the ion-solvent interaction may have differing effects upon the proton linewidth, first because the solvent molecules are different, and second because the ion-ion interaction occurs to a differing extent in the two cases.

The labeling of linewidth variations as due to ion-ion or ion-solvent effects is misleading unless these terms are rigorously defined. In addition, the distinction is superfluous until more detailed theories of these interactions are available for test. So far a small number of tests of a simple ion-solvent model of quadrupolar broadening has been made and the model has not been found entirely adequate (*3*). We are unaware that any reasonable ion-ion or more comprehensive model exists for testing.

The extension of our linewidth studies has been hindered by lack of suitable solvents. Important considerations are good solubility and an absence of interfering proton resonance signals in the solvent. A wide range of χ values for the solution is desirable. The ion containing the quadrupolar nucleus must therefore be sensitive to environmental influences, i.e., the nucleus must have a large antishielding factor. The case of the hexafluoroarsenate ion, AsF_6^-, seems far more promising from these points of view (*4*). An important factor here is the large coupling constant, $J_{(As\text{-}F)}$ = 930 c/s (*17*), together with the lack of more than one coupling in the molecule.

Proton Chemical Shifts

Effect of a Change in X. The internal chemical shift, Δ, between the CH_3 and CH_2 resonance in ethyl compounds $(CH_3CH_2)_nX$ has been used to study the nature of the group attached to the ethyl radical. Correlations between Δ and the electronegativity of X have been extensively discussed (*18–23*). Reasonable linear correlations have been found either with (*2*) or without (*18*) allowances made for magnetic anisotropy effects. Changes in Δ due to coordination of an ethyl group have been noted (*24*) and also due to change both in coordination number and the electric charge of the compound (*1*). For example, Δ was larger for the cationic species XEt_4 ($X = N^+$ and As^+) than for the corresponding compounds XEt_3 ($X = N$ and As). We find a similar behavior in the cases both of phosphorus (*7*) and of antimony. If the effect of magnetic anisotropy is ignored in the neutral compounds (it is zero in the symmetrical cations) the change in Δ can be ascribed to the larger electronegativity of the cation, X^+, compared with X. The change in Δ for the anion Al^- compared with Al is in the opposite sense, since the ion in this case is less electronegative than the neutral atom. This behavior was reported previously for $X = B^-$ and B (*1*).

The electronegativity sequence for X in the XEt_3 compounds, again ignoring anisotropy effects, is $N > P \geq As > Sb$. This is in agreement with the sequence arrived at by Allred and Hensley from proton shifts in XMe_3 compounds (*25*). This sequence is repeated for X^+ in the cations $X\overset{+}{E}t_4$.

Solvent Effects. For each of the elements of group V, Δ for $X\overset{+}{E}t_4$ is larger in chloroform solution than in water solution. Moreover, in the case of $X = As^+$, Δ is the same within experimental error (± 0.02 ppm) for the solvents water, arsenic trichloride, and liquid sulfur dioxide, each of which is an ionizing solvent of high dielectric constant. These facts are consistent with the hypothesis that ion pair formation occurs in the solvent of low dielectric constant and affects Δ.

In the case of 1,4-diethylpyridinium ion, changing the solvent affects the Δ value to a larger extent for the 4-ethyl group than for the 1-ethyl group. This differential effect may be rationalized on the above hypothesis on the grounds of the closer proximity of the 1-ethyl than the 4-ethyl group to the main center of ionic charge.

Effects of Anion. Changes in the anion for aqueous solutions of tetraethylarsonium salts had no effect upon Δ. This invariance would be expected on the above hypothesis since in aqueous solutions the ions are essentially independent at the concentrations used. On the other hand, for chloroform solutions where ion pair formation does occur the importance of anion effects should be greater and indeed Δ does vary.

A very similar behavior is found for the 1-ethyl group of 1,4-diethylpyridinium. Δ is changed by variation of solvent, or by variation of anion in the case of chloroform solution, whereas it is constant in aqueous solution at a fixed concentration. An additional feature here is the ethyl group in the 4 position. This is remote from the ionic center and would be expected, on the ion-pairing model, to be less susceptible to the environmental influences discussed. It can be seen from Table IV that this is so.

Concentration. The spectrum of the concentrated solution of $As\overset{+}{E}t_4I^-$ in water not only differed from more dilute solutions in linewidth, as mentioned previously, but also differed in Δ. Moreover, the change in Δ

was toward the value found in the chloroform, ion-pairing, case. More detailed concentration studies are in hand.

No other chemical shifts were measured. Solvent and anion effects on the X resonances are anticipated, however, both on the basis of this investigation and the work of Richards and Yorke. They found concentration shifts of ^{133}Cs resonances in aqueous solutions of cesium bromide, which possibly arise from ion-ion interaction (*3*).

Acknowledgments

We are indebted to the U. S. Department of the Army through its European Research Office for support of this research and to Mr. D. G. Gillies for the decoupling experiments. We thank Professor Gillespie, his coauthors, and the *Canadian Journal of Chemistry* for permission to reproduce Fig. 2 (*10*).

References

1. A. G. Massey, E. W. Randall, and D. Shaw, *Spectrochim Acta*. **20**, 379 (1964).
2. J. Itoh and Y. Yamagata, *J. Phys. Soc. Japan*, **13**, 1182 (1958).
3. R. E. Richards and B. A. Yorke, *Mol. Phys*. **6**, 289 (1963).
4. K. J. Packer and E. L. Muetterties, *Proc. Chem. Soc*., p. 145 (1964).
5. J. T. Arnold and M. E. Packard, *J. Chem. Phys*. **19**, 1608 (1951).
6. D. G. Gillies, Doctoral Thesis, University of London (1965).
7. A. G. Massey, E. W. Randall, and D. Shaw, *Spectrochim. Acta* **21**, 263 (1965).
8. R. J. Chuck and E. W. Randall, unpublished results.
9. J. A. Pople, *Mol. Phys*. **1**, 168 (1958).
10. J. Bacon, R. J. Gillespie, and J. W. Quail, *Can. J. Chem*. **41**, 3063 (1963).
11. K. A. Valiev, *Soviet Phys. JETP* **11**, 883 (1960).
11a. H. Dreeskamp, personal communication.
12. M. M. Cohen and F. Reif, *Solid State Phys*. **5**, 321 (1957); also references therein.
13. R. M. Fuoss and C. A. Kraus, *J. Am. Chem. Soc*. **55**, 1019 (1933).
14. H. G. Hertz, *Z. Electrochem*. **65**, 20 (1961).
15. J. E. Wertz and O. Jarditzky, *J. Am. Chem. Soc*. **82**, 318 (1960).
16. H. G. Hertz, *Mol. Phys*. **7**, 499 (1964).
17. E. L. Muetterties and W. D. Philips, *J. Am. Chem. Soc*. **82**, 248 (1960).
18. B. P. Dailey and J. N. Shoolery, *J. Am. Chem. Soc*. **77**, 3977 (1955).
19. E. B. Baker, *J. Chem. Phys*. **26**, 960 (1957).
20. H. Spiesecke and W. G. Schneider, *J. Chem. Phys*. **35**, 722 (1961).
21. J. R. Cavanaugh and B. P. Dailey, *J. Chem. Phys*. **34**, 1099 (1961).
22. A. L. Allred and E. G. Rochow, *J. Inorg. Nucl. Chem*. **5**, 269 (1958).
23. M. P. Brown and D. E. Webster, *J. Phys. Chem*. **64**, 689 (1960).
24. S. Brownstein, B. C. Smith, G. Ehrlich, and A. W. Laubengayer, *J. Am. Chem. Soc*. **81**, 3826 (1959).
25. A. L. Allred and A. L. Hensley, *J. Inorg. Nucl. Chem*. **17**, 43 (1961).

NMR Longitudinal Relaxation Time and Structure of Aqueous Solutions

H. G. HERTZ and G. KELLER

Institut für physikalische Chemie der Universität Münster, Münster, Germany

Introduction: Negative Hydration of Univalent Ions, Positive Hydration of Nonpolar Solutes, and Stability Constants of Complexes as Studied by Relaxation Time Measurements

Nuclear magnetic relaxation time measurements form a powerful tool for the investigation of the structure of liquids, especially liquid mixtures. The designation "structure of a liquid" is generally used in a rather broad sense. It includes the description of microscopic motions by giving certain time constants for jumping, reorientation, and exchange processes of molecules. These processes are considered to be partly the subject of reaction kinetics or chemical kinetics. But the borderline between a general microdynamical process and a chemical reaction is not very clearly established. Other properties, which are also included in the concept of the structure of a liquid, do not contain the time; these properties are most generally described by molecular distribution functions. A special way of treating this second class of structural information—appropriate for certain types of problems—is to determine equilibrium constants, e.g., dissociation constants or stability constants, etc.

Many of the investigations which concern the first type of structural study, i.e., which lead to one or more parameters of the microdynamical structure of the liquid, use the transverse NMR relaxation time or both the transverse and the longitudinal relaxation time (*1–3*).

Here we are concerned with the results of longitudinal relaxation time measurements alone. Longitudinal relaxation times are easier to measure than transverse relaxation times if they are relatively long, i.e., of the order of magnitude of 1 second or more. Furthermore, the two relaxation times are probably almost equal (*4*) in the cases mentioned here as examples of the study of the microdynamical structure of aqueous solutions.

The first example is the nature of hydration of big univalent ions, especially noble gas ions like Cl^-, Br^-, I^-, Rb^+, and Cs^+. As shown by the hydration enthalpies, there exists a rather strong ion-dipole interaction between these ions and the water molecules in aqueous solution. As a consequence, one would expect a rigid hydration shell with definite dipole orientation to exist for these ions as for other ions with smaller radii. But other properties are known, e.g., the viscosity of aqueous solutions of the respective salts (*5*, *6*) as well as the self-diffusion coefficient of the water molecules in these solutions (*7*), which show that the liquid is more fluid in these solutions compared with pure water. The question arises whether the region of enhanced fluidity extends to the surface of the ion or whether the inner rigid hydration sphere is still present, the average effect of all regions in the solution being a decrease of structure, i.e., an increase in fluidity. The first view has been proposed by Gurney (*6*) and by Samoilow (*8*), the second one by Frank and Wen (*9*). The first type of hydration has been called "negative hydration" by Samoilow. Its difficulty is to account for the negative hydration enthalpy. Measurements of the longitudinal NMR relaxation time (called T_1 in the following) of the solvent water have proved that Gurney and Samoilow's picture is the correct one (*10–12*). These investigations concern proton (*10*, *11a*, *12*) as well as deuteron (*11b*) relaxation times. In addition, transverse relaxation times (T_2) of the ion nuclei corroborate the hypothesis of negative hydration. This additional information is partly based on the concentration dependence of the ion resonances (*11b*) and partly on their temperature dependence, which give rather low activation energies of the reorientation processes in the first hydration spheres (*13–15*) around the ion considered.

In view of the fact that there is an appreciable negative hydration enthalpy of the above-mentioned ions, the effect of "negative hydration" is unexpected, and in the same way it is quite surprising that nonpolar particles with only weak van der Waals interaction with the water molecules are positively hydrated in aqueous solution. Frank and Evans (*16*) concluded from the interpretation of entropy data that there should exist such a hydration effect. They called this special type of hydration "iceberg formation." To avoid misinterpretation, another name, "hydration of the second kind," was proposed later (*17*, *18*). Again it was possible to show (*17*, *19*, *20*) unambiguously by NMR T_1 measurements that hydration of the second kind exists. These measurements were performed with aqueous solutions of charged or polar solutes which possess shorter or longer hydrocarbon chains (alkyl groups). Here the evidence for positive hydration comes from the observation that the average reorientation time of the solvent (water) molecules is greater in the presence of the partly nonpolar dissolved particles than in pure water. Essentially, the effect gets the more pronounced the greater the number of CH_2 groups in a dissolved particle. The interpretation of these findings may be carried one step further. The water molecules with increased tumbling times are located around the nonpolar groups; the molecules further apart are unaffected. Thus it follows that hydration of the second kind should exist.

The connection between the NMR relaxation time and structural changes in aqueous solutions of nonpolar particles as described by molecular distribution functions has been discussed elsewhere by one of the authors (*18*).

The present paper concerns the second class of structural studies mentioned above, namely, the determination of stability constants of complexes. The complexes treated are simple systems like MA_n where M is the central ion and A is a ligand which does not contain protons. The ions M are the paramagnetic ions Mn^{2+}, Cu^{2+}, Ni^{2+}, and Fe^{3+}; some results for the diamagnetic ions Cd^{2+} and Zn^{2+} are also reported. Only some of the solutions investigated are evaluated quantitatively; other examples, especially those with diamagnetic central ions, are discussed qualitatively.

Studies of T_1 of aqueous solutions containing complex-forming paramagnetic particles have been made by several authors (*21–27*). The most extensive work is due to Russian workers (*23, 25*). The present paper is related to the one of Rivkind (*23*) in the following way: Rivkind has reported the very striking effects of typical strong complex formers, often with complicated organic ligands. Our paper—like part of the work of Wdowenko and co-workers (*25*)—regards the other end of the scale of stability constants and it is shown that the T_1 method is also applicable to the study of very weak complexes for which it has been doubtful whether there would be any complex formation at all.

Theoretical Background

In all cases to be described here the general nature of the solution is the following: The solvent is water. One of the solutes is the salt ML_2, where $L = Cl^-$, NO_3^-, ClO_4^- (ML_3 in the case of $FeCl_3$). The concentration of ML_2 is generally small, of the order of magnitude of 10^{-3} to 10^{-2} m (m = moles/kg H_2O). Only when M is the ion Cd^{2+} or Zn^{2+} the concentration is greater (see below). To this solution is added another salt: K_mA ($m = 1$ in most cases; $m = 2$ only for Na_2SO_4). The concentration c of this salt is much greater; it is of the order of magnitude of some moles per kilogram H_2O. At this stage of description of the model any incomplete dissociation or complexing will be disregarded. Actually, however, the presence of the anion A in lower or higher concentration causes the formation of complexes MA_n, $n = 1, 2 \ldots$, the stability constants of which are to be determined. This complication will be introduced later. Up to concentrations $\lesssim 5$ m the solution thus defined can be divided into microscopic regions of five different types:

(1) the hydration shell of the cation M;
(2) the hydration shell of the anion L;

(3) the hydration shell of the cation K;
(4) the hydration shell of the anion A;
(5) the free water.

Any individual water molecule travels from one microscopic region to the other in the course of time; it forms a part of such a region for a certain time and leaves this region after the residence time to become a member of another region. In the five different microscopic regions the protons of the water molecule would have different NMR relaxation times if these regions existed as pure bulk phases. Of course, this is possible only for the free water region; we have T_1 of pure water at the same temperature. But still it is sensible to correlate relaxation times T_{1M}, T_{1L}, T_{1K}, T_{1A}, and T_{10} with the respective hydration regions of the ions M, L, K, A, and with pure water. The relaxation time T_{1M} is of the order of magnitude of 10^{-4} sec if M is a paramagnetic ion; all other relaxation times are of the order of 1 sec. If the residence time in the different regions is very much smaller than the respective relaxation time, the measured relaxation rate $1/T_1$ of the water proton in the solution equals the weighted average of the individual relaxation rates $1/T_{1M}$, $1/T_{1L}$, $1/T_{1K}$, ... of the various regions (*28*):

$$\frac{1}{T_1} = \frac{p_M}{T_{1M}} + \frac{c_M^0}{T_{1t}} + \frac{p_L}{T_{1L}} + \frac{p_K}{T_{1K}} + \frac{p_A}{T_{1A}} + \frac{1 - \Sigma p_\nu}{T_{10}}, \tag{1}$$

where ν = M, L, K, A and p_M, p_L, p_K, p_A are the mole fractions of water molecules in the different regions of the solution. The term c_M^0/T_{1t} takes account of the fact that in the regions 3–5 there is a certain contribution to $1/T_1$ caused by the translational motion of the water molecules in regions 3–5 relative to the paramagnetic ions (*29*).

As already mentioned, when ML_2 is a paramagnetic salt, $1/T_{1M} \gg 1/T_{1L}$ and the solutions of interest have the property $p_M \ll p_K, p_A$, since the concentration of the paramagnetic salt is always small (see above). Thus the third term on the right-hand side of Eq. (1) can be neglected compared with the other terms and in the last term of the right-hand side p_M and p_L can be dropped. It follows:

$$\frac{1}{T_1} = \frac{p_M}{T_{1M}} + \frac{c_M^0}{T_{1t}} + \left\{\frac{p_K}{T_{1K}} + \frac{p_A}{T_{1A}} + \frac{1 - p_K - p_A}{T_{10}}\right\}. \tag{2}$$

The expression in parentheses in Eq. (2) is just the relaxation rate of the solution under consideration but *without* the (paramagnetic) salt ML_2. This relaxation rate can be measured independently. The necessary data are partly available in the literature (*11a, 12, 19*) and have been obtained partly in this work. If we call the expression in parentheses in Eq. (2) $(1/T_1)_{dia}$ (relaxation rate of the corresponding diamagnetic solution) we have

$$\Delta(1/T_1) = (1/T_1)_{par} = (1/T_1) - (1/T_1)_{dia} = \frac{p_M}{T_{1M}} + \frac{c_M^0}{T_{1t}}, \tag{3}$$

where $\Delta(1/T_1)$ gives the contribution of the paramagnetic salt to the measured relaxation rate. If M is one of the paramagnetic ions Mn^{2+}, Fe^{3+} (or Cu^{2+}), $1/T_{1M}$ is given by the following formula (*30, 31*):

$$\frac{1}{T_{1M}} = \frac{2}{15} b^{-6} \gamma_n^2 \gamma_e^2 \hbar^2 S(S+1) \left\{ \tau + \frac{7\tau/3}{1 + \omega_s^2 \tau^2/9} \right\}. \tag{4}$$

T_{1M} according to this equation is the relaxation time of a proton in a pair of unlike spins undergoing statistical rotational motions. The other spin of the pair is the spin of the unpaired electrons of the paramagnetic ion. Both spins interact by magnetic dipole-dipole interaction. b is the distance between the proton in the rigid hydration sphere and the center of the ion M; b is not very different from the radius a of the fully hydrated ion $M(H_2O)_6$. γ_n and γ_e are the gyromagnetic ratios of the proton and the electron, respectively; S is the quantum number characterizing the total spin of the unpaired electrons in M; τ is the reorientation or tumbling time of the fully hydrated complex $M(H_2O)_6$; ω_s is the angular frequency of electron spin resonance at the magnetic field which acts during the T_1 measurement. A rough approximation to the tumbling time τ is the Debye formula

$$\tau = \frac{4\pi\eta a^3}{kT} \tag{5}$$

(η = viscosity, k = Boltzmann constant, T = absolute temperature). In the case of Cu^{2+} the magnetic moment of the ion contains some orbital contribution. We can still use Eq. (4) but have to replace $\hbar^2 \gamma_e^2 S(S+1)$ by $\langle\mu^2\rangle$, the mean square of the magnetic moment of the ion. Formally it can be written $\langle\mu^2\rangle = \hbar^2 \gamma_e^2 S'(S'+1)$, defining thus the number S' ($S' \approx \frac{1}{2}$ for Cu^{2+}). The residence time τ_h of a water molecule in the hydration shell is very long compared with the tumbling time (*1–3, 32*), which is of the order of 10^{-10} sec (*2, 3, 29*). The electron spin relaxation time τ_e is longer than the tumbling time for the ions mentioned (*1–3, 33*). For Ni^{2+} this is no longer valid (*1, 3*) and Eq. (4) is not applicable.

In the ideal solution considered so far p_M, the mole fraction of water molecules in the region of the solution which corresponds to the hydration sphere of the ion M, is given by

$$p_M = \frac{n_h^0 c_M^0}{55.5}, \tag{6}$$

with $n_h^0 = 6$ (n_h^0 is the hydration number or coordination number of the ion M). c_M^0 is the concentration of M (moles/kg H_2O). From Eqs. (3), (4), and (6) one obtains (for S-state ions):

$$\Delta(1/T_1) = \frac{2}{15} b^{-6} \gamma_n^2 \gamma_e^2 \hbar^2 S(S+1) \frac{n_h^0 c_M^0}{55.5} \left\{ \tau + \frac{7\tau/3}{1 + \omega_s^2 \tau^2/9} \right\} + \frac{c_M^0}{T_{1t}}. \tag{7}$$

$\Delta(1/T_1)$ increases for increasing reorientation time τ. The first term on the right-hand side of Eq. (7) is the dominant one: the second one amounts to about 10% of the first (*29*).

Apart from a small variation of $1/T_{1t}$ in the ideal solution represented by the model and leading to Eq. (7), $\Delta(1/T_1)$ should vary linearly with the concentration c_M^0 of the salt ML_2, but should be independent of the concentration K_mA. Generally, however, $\Delta(1/T_1)$ depends on the concentration c of the diamagnetic salt K_mA as will be explained now. The reason for this behavior is the formation of complexes MA_n, $n = 1, 2, \ldots$. Thus the study of the concentration dependence on c allows the determination of stability constants in favorable cases.

The effects of complexing on $\Delta(1/T_1)$ will now be discussed. First of all, we have no reason to suspect that the quantum number S changes under complex formation with any of the ligands to be considered here. For the number $S'(Cu^{2+})$ the situation is a bit more critical. Having no knowledge about this question, we felt it to be most reasonable to take $S' =$ const. in all cases, too. In a certain way the very satisfactory agreement of the calculated stability constants with literature data derived from other sources justifies this assumption.

But there are two other effects of complexing, acting in opposite directions, which are important. The first one is the decrease of the hydration number, because one or more of the sites *at the surface* of the cation M are occupied by a ligand A (which in the systems considered here does not contain a proton). The type of complex with A where this happens is generally called an "inner sphere complex." In Eq. (7) n_h^0 has to be replaced by n_h, the mean hydration number[1] at a given concentration c. Thus $n_h = n_h(c)$, and n_h is no longer equal to the coordination number n_h^0 but decreases if inner sphere complexes appear with increasing concentration c. It follows from Eq. (7) after n_h^0 is replaced by $n_h(c)$ that one effect of inner sphere complexes is to make $\Delta(1/T_1)$ smaller.

[1]The name hydration number is generally used in a very wide sense; here it is the number of water molecules in the *first coordination* sphere.

The other effect of complex formation with A is the change of the geometrical dimensions of the species which undergoes the rotational Brownian movement. Generally, the ligand A will have a radius different from that of the water molecule. The hexaquo complex $M(H_2O)_6$ may be considered as approximately of spherical shape. If one or two of the ligands have a radius which is only slightly greater than that of the water molecule, the deviation from the quasi-spherical shape will be small and there should be practically no change in reorientation time. This situation will be valid for Cl^-; for Br^- and I^- the deviation from the quasi-spherical shape will be greater and the complex has different reorientation times about different axes of the complex. The anisotropy of rotational motion is still more pronounced with the big ions SO_4^{2-} or ClO_4^- as ligands, whose radii are about 3 Å. For these ions Eq. (5) would give a drastic increase in reorientation time if one put the quantity a equal to $\frac{1}{2}$ of the greatest extension of the complex. But as a consequence of the nonspherical shape the average or effective reorientation time of the vector connecting the water protons with the center of the ion M will be much shorter. Moreover, it can easily be imagined that by the introduction of the ligand A the water molecules in the first hydration sphere regain their free rotation to a certain amount, which causes an additional modulation of the magnetic dipole-dipole interaction with the electrons, and thus an increase of τ is partly compensated. For reasons of simplicity we assume that the effective reorientation time of an inner sphere complex with the ligand A is equal to the reorientation time of the hexaquo complex. This approximation may be uncertain in the case of large ligands (NO_3^-, SO_4^{2-}, ClO_4^-) but for these ligands no quantitative evaluations will be given in this paper (with two exceptions, $MnClO_4^+$ and $CuClO_4^+$). Thus we state the result that if a decrease of $\Delta(1/T_1)$ with increasing concentration of the ligand A is observed, this decrease is interpreted as a consequence of inner sphere complex formation. Writing mathematically, we have (in the case of inner sphere complex formation, M paramagnetic)

$$\Delta(1/T_1) = \frac{2}{15}\, b^{-6} \gamma_n^2 \gamma_e^2 \hbar^2 S(S+1)\, \frac{c_M^0}{55.5} \left\{ \tau + \frac{7\tau/3}{1 + \omega_s^2 \tau^2/9} \right\} \cdot n_h(c) + \frac{c_M^0}{T_{1t}}, \quad (7a)$$

with

$$n_h(c) \longrightarrow n_h^0 \text{ as } c \longrightarrow 0, \quad \text{and} \quad dn_h(c)/dc \leqslant 0,$$

the term c_M^0/T_{1t} not being affected.

The other type of complex with the ligand A which is of interest here is the so-called "outer sphere complex." It is often symbolically written MH_2OA. Here the symbol H_2O stands for a complete hydration sphere of six water molecules, and the ligand A is a more or less firmly bound member of the second hydration shell. In an outer sphere complex the hydration number n_h is always 6 and does not change as long as the equilibrium concentration of the inner sphere complex MA, which is formed by the reaction $MH_2OA \rightleftharpoons MA + H_2O$, is negligible small. Thus the continuously growing concentration of an outer sphere complex at increasing concentration of K_mA would have no influence on $\Delta(1/T_1)$, were it not for the fact that the reorientation time is different for the hexaquo and the outer sphere complex. It is a difficult task to give an appropriate number for the reorientation time of the outer sphere complex. If one were to use Eq. (5) and take a as the sum of the radii of the hexaquo complex (~ 3 Å) and the ligand (~ 3 Å for ClO_4^-, SO_4^{2-}), the resulting reorientation time certainly would be much too great. Only the unreasonable assumption that the second hydration sphere of the complex also is rigidly bound would justify this value. In a model in which merely the ligand A is firmly attached to the hexaquo complex, the complex will not be of spherical shape, and the effective reorientation time of the vector (water proton) $\longrightarrow$ (central ion) will be shorter than the Debye time with a as mentioned above, and longer than the reorientation time of the unburdened hexaquo complex. But it is not even sure that in an outer sphere complex the ligand A is rigidly connected to the hydrated ion (*26*), since the definition of the outer sphere complex as given in the literature generally does not imply such dynamical details. For the present considerations it must be sufficient to assert that the effective reorientation time of an outer sphere complex is in any case longer than that of the fully symmetrical hexaquo complex. Since in considering the inner sphere complex it was assumed that its reorientation time equals that of the hexaquo complex, there follows the simple and approximate rule: Whenever one observes an increase of $\Delta(1/T_1)$ with increasing c this means that an outer sphere complex is formed. The possibility remains that on addition of the diamagnetic salt, $\Delta(1/T_1)$ increases without any complex formation with the ligand A at all. In the picture of the Debye equation (5) this can happen if the local or microviscosity η becomes greater with increasing concentration of K_mA. In fact, in many cases the macroscopic viscosity of the solution increases, but this does not mean that the microviscosity in Eq. (5) must become greater too. Obviously the change of the microviscosity is accompanied by some rearrangement of the water molecules in the second (and third) hydration shell. In the event that this rearrangement occurs without an anion A entering the second hydration sphere, there would really exist another reason for the increase of $\Delta(1/T_1)$ besides formation of outer sphere complexes. It follows that the statement concerning the connection between increase of $\Delta(1/T_1)$ and the presence of outer sphere complexes is not quite unambiguous. In the following this has to be kept in mind when interpretations of experimental results in terms of outer complexes are given. On the other hand the statement that microviscosity increases because an anion enters the second hydration shell is another way of stating that outer sphere complexes are formed.

Finally it should be mentioned that inner sphere or outer sphere complexes ML and MH_2OL can be formed with increasing concentration of K_mA because the activity coefficients of the ions M and L change with increasing concentration of the added diamagnetic salt. This influence on $\Delta(1/T_1)$ will be neglected in all cases where the ligand L differs from the ligand A.

The next step is to present the relation between the average hydration number $n_h = n_h(c)$ and the equilibrium constants. Consider the reaction

$$\mathrm{M} + n\mathrm{A} \rightleftharpoons \mathrm{MA}_n$$

with the cumulative or gross stability constants

$$\beta_n^0 = \frac{\{\mathrm{MA}_n\}}{\{\mathrm{M}\}\{\mathrm{A}\}^n} \tag{8}$$

or the consecutive or stepwise stability constants

$$K_1^0 = \frac{\{\mathrm{MA}\}}{\{\mathrm{M}\}\{\mathrm{A}\}}, \qquad K_2^0 = \frac{\{\mathrm{MA}_2\}}{\{\mathrm{MA}\}\{\mathrm{A}\}}, \qquad K_3^0 = \frac{\{\mathrm{MA}_3\}}{\{\mathrm{MA}_2\}\{\mathrm{A}\}}, \ldots . \tag{9}$$

The quantities in parentheses are activities. From Eqs. (8) and (9) we have

$$\beta_n^0 = K_1^0 \cdot K_2^0 \ldots K_n^0. \tag{10}$$

We now define the concentration-dependent quantities

$$\beta_n = \frac{c_{\mathrm{MA}_n}}{c_\mathrm{M}\, c_\mathrm{A}^n} = \beta_n^0 \frac{f_\mathrm{M}(f_\mathrm{A})^n}{f_{\mathrm{MA}_n}} \tag{11}$$

and

$$K_1 = \frac{c_\mathrm{MA}}{c_\mathrm{M}\, c_\mathrm{A}} = K_1^0 \frac{f_\mathrm{M}\, f_\mathrm{A}}{f_\mathrm{AM}}, \qquad K_2 = \frac{c_{\mathrm{MA}_2}}{c_\mathrm{MA}\, c_\mathrm{A}} = K_2^0 \frac{f_\mathrm{MA}\, f_\mathrm{A}}{f_{\mathrm{MA}_2}}, \ldots . \tag{12}$$

$c_{\mathrm{M,A},\ldots}$ = concentration of the respective particles; $f_{\mathrm{M,A},\ldots}$ = activity coefficient of the respective particles. Since the concentration c_{MA_n} is always very small in the cases of interest we may put $c_\mathrm{A} = c_\mathrm{A}^0$ where c_A^0 is the analytical concentration of the anion A. Furthermore, for the salts used in this work $c_A = c_A^0 = c$ (c = concentration of K_mA). It follows that

$$\frac{c_M}{c_M^0} = \frac{1}{1 + \beta_1 c + \beta_2 c^2 + \ldots + \beta_n c^n}; \tag{13}$$

$$\frac{c_{MA_n}}{c_M^0} = \frac{\beta_n c^n}{1 + \beta_1 c + \beta_2 c^2 + \ldots + \beta_n c^n} \tag{14}$$

(c_M^0 is the analytical concentration of the cation M).

From Eqs. (13) and (14) we have

$$n_h = \frac{6 c_M + 5 c_{MA} + 4 c_{MA_2} + \ldots}{c_M^0}$$

$$= \frac{6 + 5\beta_1 c + 4\beta_2 c^2 + \ldots}{1 + \beta_1 c + \beta_2 c^2 + \ldots}. \tag{15}$$

This relation is to be introduced in Eq. (7a). It is convenient to express the $\Delta(1/T_1)$ values as relative quantities

$$\Delta(1/T_1)_{\text{rel}} = g(c) = \frac{\Delta(1/T_1)_c}{\Delta(1/T_1)_0} \cdot 100\%. \tag{16}$$

$\Delta(1/T_1)_c$ is the value of $\Delta(1/T_1)$ at the concentration c of K_mA; $\Delta(1/T_1)_0$ is the value of $\Delta(1/T_1)$ at the concentration 0 of K_mA; the amount of ML_2 is always kept constant (and generally of the order of 10^{-3} to 10^{-2} *m*). In order to obtain a more compact final expression the term c_M^0/T_{1t} may be canceled from Eq. (7a) without appreciable error as long as one considers only the first steps of complex formation ($\Delta(1/T_1)_{\text{rel}} \geqslant 0.5$). With Eqs. (7), (7a), (15), and (16) we write

$$\Delta(1/T_1)_{\text{rel}} = \frac{1}{6} \cdot \frac{6 + 5\beta_1 c + 4\beta_2 c^2 + \ldots}{1 + \beta_1 c + \beta_2 c^2 + \ldots} \cdot 100\% \tag{17}$$

(valid for $100\% \geqslant \Delta(1/T_1)_{\text{rel}} \gtrsim 50\%$; M is paramagnetic).

Since all the β_n depend on c, the decrease of the relative paramagnetic relaxation rate is a complicated function of the concentration of the added diamagnetic salt. If the inner sphere complex MA_n is the highest one which forms within a certain concentration range, then $\Delta(1/T_1)$ decreases from 100% to about $(100 - n \cdot 16.7)\%$ in this range ($n \leqslant 3$). This holds only if at c = 0, M is present as an undisturbed hexaquo complex (for complications see later).

The following types of activity coefficients occur in the evaluation of Eq. (17).

(a) f_M for divalent ions: This is the individual ionic activity coefficient of the doubly charged ion M at low concentration in a solution of K_mA with high concentration. A method of estimating this activity coefficient for a sufficiently extended concentration range has been given by one of the authors elsewhere (*34*). The starting point for this calculation is the activity coefficient of a hypothetical completely dissociated salt MA_2 derived from the known activity coefficients of MgA_2, CaA_2, SrA_2, BaA_2, by interpolation of the ionic radii. The results of these calculations are given in Table I. The UO_2^{2+} ion is taken as a model for the $FeCl^{2+}$ ion, and its activity coefficients in the desired solution are also included in Table I. At sufficiently low concentrations the formula

TABLE I
APPROXIMATE ACTIVITY COEFFICIENTS OF SOME IONS AT LOW CONCENTRATION (IONIC STRENGTH $J \ll 1\ m$) IN CONCENTRATED SOLUTIONS OF THE SALTS KA

Concentration KA (m)	Mn^{2+} in KCl	Mn^{2+} in KBr	Mn^{2+} in KI	Mn^{2+} in $NaClO_4$	Cu^{2+} in KCl	Cu^{2+} in KBr	Ni^{2+} in KBr	UO_2^{2+} in KCl
0.5	0.21	0.22	0.25	0.26	0.22	0.24	0.23	0.23
1.0	0.18	0.21	0.25	0.25	0.19	0.23	0.22	0.21
1.5	0.19	0.23	0.29	0.29	0.20	0.26	0.25	0.21
2.0	0.20	0.27	0.36	0.37	0.23	0.30	0.30	0.23
3.0	0.29	0.44	—	0.70	0.34	0.53	0.51	0.27
4.0	0.46	0.82	—	1.45	0.58	1.03	0.99	0.35

$$-\log f^{\pm} = 0.5\, z_+ \mid z_- \mid \left\{ \frac{J^{1/2}}{1 + J^{1/2}} - 0.30\, J \right\}$$

proposed by Davies (*35*) was used. The $z_{+,-}$ are ionic charges, and J = ionic strength $[J = \frac{1}{2}(\Sigma z_i^2 c_i)]$.

(b) f_A: The activity coefficient of the anion A is equal to the mean activity coefficient $f^{\pm}$ of the diamagnetic salt added. Their values are taken from the literature (*36*).

(c) f_{MA}: The activity coefficient of the ion MA is assumed to be equal to f_A.

(d) f_{MA_2}: All uncharged particles are assumed to have the activity coefficient 1 over the whole concentration range.

Regarding the treatment of Fe^{3+}, where an activity coefficient f^{3+} would occur, see below.

So far we have presumed that the ion M is a paramagnetic one. We shall conclude this section with some remarks concerning diamagnetic cations M (Cd^{2+} and Zn^{2+} are the examples studied here). The fundamental difference is that now T_{1M} is of the order 1 sec whilst with paramagnetic cations it was of the order 10^{-4} sec. Thus, to observe any effect of the ion M at all, the concentration of M has also to be of the order 1 *m*. It follows that the simplifications introduced between Eqs. (1) and (2) are no longer justified. Instead, we have now

$$\Delta(1/T_1) = 1/T_1 - (1/T_1)_{\text{dia}}$$
$$= p_{\text{M}}\left(\frac{1}{T_{1\text{M}}} - \frac{1}{T_{10}}\right) + p_{\text{L}}\left(\frac{1}{T_{1\text{L}}} - \frac{1}{T_{10}}\right). \tag{18}$$

T_{1M} now is caused by the magnetic dipole-dipole interaction between the protons of the water molecules. In a fair approximation we may consider only the rotational part of the Brownian motion, for which the following equation holds (*11b, 31*):

$$\left(\frac{1}{T_1}\right)_{\text{rot}} = \frac{1}{2}\frac{\gamma_n^4 \hbar^2}{d^6}\tau_x, \tag{19}$$

where d is the proton-proton distance in the water molecule, and τ_x is the reorientation time of a pair of identical spins, i.e., the water molecule. From Eqs. (18), (19), and (6) it follows that

$$\Delta(1/T_1) = \frac{1}{2}\cdot\frac{c_{\text{M}}^0}{55.5}\cdot\frac{\gamma_n^4\hbar^2}{d^6}\,\xi\,\{n_{h\text{M}}(\tau_{\text{M}} - \tau_0) + 2n_{hL}(\tau_{\text{L}} - \tau_0)\}. \tag{20}$$

The factor ξ is of order unity and takes care of the fact that here only the rotational part of the relaxation rate is considered. The indices M, L, 0, as usual, refer to the ions M, L, and pure water, respectively. In the corresponding equation (7a) n_h is defined as the number of water molecules on the surface of the ion M. This definition results from the very rapid decrease of $(1/T_1)_{\text{par}}$ [Eq. (4)] with b, the distance (water molecule)–(center of the ion). But in Eq. (20) d has the same value everywhere in the solution. Thus n_{hM} in Eq. (20) is not at all clearly defined by the interaction with the ion. The only and more indirect connection to an interaction is the fact that if the water molecule is close to the surface of the divalent ion M, the electrostatic field is strong, and the water molecule is impeded from rotating freely and has to take part in the reorientation of the whole hydration complex.

If a ligand enters in the first hydration shell the water molecule is pushed away from the surface of M and finds itself further away where the orientating electric field is weaker and the water molecule has a greater chance to rotate freely. The effect is that we have to define n_{hM} *ad hoc* as a more or less constant number given by the geometry of the complex, and that τ_M depends on the state of the complex of the species under consideration; τ_M is the average reorientation time of a water molecule in a geometrically defined hydration sphere of the complex MA_n, $n = 0, 1, 2, \ldots$. If the hydration sphere is located close to the ion M, i.e., the center of the electric field, τ_M will be relatively long; if the hydration shell is far away because it has been pushed outward by anions sitting on the surface of M, τ_M will be short. Clearly, the first situation applies to the hexaquo and outer sphere complexes, the second situation to the inner sphere complex. It follows that $\tau_M = \tau_M(c)$ and $d\tau_M(c)/dc < 0$ or > 0 indicates the formation of inner or outer sphere complexes, respectively. The second term in the parentheses of Eq. (20) is relatively small and in first approximation independent of c. Transforming Eq. (20) to the corresponding relative quantity gives [see Eq. (16)]

$$\Delta(1/T_1)_{\text{rel}} \approx \frac{n_{hM}(\tau_M(c) - \tau_0) + 2n_{hL}(\tau_L - \tau_0)}{n_{hM}(\tau_M^0 - \tau_0) + 2n_{hL}(\tau_L - \tau_0)} \cdot 100\%, \tag{21}$$

τ_M^0 is the reorientation time in the absence of the complex-forming ligand A $(c = 0)$.

As the symbol $\approx$ indicates, this relation generally holds only approximately, because under conditions of strong complex formation with the ligand A the term $(1/T_1)_{\text{dia}}$ in Eq. (18) implies too great a number of anions A. An appreciable amount of these anions has been trapped by the cations M. Often, however, $(1/T_1)_{\text{dia}}$ does not depend strongly on c.

For evident reasons no quantitative evaluation of Eq. (21) has been undertaken.

Experimental

The T_1 measurements have been performed with a spin echo apparatus of conventional construction (*11b, 37*). The method used was partly the 90°–90° and partly the 180°–90° pulse sequence. The samples which did not contain paramagnetic salts were freed from oxygen in the usual way (*11b*). All relaxation times measured are those of protons in the water molecules of the solution. Each experimental point in the Figs. 1–14 is generally an average of several single measurements. For the solutions with diamagnetic divalent cations the scatter is greater, which follows from the definition of $\Delta(1/T_1)_{\text{rel}}$ and the magnitudes of the two terms in the first line of Eq. (18). In all cases the resonance frequency was 20 Mc.

The solutions studied were the following:
(A) 10^{-3} *m* $MnCl_2$ and diamagnetic salt added.
(B) 10^{-2} *m* $CuCl_2$ and diamagnetic salt added.
(C) $6.4 \cdot 10^{-4}$ *m* $FeCl_3$ + 0.05 *m* HCl and diamagnetic salt added.
(D) $2.3 \cdot 10^{-2}$ *m* $Ni(NO_3)_2$ and diamagnetic salt added.
(E) 1.59 *m* $Zn(ClO_4)_2$ and diamagnetic salt added.
(F) 1.50 *m* $Cd(NO_3)_2$ and diamagnetic salt added.
In the last two solutions small amounts of $HClO_4$ and HNO_3 were added to produce a pH of ≈ 2. The diamagnetic salts were KCl, NaCl, NaBr, KI, NaI, Na_2SO_4, $NaNO_3$, $NaClO_4$. The concentration ranges of these salts may be seen from Figs. 1–6. The chemicals were commercial ones; they were as pure as possible, mostly p.a. (*pro analysi*); they were used without further purification. All measurements were made at 25°C. The results of T_1 measurements of pure water (3.6 sec) and of the solutions of paramagnetic ions without added diamagnetic salts were in satisfactory agreement with the data given in the literature (*38–40*). The measured relaxation rates for the solutions quoted above *without* added diamagnetic salts are (A) 9.20; (B) 9.76; (C) 5.89; (D) 14.65; (E) 0.79; (F) 0.52 sec^{-1}. The relaxation rates of Na_2SO_4 and $NaClO_4$ solutions without paramagnetic admixtures, which are not available in the literature, have been measured and are given in Table II.

TABLE II
RELAXATION RATES OF THE WATER PROTONS IN SOLUTIONS OF $NaClO_4$ AND Na_2SO_4 ($\pm 5\%$, $\vartheta = 25°C$)

$NaClO_4$		Na_2SO_4	
Concentration, *m*	$1/T_1$, sec^{-1}	Concentration, *m*	$1/T_1$, sec^{-1}
0	0.28	0	0.28
2.5	0.27	0.25	0.29
4.9	0.30	0.49	0.31
7.3	0.33	0.70	0.34
12.3	0.48	0.95	0.37
14.4	0.54	1.19	0.39
		1.38	0.44

Results[2]

Figures 1–6 show the behavior of the relative (paramagnetic) relaxation rate $\Delta(1/T_1)_{rel}$ [definition see Eq. (16)] of the water protons as a

[2]We wish to thank Mr. H. Küper for his assistance with some measurements.

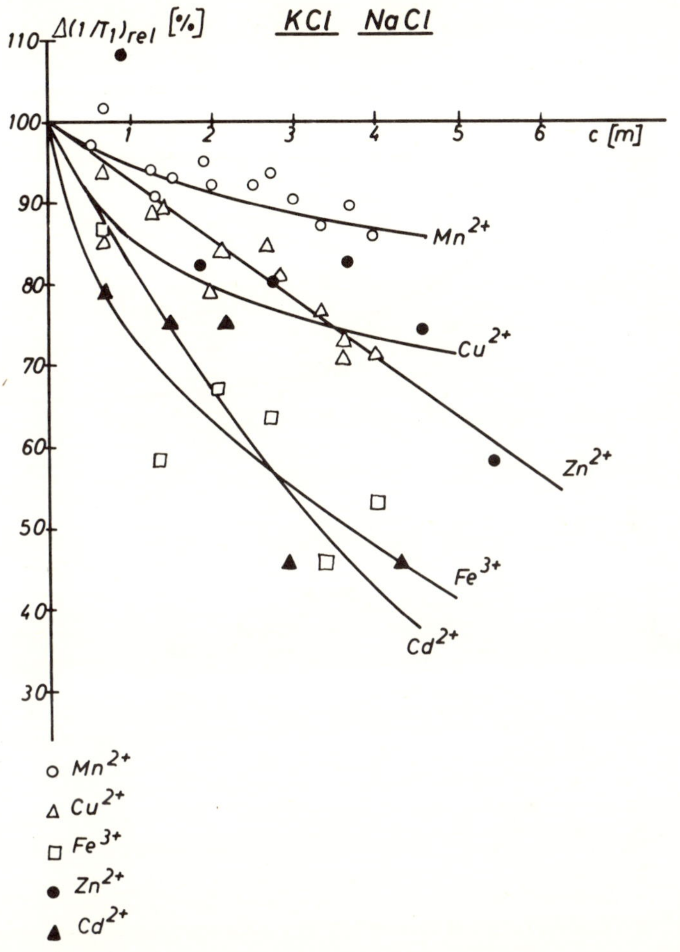

FIG. 1. Relative relaxation rate $\Delta(1/T_1)_{rel}$ of the water protons as a function of the concentration c of added chlorides. The concentrations of di- and trivalent cations are quoted in the third section (A, B, C, . . .). For the paramagnetic solutions the quantity $\Delta(1/T_1)_{rel}$ is practically equal to $[1/T_1(c)]/[1/T_1(0)]$, $T_1(c)$ and $T_1(0)$ being the relaxation times at c and c = 0, respectively. For a more precise definition of $\Delta(1/T_1)_{rel}$ see Eqs. (16) and (3). The Zn^{2+} solution is the only one where NaCl has been used as added salt; all other solutions contain KCl ($\vartheta = 25°C$).

function of the added diamagnetic salts K_mA. The kind and concentration range of these salts are also indicated in the figures and their legends. The concentrations of the constant paramagnetic admixtures are quoted in the preceding section. For the paramagnetic solutions

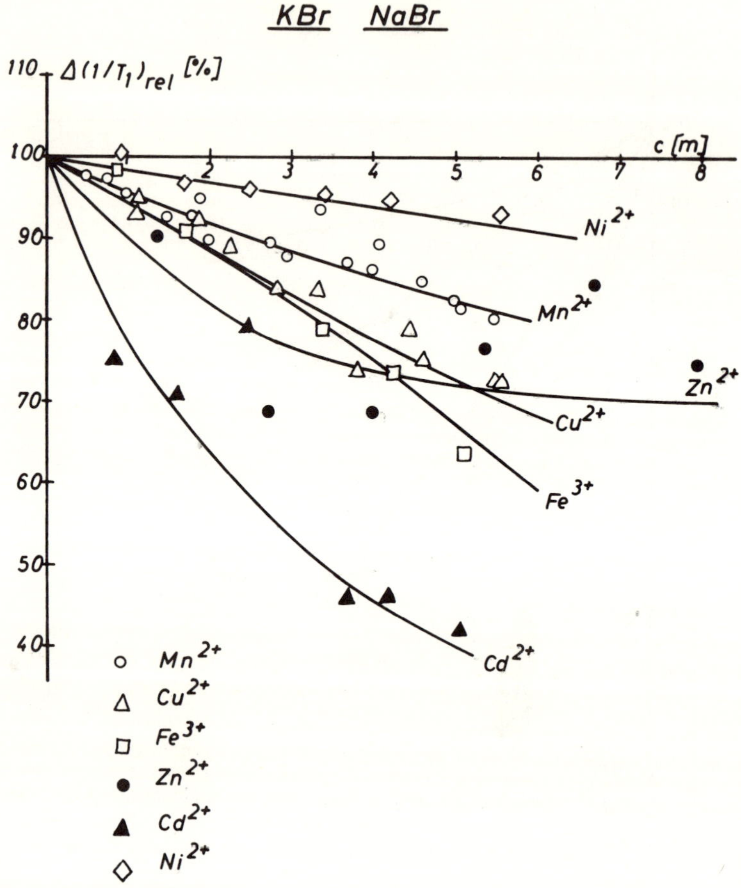

FIG. 2. Relative relaxation rate $\Delta(1/T_1)_{rel}$ as a function of the concentration c of added bromides. The Zn^{2+} solution is the only one where NaBr has been used as added salt; all other solutions contain KBr. For other details see the legend of Fig. 1 ($\vartheta = 25°C$).

$\Delta(1/T_1)_{rel}$ is practically equal to $[1/T_1(c)]/[1/T_1(0)]$, $T_1(c)$ and $T_1(0)$ being the relaxation times at the concentration c and 0, respectively, of the added diamagnetic salt. The effects of the latter salts on the relaxation rates in solutions of the diamagnetic divalent cations Cd^{2+} and Zn^{2+} are included in the figures. As may be seen from the figures, the relative (paramagnetic) relaxation rate $\Delta(1/T_1)_{rel}$ goes down or goes up with increasing concentration for different salts added (K_mA), which according to our interpretation means that inner sphere as well as outer sphere complexes are formed.

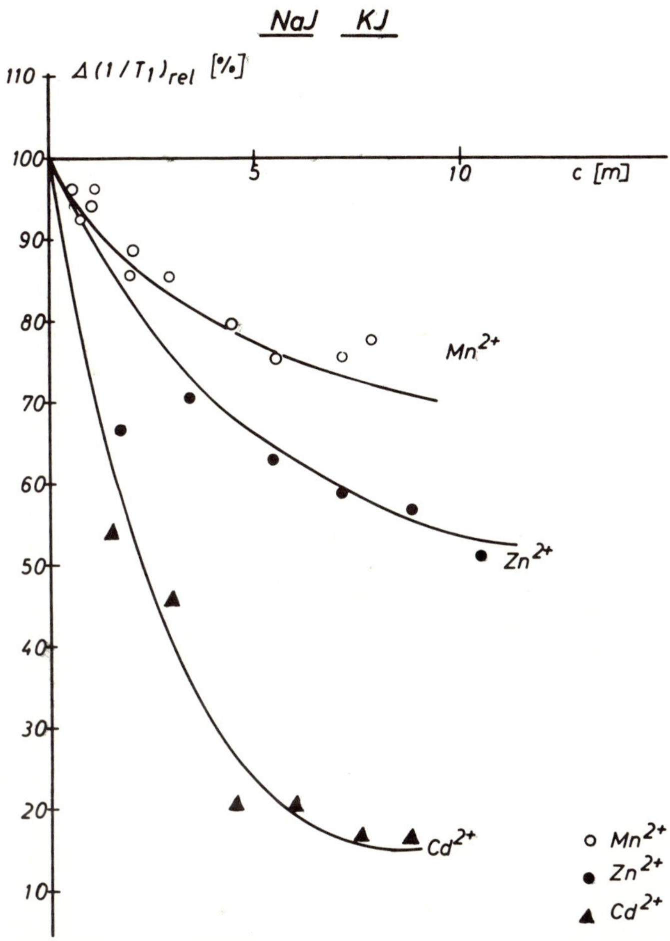

FIG. 3. Relative relaxation rate $\Delta(1/T_1)_{rel}$ as a function of the concentration c of added iodides. In the case of Zn^{2+} the added salt is NaI; in the two other cases KI was used. For other details see the legend of Fig. 1 ($\vartheta = 25°C$).

In order to check the method applied, the following experiment was performed. Mg^{2+} is known to give no complexes with Cl^-. Two series of solutions were prepared: (1) $Mg(NO_3)_2$, 1.5 m + increasing concentration of KCl up to 2.3 m; (2) $Mg(NO_3)_2$, 0.5 m + increasing concentration of KCl up to 3.6 m. Because of limited solubility solutions (1) had to be measured at 40°C. In both cases the experimental accuracy was not sufficient to detect with certainty an increase or decrease of the relaxation rate with increasing concentration of KCl.

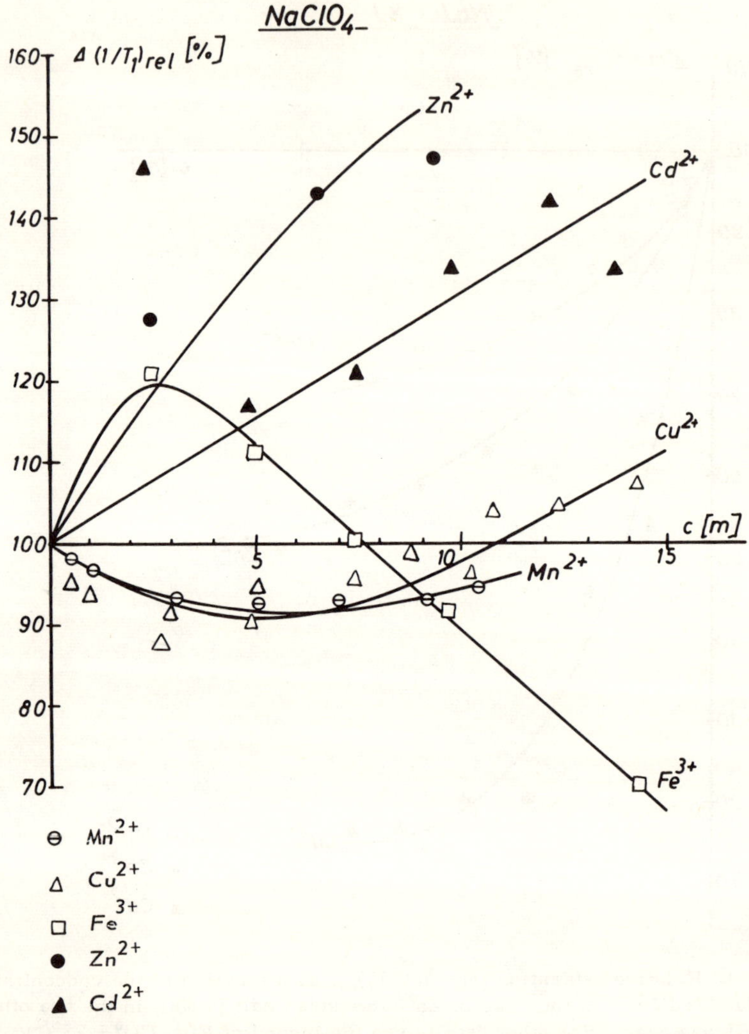

FIG. 4. Relative relaxation rate $\Delta(1/T_1)_{rel}$ as a function of the concentration c of added $NaClO_4$. For other details see the legend of Fig. 1 ($\vartheta = 25°C$).

We give now the detailed discussion of the various systems investigated.

(a) $CuCl_2 + KCl$

With the activity coefficients given according to the rules of the second section and in Table I, the measured values $\Delta(1/T_1)_{rel}$ have to be

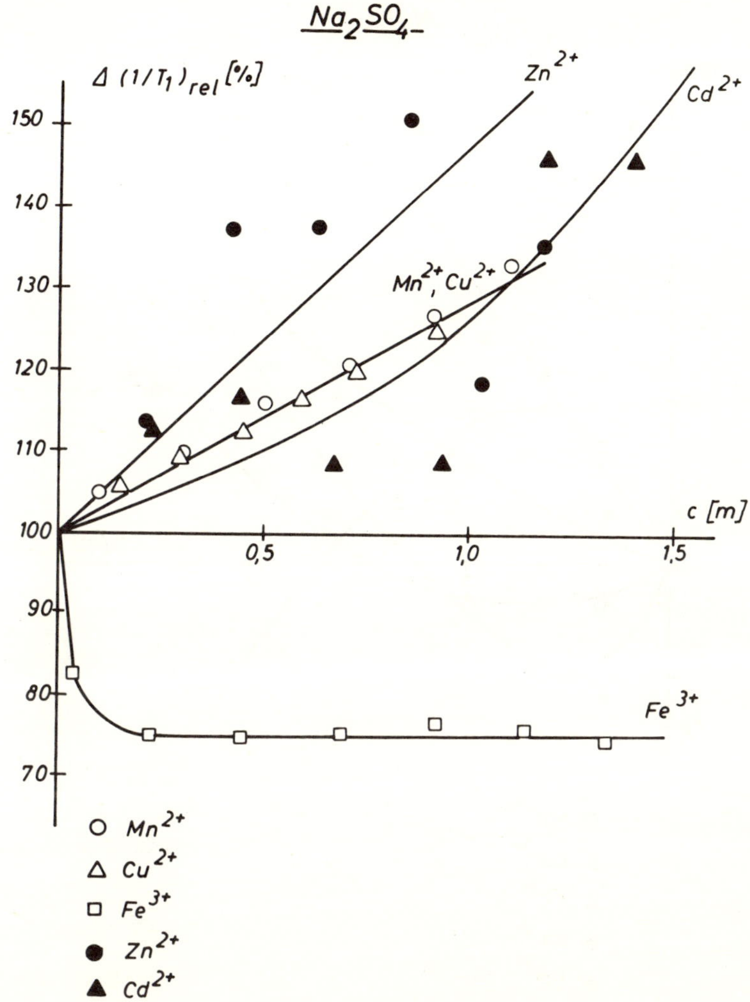

FIG. 5. Relative relaxation rate $\Delta(1/T_1)_{rel}$ as a function of the concentration c of added Na_2SO_4. For other details see the legend of Fig. 1 ($\vartheta = 25°C$).

fitted by the parameters β_1^0 and β_2^0 or K_1^0 and K_2^0 [Eq. (*17*)], higher complex constants being neglected. The following results were obtained:

$$K_1^0 = 7 \text{ m}^{-1}; \qquad K_2^0 = 0.6 \text{ m}^{-1}.$$

If we apply the activity coefficients as indicated in the second section, the nonthermodynamical equilibrium constants K_n [see Eq. (12)] at ionic strength $J = 1$ are

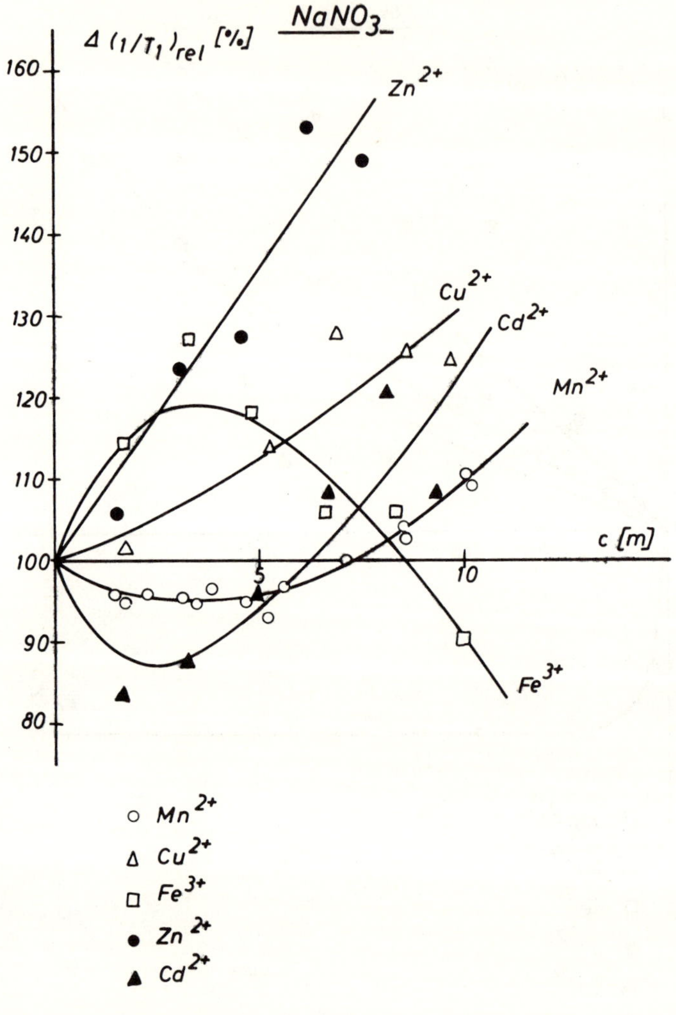

FIG. 6. Relative relaxation rate $\Delta(1/T_1)_{\text{rel}}$ as a function of added $NaNO_3$. For other details see the legend of Fig. 1 ($\vartheta = 25°C$).

$$K_M = 1.3 \text{ m}^{-1}; \qquad K_2 = 0.2 \text{ m}^{-1}.$$

These figures compare excellently—of course, fortuitously—with the values of McConnell and Davidson (*41*) at $J = 1$:

$$K_1 = 1.29 \ m^{-1}; \qquad K_2 = 0.23 \text{ m}^{-1}.$$

Other values stated in the literature for $J = 0.42$ are in reasonable agreement with our data (K_1^0 between 1 and 2.5 m^{-1}; $K_2^0 = 0.2 \text{ m}^{-1}$).

Figure 7 shows the measured $\Delta(1/T_1)_{rel}$ values in comparison with the ones calculated by using various parameters β_1^0 and β_2^0.

(b) $CuCl_2 + KBr$

Neglecting higher constants than K_2^0 we find (see Fig. 8)

$$K_1^0 = 2 \text{ m}^{-1}; \qquad K_2^0 = 0.4 \text{ m}^{-1},$$

and at $J = 1$,

$$K_1 = 0.5 \text{ m}^{-1}; \qquad K_2 = 0.15 \text{ m}^{-1}.$$

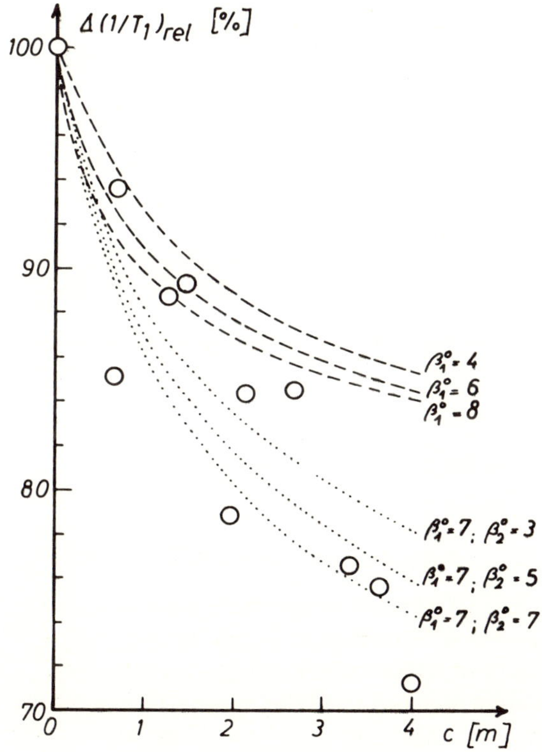

FIG. 7. $CuCl_2 + KCl$.

FIGS. 7–14. Comparison between calculated (dotted or dashed curves) and measured relative relaxation rates (circles) for different solutions. The measured values are identical with the corresponding ones of Figs. 1–4. The β_n^0 are the cumulative or gross stability constants (m^{-n}) used for the calculation [see Eq. (8)] ($\vartheta = 25°C$).

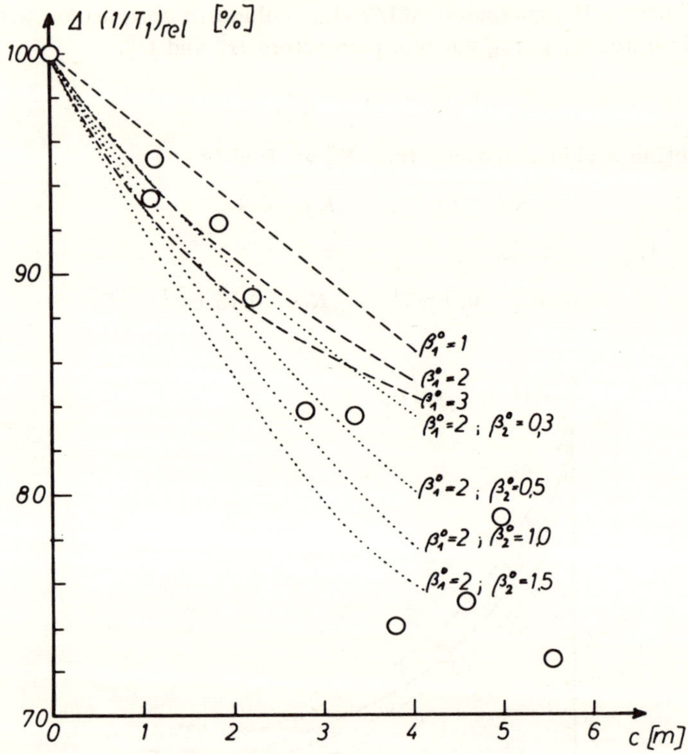

FIG. 8. $CuCl_2$ + KBr.

The agreement with the literature values is again quite satisfactory ($K_1^0 = 0.93\ m^{-1}$, $K_1(J = 1) = 0.4\ m^{-1}$, $K_2 = 0.25\ m^{-1}$) (*42*).

(c) $FeCl_3$ + HCl + KCl

The evaluations show that it is not sufficient to consider only a two-stage equilibrium here. The stability constant $K_1^0 \equiv \beta_1^0 = 30\ m^{-1}$ was taken from the work of Rabinowitch and Stockmayer (*43*). These authors determined spectrophotometrically the concentration dependence of β_1 over the concentration range up to $\approx 4\ m$. Their results were used here without modification. For β_2^0 we used the value 228 m^{-2} which follows from the literature values of Rabinowitch and Stockmayer ($\beta_2 = 162\ m^{-2}$ at $J = 1\ m$) (*43*) and Lister and Rivington ($\beta_2 = 185\ m^{-2}$ at $J = 1.2\ m$) (*44*) by applying our method of approximating the activity coefficient. The big doubly charged cation UO_2^{2+} was taken as a model for $FeCl^{2+}$. Curve fitting yielded $K_3^0 = 2.4\ m^{-1}$ ($\beta_3^0 = 547\ m^{-3}$) or $K_3 = 0.8\ m^{-1}$ at $J = 2\ m$ (Fig. 9). This is in agreement with the values of Gamlen and

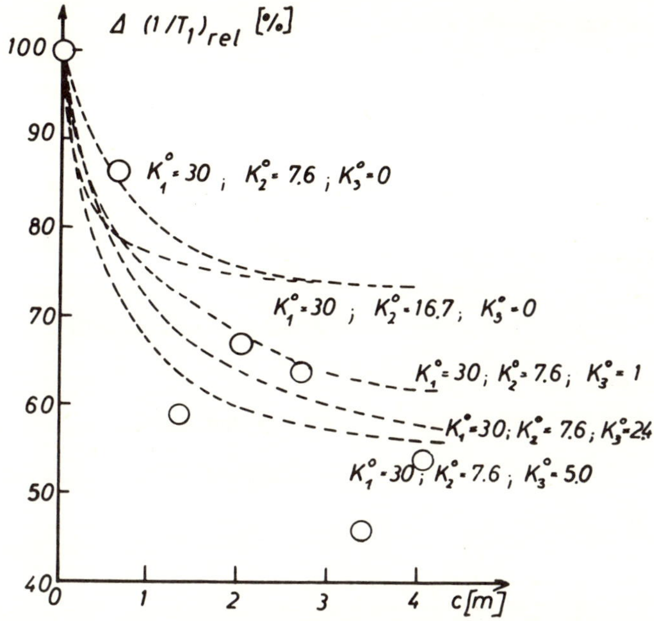

FIG. 9. $FeCl_3 + HCl + KCl$.

Jordan (*45*) ($K_3 = 0.72\ m^{-1}$ at $J = 2\ m$) and Olerup (*46*) ($K_3 = 0.87\ m^{-1}$ at $J = 2\ m$). Rabinowitch and Stockmayer, however, find $K_3^0 = 0.1\ m^{-1}$ and Marcus gives the value $K_3^0 = 0.04\ m^{-1}$ (cited after *42*). In order to avoid effects of hydrolysis 0.05 *m* HCl was added to the stock solution of $FeCl_3$. This has the consequence that it is not correct to consider the Fe^{3+} as hexaquo complex at the concentration 0 of added diamagnetic salt. The values of $\Delta(1/T_1)_{rel}$ are corrected to give $\Delta(1/T_1)_{rel} = 100\%$ at zero concentration of the diamagnetic salt.

The solution $FeCl_3 + HCl + KBr$ behaves similarly to that with KCl (Fig. 2). Complex formation is probably somewhat weaker here. We refrained from calculating stability constants; to avoid mixed complexes, solutions with HBr would have to be measured.

(d) $MnCl_2 + KCl$

There is little known about the complex formation of Mn^{2+} with halide ions. Stokes (*47*) concludes from the behavior of the activity coefficient in aqueous solutions of $MnCl_2$ that complex formation should occur. The activity coefficient is lower than it should be for a completely dissociated 1-2 electrolyte in the concentration range $> 0.1\ m$. The only literature value known to the authors is $K_1^0 \approx 1\ m^{-1}$ (*48*).

In the present work the values

$$K_1^0 = 2\ m^{-1}, \qquad K_2^0 = 0$$

were found. (See Fig. 10.)

(e) $MnCl_2 + KBr$; $MnCl_2 + KI$

No literature data are available for these equilibria. Figure 11 and 12 yield

$$K_1^0 = 1.5\ m^{-1}, \qquad K_2^0 \approx 0.2\ m^{-1} \text{ (estimated)}$$

for $MnBr_2$, and

$$K_1^0 = 3\ m^{-1}, \qquad K_2^0 > 0, \text{ but small,}$$

for MnI_2.

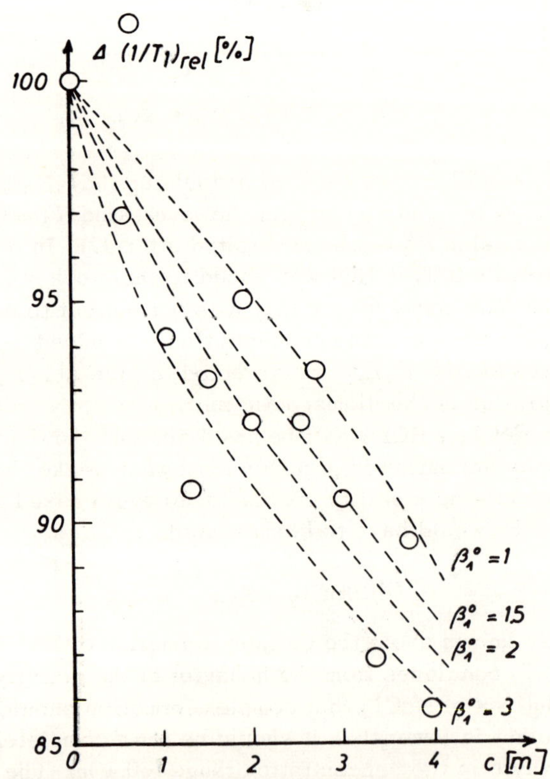

FIG. 10. $MnCl_2 + KCl$.

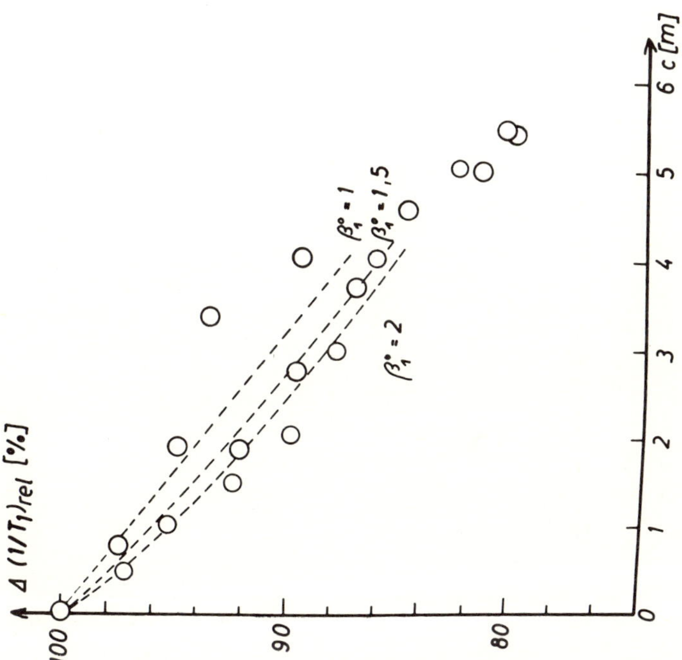

FIG. 11. $MnCl_2 + KBr$.

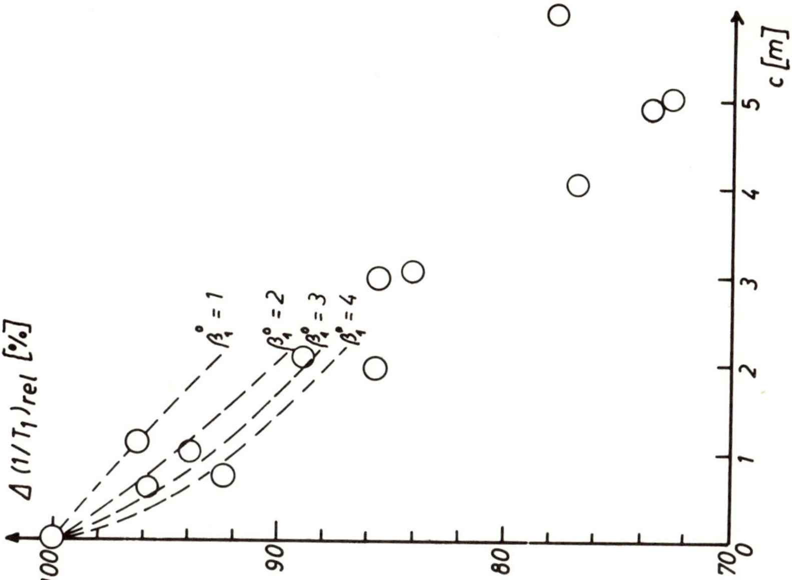

FIG. 12. $MnCl_2 + KI$.

Because of the limited solubility of CaI_2 the evaluation of the activity coefficient of MnI_2 by ionic radius interpolation could not be extended to higher concentrations (see Table I).

(f) $MnCl_2 + NaClO_4$; $CuCl_2 + NaClO_4$

It is generally accepted that the perchlorate anion shows only a weak tendency to form complexes. But for the equilibrium

$$Fe^{3+} + ClO_4^- \rightleftharpoons FeClO_4^{2+}$$

stability constants of the order of magnitude 10 m^{-1} have been reported (*42*). Moreover, according to Murray and Cleveland (*49*), one has to assume that also divalent metal ions are incompletely dissociated in the second stage. We find (Fig. 13):

$$K_1^0 \approx 2m^{-1};$$

for both cations Mn^{2+} and Cu^{2+}. At higher concentrations outer sphere complex formation gets predominant.

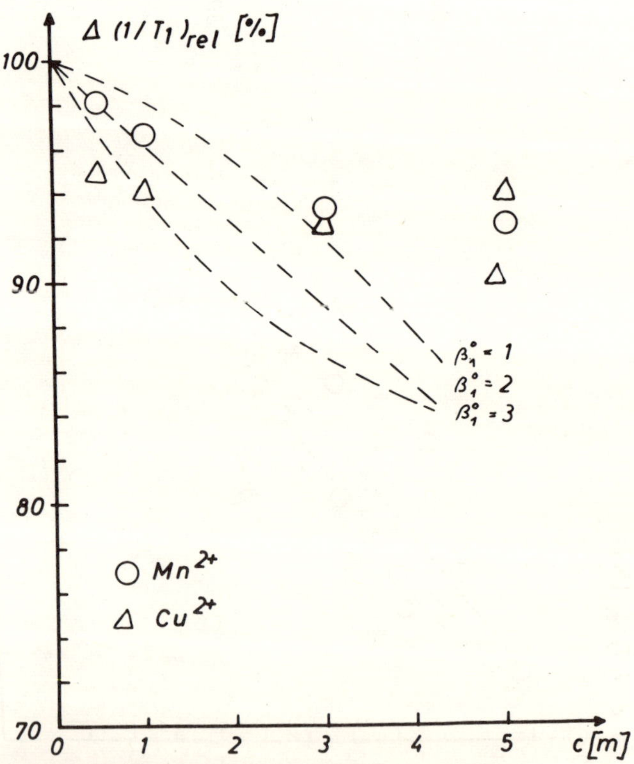

FIG. 13. $CuCl_2 + NaClO_4$; $MnCl_2 + NaClO_4$.

(g) $Ni(NO_3)_2 + KBr$

There has been observed a weak complex formation between Ni^{2+} and Br^- (*50*), K_1 being of the order of 1 m^{-1}. Figure 14 shows, however, that the measured $\Delta(1/T_1)_{rel}$ values cannot be fitted by any curve representing an association equilibrium. It is concluded that the electron spin relaxation time contributes in a still unknown manner to the relaxation mechanism of the protons in the hydration shell and thus the method used in this work cannot be applied to ions like Ni^{2+}, Co^{2+}, and Fe^{2+}.

(h) *Solutions with* Na_2SO_4 *as Added Salt*

The stability constants of complexes between SO_4^{2-} and the cations Mn^{2+}, Cu^{2+}, Cd^{2+}, and Zn^{2+} as available in the literature are all of the same order of about $K_1^0 \approx 200\ m^{-1}$ (*42*). According to Eigen and Tamm

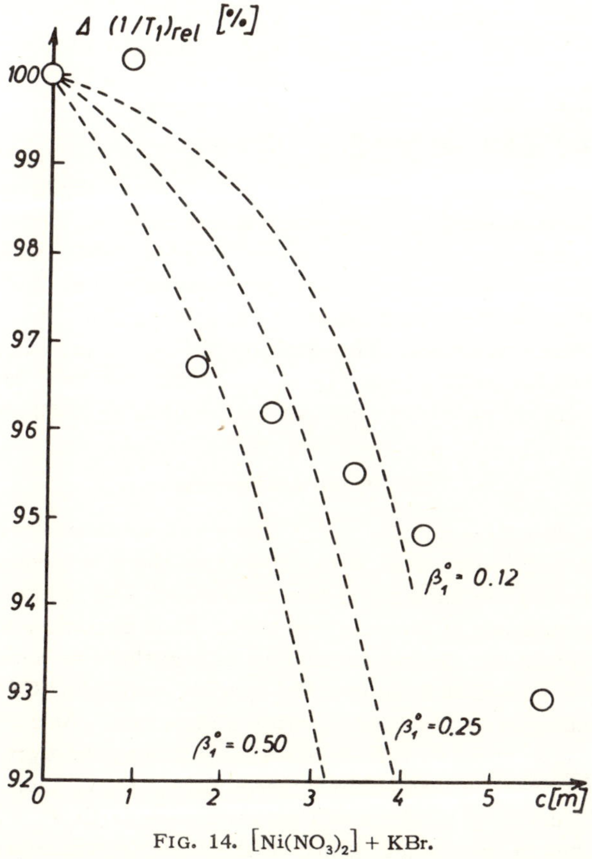

FIG. 14. $[Ni(NO_3)_2]$ + KBr.

(*51*) by far the greatest fraction of the divalent metal ions in solutions of their sulfates is fully surrounded by water molecules. For the Mn^{2+} ion this was confirmed by proton relaxation time measurements which were performed with $MnSO_4 + MgSO_4$ solutions and yielded results similar to the present ones (Lindner and Pfeifer, *26*). Figure 5 shows that Cu^{2+}, Cd^{2+}, and Zn^{2+} form outer sphere complexes as well. The conclusion of Wendt and Strehlow (*52*), that in aqueous solutions of iron(III)-sulfates inner sphere complex formation should predominate over outer sphere complex formation, is confirmed by this work (Fig. 5). Calculation of stability constants was not carried out because the equilibrium situation is rather complicated (*52*).

(i) *Solutions with* $NaNO_3$ *and* $NaClO_4$ *as Added Salts*

The following stability constants are reported in the literature (*42*):

$$Fe^{3+} + NO_3^- \rightleftharpoons FeNO_3^{2+}, \qquad K_1^0 \approx 10\ m^{-1}$$
$$K_1(J = 0.6\ m) = 0.6\ m^{-1},$$
$$K_1(J = 1.0\ m) = 0.3\ m^{-1};$$
$$Cd^{2+} + NO_3^- \rightleftharpoons CdNO_3^+, \qquad K_1^0 \approx 2.5\ m^{-1};$$
$$Zn^{2+} + NO_3^- \rightleftharpoons ZnNO_3^+, \qquad K_1^0 \approx 0.$$

At lower concentrations the presence of inner sphere complexes with Fe^{3+} seems to be masked by some outer sphere complex formation (see Fig. 6). The existence of an inner sphere complex $CdNO_3^+$ at moderate concentrations is confirmed by the present measurements. Mn^{2+} shows some competition between both types of complexing; Cu^{2+} and Zn^{2+} both form outer sphere complexes. The addition of $NaClO_4$ produces similar effects as found for $NaNO_3$ (see Fig. 4). The data for Mn^{2+} and Cu^{2+} have been evaluated quantitatively (see subsection on $MnCl_2 + NaClO_4$).

(j) Cd *and* Zn *Halides*

Cd^{2+} forms strong complexes with halides probably up to the stage CdA_4^{2-} (*42*) ($A = Cl^-, Br^-, I^-$). It is known that there is covalent bonding between the halide anions and the cation Cd^{2+} (*34*) which indicates that the complexes are of inner sphere type. Thus according to the arguments given in the second section we expect that the mean reorientation time in the hydration sphere of the species containing Cd^{2+} in its center will decrease as the concentration c increases. At $c = 0$ essentially the hexaquo complex will be present. As more and more halide ions are added the complexes CdA^+, CdA_2, CdA_3^-, CdA_4^{2-} appear. The first, second, and third members of this series are particles with small or no charge and of increasing geometrical extension; the reorientation time of the hydrate water will decrease along the series until eventually in the last complex CdA_4^{2-}, where the charge is high but the radius is large, one would expect τ_M to rise again moderately. As seen from Figs. 1–3 the observed behavior is qualitatively as it should be according to this interpretation. Zn^{2+} shows a weaker tendency to form complexes with the ions Cl^-, Br^-, and I^- (*42*), but the existence of complexes up to the stage ZnA_4^{2-} is well established (*34*). Covalent bonding has been observed (*34*) and the measured relative relaxation rates again show the correct qualitative behavior (Figs. 1–3).

References

1. T. J. Swift and R. E. Connick, *J. Chem. Phys.* **37**, 307 (1962).

2. R. A. Bernheim, T. H. Brown, H. S. Gutowsky, and D. E. Woessner, *J. Chem. Phys.* **30**, 950 (1959).

3. N. Bloembergen and L. O. Morgan, *J. Chem. Phys.* **34**, 842 (1961).

4. W. Lohmann, *Z. Naturforsch.* **19a**, 814 (1964).

5. M. Kaminsky, *Discussions Farad. Soc.* **24**, 171 (1957).

6. R. W. Gurney, "Ionic Processes in Solution." McGraw-Hill, New York, 1953.

7. J. H. Wang, *J. Phys. Chem.* **58**, 686 (1954).

8. O. J. Samoilow, *Discussions Farad. Soc.* **24**, 141 (1957); "Die Struktur wässriger Elektrolytlösungen und die Hydratation von Ionen," Teubner, Leipzig, 1961.

9. H. S. Frank and W. Y. Wen, *Discussions Farad. Soc.* **24**, 133 (1957).

10. W. M. Wdowenko and W. A. Tscherbakow, *Zh. Struct. Khim* **1**, 122 (1960).

11a. H. G. Hertz, *Ber. Bunsenges. Physik. Chem.* **67**, 311 (1963).

11b. H. G. Hertz and M. D. Zeidler, *Ber. Bunsenges. Physik. Chem.* **67**, 774 (1963).

12. B. P. Fabricand, S. S. Goldberg, R. Leifer, and S. G. Ungar, *Mol. Phys.* **7**, 425 (1964).

13. J. Itoh and Y. Yamagata, *J. Phys. Soc. (Japan)* **13**, 1182 (1958).

14. D. E. O'Reilly, G. E. Schacher, and K. Schug, *J. Chem. Phys.* **39**, 1756 (1963).

15. L. Endom and H. G. Hertz (to be published).

16. H. S. Frank and M. W. Evans, *J. Chem. Phys.* **13**, 507 (1945).

17. H. G. Hertz and M. D. Zeidler, *Ber. Bunsenges. Physik. Chem.* **68**, 821 (1964).

18. H. G. Hertz, *Ber. Bunsenges. Physik. Chem.* **68**, 907 (1964).

19. M. D. Zeidler, Dissertation, University of Münster, 1963.

20. S. S. Danyluk and E. S. Gore, *Nature* **203**, 748 (1964).

21. B. M. Kozyrev and A. I. Rivkind, *Zh. Eksper. i Teor. Fiz.* **27**, 69 (1954).

22. L. O. Morgan, A. W. Nolle, R. L. Hull, and J. Murphy, *J. Chem. Phys.* **25**, 206 (1956).

23. A. I. Rivkind, *Zh. Neorgan. Khim.* **2**, 1263 (1957); **4**, 1291 (1959).

24. J. King and N. Davidson, *J. Chem. Phys.* **29**, 787 (1958).

25. W. M. Wdowenko, G. A. Romanow, and W. A. Tscherbakow, *Radiokhimiya* **5**, 137, 581, 664 (1963).

26. U. Lindner and H. Pfeifer, *Z. Naturforsch.* **18a**, 773 (1963).

27. P. A. Zagorets, V. I. Ermakov, and A. P. Grunau, *Russ. J. Phys. Chem.* **37**, 758 (1963).
28. J. R. Zimmerman and W. E. Brittin, *J. Phys. Chem.* **61**, 1328 (1957).
29. H. Pfeifer, *Ann.* Physik 7. Folge 8, 1 (1961); *Z. Naturforsch.* **17a**, 279 (1962).
30. I. Solomon, *Phys. Rev.* **99**, 559 (1955).
31. A. Abragam, "The Principles of Nuclear Magnetism." Oxford Univ. Press (Clarendon), London and New York, 1961.
32. E. E. Genser, Lawrence Radiation Laboratory, Univ. of California, Berkeley, UCRL-9846, January, 1962.
33. R. G. Hayes and R. J. Myers, *J. Chem. Phys.* **40**, 877 (1964).
34. H. G. Hertz, *Z. Elektrochem.* **65**, 36 (1961).
35. C. W. Davies, "Ion Association." Butterworth, London, 1962.
36. B. E. Conway, "Electrochemical Data," Elsevier, Amsterdam, 1952.
37. J. Buchta, H. S. Gutowsky, and D. E. Woessner, *Rev. Sci. Instr.* **29**, 55 (1958).
38. J. H. Simpson and H. Y. Carr, *Phys. Rev.* **111**, 1201 (1958).
39. J. W. Hennel, A. Z. Hrynkiewicz, K. Krynicki, T. Waluga, and G. Zapalsky, *Arch. Sci.* (*Geneva*): *Colloq. Ampère* **11**, 243 (1958).
J. W. Hennel, K. Krynicki, T. Waluga, and G. Zapalsky, *Acta Phys. Polon.* **20**, 77 (1961).
40. L. O. Morgan and A. W. Nolle, *J. Chem. Phys.* **31**, 365 (1959); also literature cited therein.
41. H. McConnell and N. Davidson, *J. Am. Chem. Soc.* **72**, 3164 (1950).
42. J. Bjerrum, G. Schwarzenbach, and L. G. Sillén, "Stability Constants," Part II. The Chemical Society, London, 1958.
43. E. Rabinowitch and W. H. Stockmayer, *J. Am. Chem. Soc.* **64**, 335 (1942).
44. M. W. Lister and D. E. Rivington, *Can. J. Chem.* **33**, 1603 (1955).
45. G. A. Gamlen and D. O. Jordan, *J. Chem. Soc.*, p. 1435 (1953).
46. H. Olerup, *Svensk Kem. Tidskr.* **55**, 324 (1943).
47. R. H. Stokes, *Trans. Farad. Soc.* **44**, 295 (1948).
48. J. C. James, Thesis, University of London, 1947 (cited after *42*).
49. M. J. Murray and F. F. Cleveland, *J. Am. Chem. Soc.* **65**, 2110 (1943).
50. H. W. Lister and D. W. Wilson, *Can. J. Chem.* **39**, 2606 (1961).
51. M. Eigen and K. Tamm, *Z. Elektrochem.* **66**, 107 (1962).
52. H. Wendt and H. Strehlow, *Z. Elektrochem.* **66**, 228 (1962).

Study of Some Cyano-Metal Complexes by NMR[1]

A. LOEWENSTEIN and M. SHPORER

Department of Chemistry, Technion—Israel Institute of Technology, Haifa, Israel,

and

G. NAVON

Department of Physical Chemistry, Hebrew University, Jerusalem, Israel

Introduction

Cyano-metal complexes, especially those of iron, have been studied thoroughly for many years. (For a comprehensive data survey on ferrocyanides, see Ref. *2*.) Still there remain various unsolved problems concerning the structure and dynamic properties of these complexes. (For example, no interatomic distances were found in the literature for $K_3Fe(CN)_6$ (*3*). This study presents an attempt to apply the nuclear magnetic resonance (NMR) technique to some of these problems. All measurements were carried out in aqueous solutions.[2]

The measurement of chemical shifts in paramagnetic materials can, in principle, provide information on the spin density distribution in the

TABLE I

Compound	χ_m(cm^3/mole) $\times 10^6$ [a]	μ_{eff}(B.M.)[b]
$K_3Cr(CN)_6$	6500	3.72
$K_3Mn(CN)_6$	4423	3.18
$K_3Fe(CN)_6$[c]	2300	2.4
$K_4Mn(CN)_6 \cdot 3H_2O$	1906	1.8

[a] Taken from Landolt-Börnstein, "Magnetic Properties I," pp. 3–103, 105. Springer-Verlag, Berlin, 1962.

[b] B. N. Figgis and J. Lewis, *in* "Modern Coordination Chemistry," p. 407 (J. Lewis and R. G. Wilkins, Ed.). Wiley (Interscience), New York, 1960.

[c] The susceptibilities of aqueous solutions of $K_3Fe(CN)_6$ are given by F. W. Gray and W. M. Birse, *J. Chem. Soc.* **105**, 2707 (1914).

[1] An extended report, including most of the results given here, was published in *Inorg. Chem.* **4**, 358, 361 (1965). For a preliminary report see Ref. (*1*).

[2] Preliminary attempts to detect the ^{14}N resonances in powders of $K_3Fe(CN)_6$ and $K_4Fe(CN)_6$ at room temperature failed.

molecules. (For similar studies on proton NMR, see, for example, Ref. *4*.) Furthermore, by measuring the ^{14}N NMR linewidths in mixtures of ferri- and ferrocyanides we were able to study the kinetics of the electron transfer reaction occurring in this system. Some of the complexes studied are paramagnetic—$K_3Fe(CN)_6$, $K_3Mn(CN)_6$, $Na_4Mn(CN)_6$, $K_3Cr(CN)_6$—whereas other are diamagnetic—$K_4Fe(CN)_6$, $K_3Co(CN)_6$, $K_2Ni(CN)_4$, $KCu(CN)_2$, and $K_2Ag(CN)_2$. The magnetic susceptibilities and the calculated effective moments of the solids are given in Table I.

The NMR of ^1H, ^{14}N, and ^{13}C were measured and are reported in the following. The instrument, materials, and measuring procedure were discussed elsewhere (*1*).

Proton NMR Measurements

The ferri- and ferrocyanide acids $H_3Fe(CN)_6$ and $H_4Fe(CN)_6$ are strong acids; 0.25-*M* aqueous solutions have pH values of about 0.35 and 0.55, respectively. $H_4Fe(CN)_6$ is assumed to be strongly complexed ($pK_{A1} \simeq 4.2$, $pK_{A2} \simeq 3.0$) whereas no evidence for complex formation exists for $H_3Fe(CN)_6$ (*5*). The following shifts, in parts per million to lower magnetic fields, were measured relative to the methyl resonance in *tert*-butyl alcohol added to the solution (to avoid susceptibility corrections): $H_4Fe(CN)_6$, 3.73; $H_3Fe(CN)_6$, 3.87; equimolar solution of $H_3Fe(CN)_6$ and $H_4Fe(CN)$, 3.80. The concentration of the acids was 0.25 *M*. The linewidth in $H_4Fe(CN)_6$ is about 2c/s (completely determined by the viscosity and the inhomogeneity of the magnetic field), whereas in $H_3Fe(CN)_6$ the width is just twice this magnitude. In the mixture thc linewidth is intermediate.

The following can be concluded from the above results:

(a) Proton exchange in aqueous $H_3Fe(CN)_6$–$H_4Fe(CN)_6$ is fast. Assuming a second-order rate equation the lower limit of the rate constant was estimated and found to be 2×10^2 sec^{-1} M^{-1}. In view of proton exchange data in other acids this result is to be expected.

(b) The chemical shift and difference in linewidth between the acids are smaller than usually found in solutions of paramagnetic materials. This indicates that the effective magnetic moment of the electron is strongly "screened." This screening is due to the ligand cyanide groups since $H_3Fe(CN)_6$ is completely dissociated. The same effect accounts for the low effective magnetic moment calculated from the longitudinal relaxation time, T_1, of the ^1H resonance in aqueous solutions of $K_3Fe(CN)_6$ and $K_3Mn(CN)_6$. The values of the magnetic moments thus calculated are 0.12 (*6*) and 0.3 B.M. (calculated from Ref. *7*), respectively, and should be compared with the corresponding values calculated from the susceptibilities as given in Table I. It might be argued that in these compounds, which are not completely dissociated, screening by the counter cations may be the cause of the low effective magnetic moments.

The features of the proton resonance spectra in aqueous solutions of $K_3Fe(CN)_6$ and $K_4Fe(CN)_6$ or their mixtures resemble those of the acids.

^{14}N and ^{13}C NMR Measurements

Line Positions and Linewidths of ^{14}N

The ^{14}N line positions in $K_3Fe(CN)_6$, $K_3Cr(CN)_6$, and $K_3Mn(CN)_6$ have been previously measured by Shulman (*8*). He found that the ^{14}N chemical shift in the Cr complex was significantly smaller than those in the Mn and Fe complexes and suggested that the latter are not contact shift but might be due to an anisotropic electronic g tensor. Positions and linewidths of ^{14}N resonance and the positions of the detectable ^{13}C resonances are given in Table II. The paramagnetic shift in $K_3Cr(CN)_6$ is the smallest but still much larger than previously reported by Shulman (*8*).

TABLE II
POSITIONS AND LINEWIDTHS (IN GAUSS) OF ^{14}N AND ^{13}C RESONANCES AT 32 °C

Compound	^{14}N position[a]	Width of ^{14}N resonance[b]	^{13}C position[c]
$K_3Mn(CN)_6$	−15.0	2[f]	—[d]
$K_3Fe(CN)_6$	−10.9	1.5 to 2.0[e]	+50
$Na_4Mn(CN)_6$	− 8.7	4[f]	—[d]
$K_3Cr(CN)_6$	− 5.4	5[f]	—[d]
$K_4Fe(CN)_6$	− 0.7	1.8 to 2.2[e]	+0.8
$K_3Co(CN)_6$	− 0.8	2 to 3	0[f]
$KCu(CN)_2$	− 0.3[f]	2 to 3	—
$K_2Ni(CN)_4$	− 0.3[f]	2[f]	0[f]
$KAg(CN)_2$	− 0.3[f]	2[f]	—

[a] Measured relative to ^{14}N resonance in CH_3CN.
[b] Full width between peaks of derivative curve.
[c] Measured relative to ^{13}C resonance in C_6H_6.
[d] Unsuccessful attempts to detect the ^{13}C resonance in ^{13}C-enriched samples.
[e] Concentration dependent.
[f] Approximate values.

The temperature dependence of the ^{14}N and ^{13}C resonance positions in $K_3Fe(CN)_6$ and of ^{14}N in $K_3Mn(CN)_6$ is shown in Fig. 1. The temperature dependence of the ^{14}N resonance in $Na_4Mn(CN)_6$ was not measured and in $K_3Cr(CN)_6$ the shift is less than 0.5 gauss in the temperature range of 25 to 83 °C.

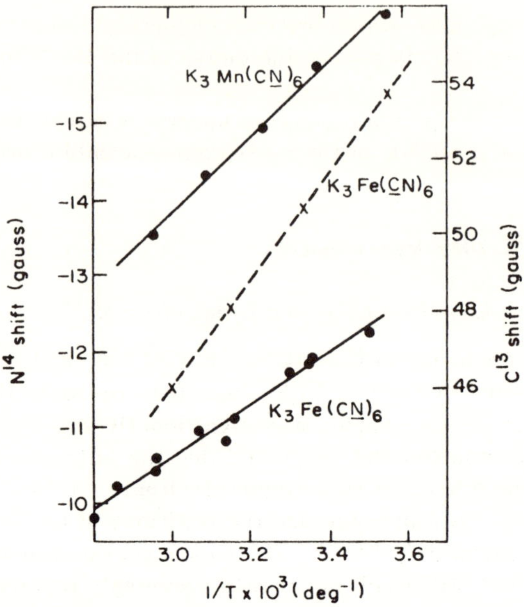

FIG. 1. Temperature dependence of ^{14}N resonances (measured relative to CH_3CN) and of ^{13}C resonance (measured relative to C_6H_6). ^{14}N resonances shift to higher field and ^{13}C resonances to lower field with increasing temperature.

The following conclusions might be drawn from these results:

(a) Line positions seem to be determined primarily by the contact interaction except for $K_3Cr(CN)_6$ and to a lesser extent for $Na_4Mn(CN)_6$ where other factors might be operative. An inverse proportionality between the lowest optical $t_{2g} \longrightarrow e_g$ transition energies [see (*9a*) for $K_3Mn(CN)_6$; (*9b*) for $K_3Cr(CN)_6$; (*9c*) for $K_3Fe(CN)_6$] and the ^{14}N chemical shifts suggests that mixing of t_{2g} and e_g orbitals in the metal might produce the contact interaction.

(b) Linewidths are determined predominantly by the quadrupolar relaxation and the contribution due to the electron spin relaxation is small. The electronic spin relaxation time might be sufficiently long in $K_3Cr(CN)_6$, and probably in $Na_4Mn(CN)_6$, to add to the quadrupolar width. Comparison with the width in the diamagnetic complexes and with EPR measurements on these systems [see (*10a*) for aqueous solutions of $Cr(CN)_6^{-3}$; (*10b*) for solid $K_4Mn(CN)_6$] support these arguments.

Measurements of ^{13}C

The ^{13}C NMR was detected in only one paramagnetic complex, $K_3Fe(CN)_6$ (Table II). The result indicates that the relative spin densities on the carbon and nitrogen atoms in this compound are *approximately* 5 to 1 with *opposite* signs. Thus a simple molecular orbitals calculation (*11*) without taking into account configuration interaction would not apply to this case. In other paramagnetic complexes, the ^{13}C resonances are probably very broad. This may be due to the relatively long electronic spin relaxation time.

The Kinetics of Electron Transfer

The electron transfer reaction between potassium ferri- and ferrocyanide in aqueous solution has been studied previously both experimentally (*12*) by isotope labeling, and theoretically (*13*). The catalytic effect of replacing the potassium by other cations has also been shown (*12*).

Solutions containing mixtures of potassium ferri- and ferrocyanide show one ^{14}N resonance. The position of the common line is a linear function of the mole fraction of the ferri- or ferrocyanide in the solution. The width of the common line depends upon the ratio of $Fe(CN)_6^{3-}$ to $Fe(CN)_6^{4-}$ in the solution and on the temperature. Using the approximate formula for fast exchange (*14*) and assuming the reaction to be described by

$$\text{rate} = k\,[Fe(CN)_6^{3-}]\,[Fe(CN)_6^{4-}],$$

the value of k was found to be $9.2 \pm 1.3 \times 10^4$ sec^{-1} M^{-1} at 32°C. The temperature dependence of the rate can be described by an Arrhenius plot with $E \simeq 4.2$ kcal/mole and a preexponential factor equal to 8.5×10^7 sec^{-1} M^{-1}.

The effect of replacing the potassium by other cations is described in Table III. The effect of Cs^+ on the rate is not included in Table III since the attempts to prepare stable solutions of $Cs_3Fe(CN)_6$ were unsuccessful. In order to measure the effect of Cs^+ on the reaction, the following two experiments were performed: (1) CsCl was added to mixtures of $K_3Fe(CN)_6$ and $K_4Fe(CN)_6$. The results are summarized in Table IV. (2) CsCl was added to equimolar (0.25 *M* of each) solutions of $H_3Fe(CN)_6$ and $H_4Fe(CN)_6$. No effect was observed on the two separate lines even when the concentration of CsCl in the solution reached 0.8 *M* and some precipitation (CsCl?) occurred.

The results led to the conclusion that there exists a marked catalytic effect of the cations on the reaction rate. The effect increases from H^+ to Cs^+ and from Ca^{2+} to Sr^{2+}. The cations may have a direct or indirect role in the reaction mechanism. The direct role implies that the electron is actually transferred by the cation which serves as a bridge between the two anions. Indirectly the cations may serve as a mean to facilitate the approach of the two anions. Otherwise the cations may differ in

TABLE III

EFFECT OF CATIONS ON THE ELECTRON TRANSFER REACTION AT 32 °C

Cation	Width (G) $Fe(CN)_6^{3-}$ [a]	Width (G) $Fe(CN)_6^{4-}$ [a]	Width in mixture (G)[b]	$\Delta\left(\frac{1}{T}\right)$ (G)[e]	Rate constant $(sec^{-1}\ M^{-1}) \times 10^{-4}$
H^+	2.1	2.2	2 Unbroadened lines	—	<0.4[c]
Li^+	2.6	2.9	2 Overlapping lines	—	<3 [c]
Na^+	2.0	2.3	4.4	2.25	5.8
K^+	1.7	2.0	3.5	1.65	7.9
Rb^+	1.7	2.2	2.9	0.95	13.1
NH_4^+	1.7	2.1[d]	3.3	1.40	8.4
Mg^{2+}	3.4	7.2	Very broad (10 to 12 G)	—	—
Ca^{2+}	3.8	7.0	Very broad (7 to 9 G)	—	—
Sr^{2+}	3.0	4.1	5.5	2.05	6.7

[a]Concentration: 0.5 *M*.

[b]Concentration: 0.25 *M* of each.

[c]Estimated from formula for slow exchange approximation.

[d]Average value between K^+ and Rb^+ since ^{14}N resonance of NH_4^+ and $Fe(CN)_6^{4-}$ overlap.

[e]$\Delta(1/T) = (1/T_2) - (P_A/T_{2A}) - (P_B/T_{2B})$.

TABLE IV

CATALYTIC EFFECT OF CsCl ON THE ELECTRON TRANSFER[a]

CsCl (*M*)	$\Delta\left(\frac{1}{T}\right)$ (G)	Rate constant $(sec^{-1}\ M^{-1})$
None	1.7	8.2×10^4
2.9×10^{-2}	1.3	10.5×10^4
8.8×10^{-2}	1.2	11.3×10^4
16.6×10^{-2}	1.0	13.7×10^4

[a]$K_3Fe(CN)_6$, 0.25 *M*; $K_4Fe(CN)_6$, 0.25 *M* (32 °C).

their ability to create the conditions required for a Franck-Condon transition between the reactants (similarity of states). Experimental evidence seems to support the first suggestion, i.e., that the electron is transferred by the cation. However, the second experiment with CsCl described above shows that matters might be further complicated. Although $H_3Fe(CN)_6$ is believed to be almost completely dissociated (*5*) Cs^+ seems to have no measurable effect on the exchange rate between $H_3Fe(CN)_6$ and $H_4Fe(CN)_6$. Hence, the pH of the solvent might also play a role in the reaction. On the other hand, the narrow lines observed in the ^{1}H NMR of mixtures indicate no participation of the solvent in the electron transfer reaction. The exact mechanism of electron transfer in the presence of cations is thus not yet fully understood.

References

1. A. Loewenstein, M. Shporer, and G. Navon, *J. Am. Chem. Soc.* **85**, 2855 (1963).
2. "The Chemistry of the Ferrocyanides," American Cyanamid Co. Edition, 1953.
3. R. W. G. Wyckoff, "Crystal Structures," Vol. III, Chapter IX. (Interscience), Wiley, New York, 1953,
4. D. R. Eaton, A. D. Josey, W. D. Phillips, and R. E. Benson, *Discussions Farad. Soc.* **34**, 77 (1962).
5. "Stability Constants," pp. 32–33. The Chemical Society (London) Special Publication No. 7, 1958.
6. N. Bloembergen "Nuclear Magnetic Relaxation," p. 97. Benjamin, New York, 1961.
7. L. O. Morgan, A. W. Nolle, R. L. Hull, and J. Murphy, *J. Chem. Phys.* **25**, 206 (1956).
8. R. G. Shulman, *J. Chem. Phys.* **29**, 945 (1958).

9a. C. I. Ballhausen, "Introduction to Ligand Field Theory," p. 250. McGraw-Hill, New York, 1962.

9b. R. Krishnamurthy and W. B. Schaap, *Inorg. Chem.* **2**, 604 (1963).

9c. C. S. Naiman, *J. Chem. Phys.* **35**, 323 (1961).

10a. R. G. Hayes, Thesis, University of California, Berkeley, 1961.

10b. J. M. Baker, B. Bleaney, and K. D. Bowers, *Proc. Phys. Soc. (London)* **B69**, 1205 (1956).

11. J. H. Van Vleck, *J. Chem. Phys.* **3**, 807 (1935).
12. A. C. Wahl, *Z. Electrochem.* **64**, 90 (1960). P. King, C. F. Deck, and A. C. Wahl, 139th Am. Chem. Soc. National Meeting, 1961, Abstract p. 30R. A. C. Wahl, R. J. Campion and P. King, 142nd Am. Chem. Soc. National Meeting, 1962, Abstract p. 32N. For a recent review on electron exchange reactions see: N. Sutin, *Ann. Revs. Nucl. Sci.* **12**, 299 (1962).
13. R. A. Marcus, *J. Chem. Phys.* **24**, 966 (1956); **26**, 867 (1957). W. F. Libby, *J. Phys. Chem.* **56**, 863 (1952).
14. A. Loewenstein and T. M. Connor, *Ber. Bunsenges. Physik. Chem.* **67**, 280 (1963). J. A. Pople, W. G. Schneider, and H. J. Bernstein, "High Resolution Nuclear Magnetic Resonance," Chapter 10. McGraw-Hill, New York, 1959.

Keto-Enol Equilibria in Derivatives of Picolylketones

G. KLOSE and E. UHLEMANN

Physical and Inorganic Chemistry Institute, University of Leipzig, Leipzig, Germany

With high-resolution NMR techniques it is often possible to get information on keto-enol tautomerism in a simple and direct way (see, for example, Refs. *1*, *2*).

We studied the keto-enol equilibria of some picolylketones and picolylthioketones in various solvents and partly at several concentrations. The measurements were made with a Trüb-Täuber spectrometer at room temperature and at a frequency of 25 Mc/s. Line separations were measured by the sideband method. The line intensities were determined by mechanical integration of several spectra and averaging.

In Fig. 1 the spectra of picolylisopropyl ketone, picolylisopropyl thioketone, and collidylphenyl ketone are plotted. The assignments of the CH_2 line of the keto form ($\tau \approx 5.5$ ppm) and of the CH line of the enol form ($\tau \approx 3$–4.5 ppm) are unique. The broad or very broad line occurring at very low magnetic fields comes from the enol H or the thioenol H, respectively.

The most important experimental results are summarized in Tables I and II. In Fig. 2 the dependence of the chemical shift of the OH line of VI O and II O (defined in Table I) on the concentration in CS_2 and acetone is represented.

The enolization takes place essentially because of a strong intramolecular hydrogen bond between the XH (X = O and S) and the picoline N. That the hydrogen bonds formed are strong can be seen from the XH resonances appearing at very low magnetic fields ($\tau_{OH} \approx -4$ ppm, $\tau_{SH} \approx -8.5$ ppm). The N . . . H—S bridges are stronger than the N . . . H—O bridges (the thioketones show a higher enol concentration than the corresponding ketones).

In a number of articles it is suggested that S—H does not form hydrogen bonds or forms only very weak bonds (see, for example, Refs. *3*, *4*). We found, however, that in the picolylthioketones there exist strong intramolecular S—H . . . N bridges. Thus the conclusion of Gordy and Stanford (*5*) is confirmed that hydrogen bonding is possible with strong nitrogen bases.

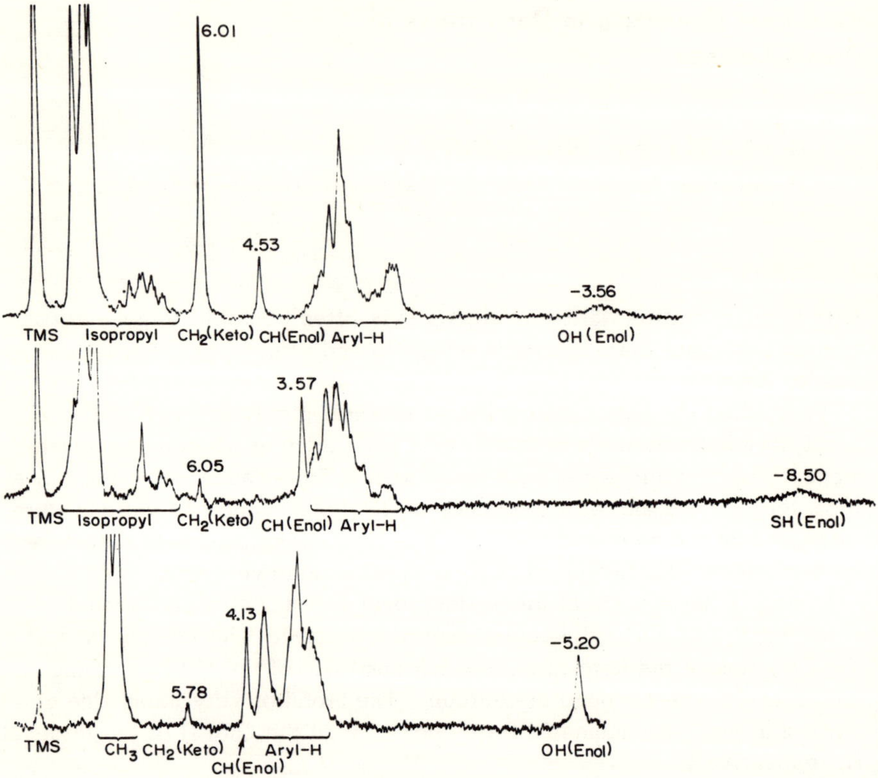

FIG. 1. The spectra of picolylisopropyl keton, picolylisopropyl thioketone, and collidylphenyl ketone at 25 Mc/s.

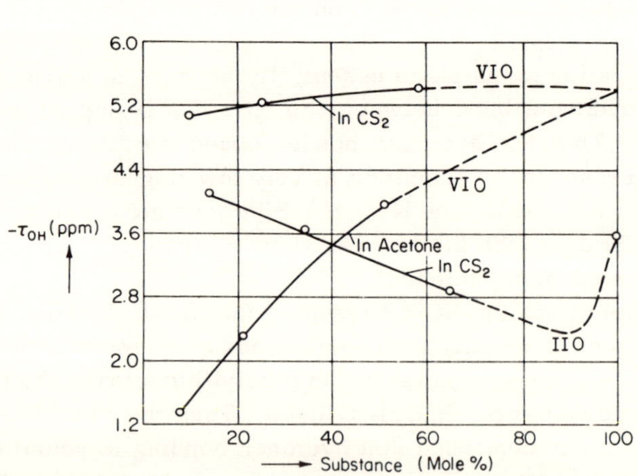

FIG. 2. The dependence of the chemical shift of the OH line of VI O and II O on the concentration in CS_2 and acetone.

TABLE I
2-PICOLYLKETONES

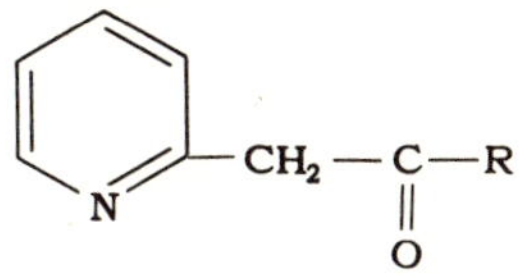

Substituent R	Solvent (mole %)[a]	Chemical shift (ppm) CH$_2$	CH	OH	Mole % enol
I O = —CH$_3$	Pure (100)	6.12	4.61	−1.00 (broad)	24
II O = —CH(CH$_3$)$_2$	Pure (100)	6.01	4.53	−3.56 (very broad)	31
	Acetone (59)	5.92	4.40	2.29 ?	22
	CS$_2$ (64)	6.03	4.63	−2.89 (very broad)	30
III O = —Phenyl	Chloroform (s)[b]	5.67	3.95	–	48
	Pyridine (s)	5.39	3.69	–	68
	CS$_2$ (s)	5.65	3.97	−4.84 (very narrow)	79
IV O = —β-Naphthyl	Chloroform (33)	5.52	3.89	–	50
	Pyridine (s)	5.24	3.54	–	70
V O = —CF$_3$	Chloroform (s)	–	4.09	−3.58 (narrow)	>90
VI O = collidyl	Acetone (51)	5.64	3.96	−3.99 (broad)	81
phenyl ketone	CS$_2$ (58)	5.78	4.13	−5.34 (narrow)	86

[a]Mole % solute.
[b]s = saturated solution.

TABLE II
2-PICOLYLTHIOKETONES

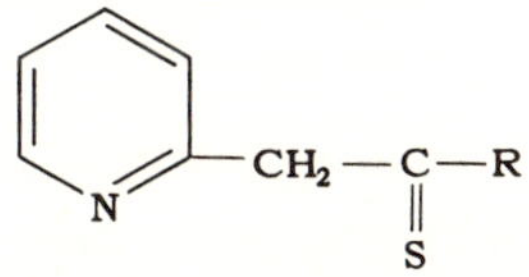

Substituent R	Solvent (mole %)[a]	Chemical shift (ppm) CH$_2$	CH	SH	Mole % enol
II S = —CH(CH$_3$)$_2$	Pure (100)	–	3.54	−8.60 (broad)	>98
	Acetone (49)	6.01	3.51	−8.56 (very broad)	89
	CS$_2$ (63)	6.05	3.57	−8.50 (broad)	91
III S = —Phenyl	Chloroform (s)[b]	–	2.91	–	>98
IV S = —α-Naphthyl	Chloroform (s)[b]	–	3.07	–	>98

[a]Mole % solute.
[b]s = saturated solution.

In the ketones the enolization is further stabilized by aromatic substituent groups, R. Burdett and Rogers (6) found the same effect for aromatic substituent groups in β-dicarbonyls. While with R = methyl and isopropyl the enol content is lower than 30 mole %, we observe with R =

aromatic substituent group an enol content higher than 48 mole % (Table I). This increasing of the enol form is a result of the effort to form conjugated double bonds between the picoline ring and the aromatic substituted group.

The enol content increases in the sequence of the solvents acetone < chloroform < CS_2 in accordance with Ref. (*7*).

Since in the studied compounds the CH_2, CH, and XH peaks appear at different τ values and the CH_2 and CH peaks are very small, it follows that the keto-enol proton exchange is slow (see, for example, Ref. *8*). At 25 Mc/s we obtained for the lifetimes

$$\begin{array}{ll} \text{ketones} & t \gg 7 \cdot 10^{-4} \text{ sec}, \\ \text{thioketones} & t \gg 5 \cdot 10^{-4} \text{ sec}. \end{array}$$

In nearly all solutions studied the OH and SH peaks are very broad or are not observable. The reason is presence of a further equilibrium, in which the protons interchange between inter- and intramolecular hydrogen bond sites. This exchange is fast or moderate, respectively. Assuming for the intermolecular bridge sites that $\tau_{XH} = 0$ ppm, one gets for the lifetimes

$$\begin{array}{ll} \text{ketones} & t \leq 2 \cdot 10^{-3} \text{ sec}, \\ \text{thioketones} & t \leq 7 \cdot 10^{-4} \text{ sec}. \end{array}$$

In II O the intermolecular hydrogen bonding between the II O molecules decreases and the intramolecular bonding increases with the dilution in CS_2 (the OH line is shifted to lower fields). Simultaneously the exchange rate decreases. The dashed part of the curve (Fig. 2) has to be studied more accurately.

In VI O dissolved in CS_2, the OH line is narrow and only a slight dependence of τ_{OH} on concentration is observable.. This proves again the strong stabilization of the intramolecular hydrogen bonding by aromatic substituent groups. We observe the same behavior also in III O, IV O, and V O in CS_2 solution. II S shows a similar effect, revealing again a strong intramolecular N . . . H—S bridge.

For II O in acetone intermolecular hydrogen bonding with acetone probably takes place ($\tau_{OH} = 2.28$ ppm?). On the other hand we find for VI O an equilibrium (stabilizing effect of aromatic substituent groups). Thereby the intermolecular bonding increases with dilution (the OH line shifts to higher fields; Fig. 2).

With increasing basicity of the picoline N (*9*), the strength of the intramolecular hydrogen bonding increases, as follows from the shift of the OH line from $\tau_{OH}(CS_2) = -4.84$ ppm (III O) to -5.34 ppm (VI O).

The CH line is shifted about 0.6 ppm downfield by aromatic substituent groups because of the diamagnetic anisotropy of R (see Tables I and II).

A further downfield shift of about 1.03 ppm for all compounds studied was observed by substituting S for O (Table III). With $\Delta = 1.03$ ppm we are able to predict for picolyl-β-naphthyl thioketone from III O τ_{CH} = 2.86 ppm and for picolyl-α-naphthyl ketone from IV S τ_{CH} = 4.10 ppm.

TABLE III
$\Delta = \tau_{CH}$ (PICOLYLKETONE) − τ_{CH} (PICOLYLTHIOKETONE)

Compounds	Solvent	
II	Pure	0.99
	Acetone	1.04
	CS_2	1.03
III	Chloroform	1.04
(IV)	Chloroform	(0.82)

This downfield shift is essentially a result of the altered diamagnetic susceptibilities of the bonds and altered electron densities.

The temperature dependence of the spectra studied in II O confirmed the presence of two equilibria, but it is not yet clear whether the one is a proton exchange between the enol form and an intermolecular site (first assumed), an exchange between a mesomeric form and an intermolecular site, or an exchange between the enol form and a mesomeric form (tautomerism) with predominance of the structure:

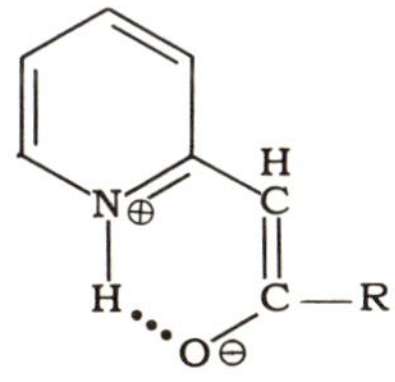

For further clarification of the equilibria, we intend to study more accurately the dependence of the spectra on temperature.

Acknowledgments

We wish to thank Mr. H. Frischleder and Mr. A. Zschunke for helpful discussions.

References

1. D. Hadzi and H. W. Thompson (eds.), "Hydrogen Bonding." Pergamon, London, 1959.
2. G. C. Pimentel, and A. L. McClellan, "The Hydrogen Bond," Freeman, San Francisco, 1960.

3. A. Di Giacomo, and C. P. Smyth, *J. Am. Chem. Soc.* **78**, 2032 (1950).
4. A. S. N. Murthy, C. N. R. Rao, B. D. N. Rao, and P. Venkateswarlu, *Trans. Faraday Soc.* **58**, 858 (1962).
5. W. Gordy, and S. C. Stanford, *J. Am. Chem. Soc.* **62**, 497 (1940).
6. J. L. Burdett and M. T. Rogers, *J. Am. Chem. Soc.* **86**, 2105 (1964).
7. J. Powling and H. J. Bernstein, *J. Am. Chem. Soc.* **73**, 4353 (1951).
8. H. M. McConnell, *J. Chem. Phys.* **28**, 430 (1958).
9. A. Gero and J. J. Markham, *J. Org. Chem.* **16**, 1835 (1952).

Tautomerism in Side-Chain Substituted 2-Methyl-*N*-Heterocycles

ROSANNA MONDELLI and LUCIO MERLINI

Istituto di Chimica, Politecnico di Milano, Milano, Italy

Availability of NMR spectroscopy has renewed interest in tautomerism, especially in prototropy of heterocyclic compounds (*1*). Promising types of substances in this respect are the side-chain derivatives of 2-methyl-*N*-heterocycles. The results obtained in such an investigation of xanthopterine derivatives (*2*) led us to extend the examination to other similar systems (I) such as 2-oxo-3-methyl-1,2-dihydroquinoxaline, 3-methyl-

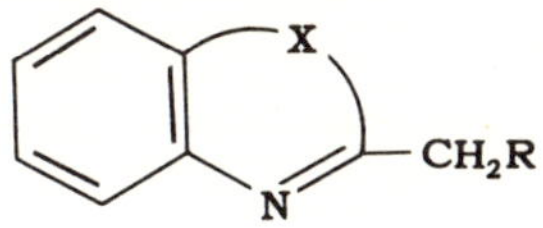

(I)

dihydro-1,4-benzoxazin-2-one, 2-methylquinoxaline, 2-methylquinoline, and 4-methylquinoline.

Irrespective of other possible equilibria, such as for instance the lactam–lactim equilibrium[1] in 2-oxo-1,2-dihydroquinoxalines, only two tautomeric structures (A) and (B) are possible, when R = H, CN, COOMe.

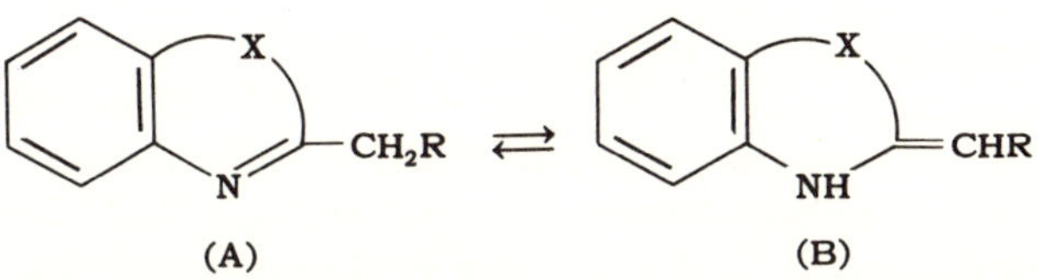

The structures become three in number (C,D,E) when R = COY (Y = CH_3, COOEt). The first series studied consists of the 2-oxo-3-methyl-1,2-di-

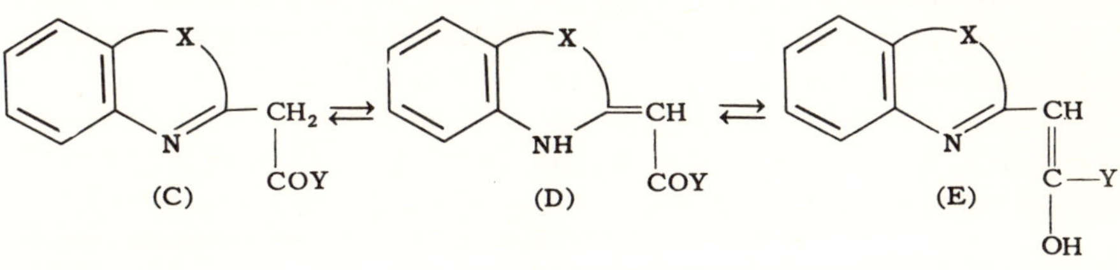

hydroquinoxaline derivatives (II) (Table I). Structures (A) and (B) are easily recognized by the shifts of the methylene and vinyl protons.

[1]This equilibrium is always on the side of the cyclic amide form; see Ref. (*3*).

TABLE I[a]

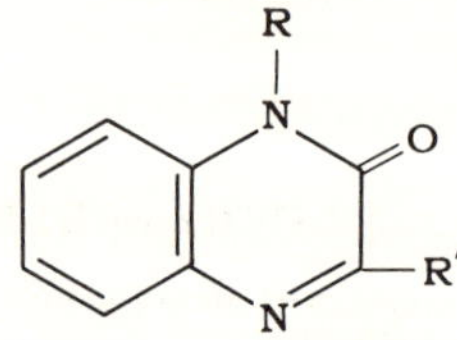

	R	R′	TFA CH$_2$	TFA CH=	DMS CH$_2$	DMS CH=	DMS NHCO	DMS NH	CDCl$_3$ CH$_2$	CDCl$_3$ CH=	CDCl$_3$ NHCO	CDCl$_3$ NH
a	H	CH$_3$	3.1	–	2.45	–	12.28	–		Insol.		
b	H	CH$_2$COOCH$_3$	4.53	–	3.83	5.52	11.85	11.00		Insol.		
c	CH$_3$	CH$_2$COOCH$_3$	4.56	–	3.84	5.54	–	11.00	–	5.83	–	11.15
d	H	CH$_2$COCH$_3$	–	6.52	–	6.01	11.82	13.0		Insol.		
e	CH$_3$	CH$_2$COCH$_3$	–	6.50	–	6.10	–	13.0	–	6.25	–	13.30
f	CH$_3$	CH$_2$COCOOEt	–	7.39	–	6.70	–	13.3	–	7.07	–	14.00
						—CH$_\beta$=CH$_\alpha$—				—CH$_\beta$=CH$_\alpha$—		
g	H	CH$_\beta$=CH$_\alpha$—COOEt				7.88	7.30			8.10	7.38	

[a]Shifts are measured in parts per million (δ) from TMS as internal standard.

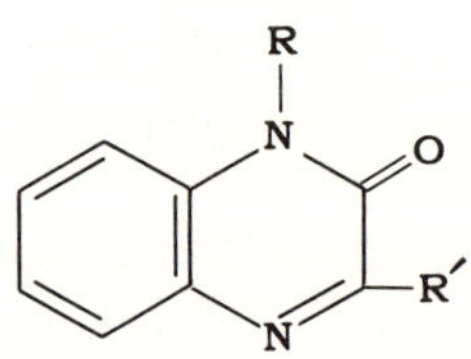

(II)

In compounds (II) (R′ = CH_2CN, CH_2COOMe), the position of the equilibrium is affected by the solvent: it shifts toward methylenic structure from $CDCl_3$ to dimethylsulfoxide (DMS) to trifluoroacetic acid (TFA). On the other hand, when R′ = CH_2COCH_3, $CH_2COCOOEt$, structure (D) with the exocyclic double bond is always stabilized. The potential enol structure (E) seems to us unlikely, on the basis of the following evidence. The chemical shift of the vinylic proton in (IIf) (Table I) is more than 1 ppm to higher field than that of β-hydrogen in (IIg);[2] in the enol structure the presence of the hydroxyl should shift the same β-hydrogen not more than 0.5–0.6 ppm to higher field.

The lack of the allylic long-range coupling between CH_3 and CH= groups in the spectra of (IId) and (IIe) is in agreement with structure (D).

[2]It should be noted that δ values of vinylic protons increase in the order =CHCOOR < =CHCOMe < =CHCOCOOR in all the series examined. See the values in xanthopterine derivatives: 5.58–6.13–6.70 (DMS); 6.39–6.79–7.58 (TFA) (*2*). This effect may be attributed to the different electronegativity of the groups —COOR, —COMe, and —COCOOR.

However, such a coupling has not been observed, as far as we know, in any substance (*4*) which contains the group

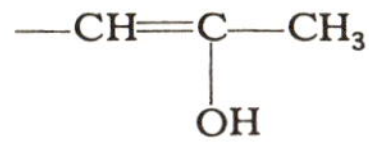

Moreover, independent evidence comes from the UV spectra. The strong difference in λ and ε values between (IIf) (λ_{max} = 405, 429, 456 mμ; ε = 13500, 19600, 17750 in $CHCl_3$) and (IIg) (λ_{max} = 319, 380 mμ; ε = 11100, 6400 in $CHCl_3$), which is assumed to have a cromophore comparable with the enolic structure (E), is consistent with the enaminic structure (D) proposed for the former. Ultraviolet absorptions of (IId), (IIe), and (IIf) are very similar to those of (IIb) and (IIc) in $CHCl_3$, the enaminic structure of which in this solvent is established by NMR.

TABLE II

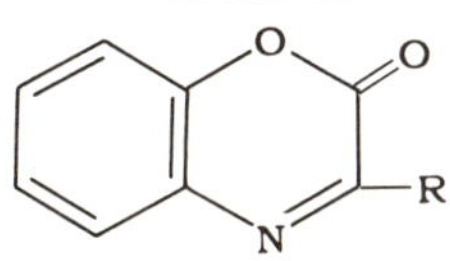

	TFA		DMS			$CDCl_3$		
R	CH_2	CH=	CH_2	CH=	NH	CH_2	CH=	NH
CH_3	2.99	–	2.42	–	–	2.56	–	–
CH_2COOEt	4.4	6.0	–	5.60	10.7	–	5.94	10.67
CH_2COCH_3	–	6.57	–	6.15	12.18	–	6.25	12.40

In the 3-methyl-dihydro-1,4-benzoxazin-2-one series (III)(Table II) we have

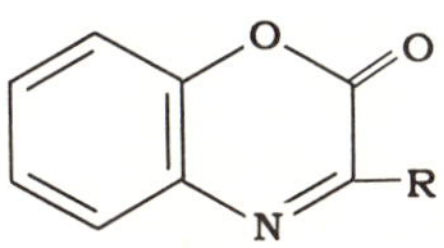

(III)

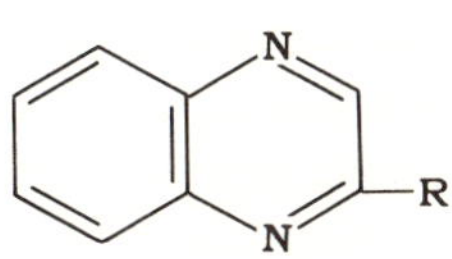

(IV)

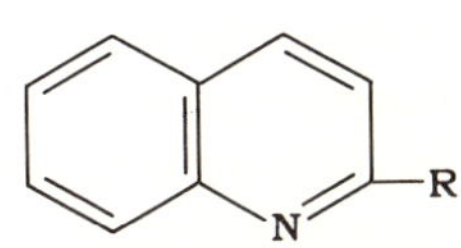

(V)

about the same behavior with a small trend toward the stabilization of the exocyclic double bond (as in pteridines (*2*)).

On the other hand 2-methylquinoxaline (IV) and especially quinaldine derivatives (V) show the opposite trend, probably because of the stronger aromaticity of the ring (Table III). These statements are supported by similar NMR and UV evidence; moreover the IR spectrum in CCl_4 of methyl quinaldylketone shows a carbonyl band at 6.1 μ and lacks the characteristic absorptions of the quinoline ring.

TABLE III

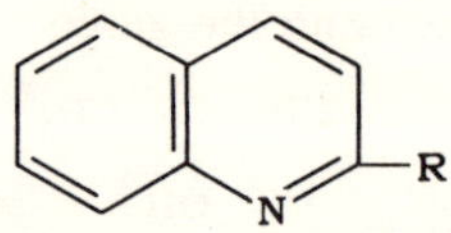

R	TFA			DMS			$CDCl_3$[a]		
	CH_2	CH=	NH^+	CH_2	CH=	NH	CH_2	CH=	NH
CH_3	3.18	–	13.5	2.73	–	–	2.68	–	–
CH_2COOEt	4.57	–	14.2	4.08	–	–	4.03	–	–
CH_2COMe	4.77	–	14.0	4.20	5.45	14.8	3.98	5.22	15.0
$CH{=}CMe_2$	–	6.70	13.0	–	6.48	–	–	6.39	–
$CH_2COCOOEt$	–	7.00	13.5	–	6.32	Not det.	–	6.37	15.7
	—CH$_\beta$=CH$_\alpha$—			—CH$_\beta$=CH$_\alpha$—					
CH$_\beta$=CH$_\alpha$—COOH	8.42	7.47	14.5	8.30	7.32			Insol.	

[a]Or in CCl_4.

The existence of a considerable number of these compounds in the enaminic form could suggest that a six-membered chelated ring strongly contributes to its stabilization. Thus the large paramagnetic shift of all

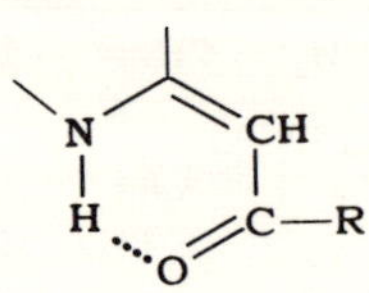

the NH absorptions indicates that they are strongly hydrogen bonded. On addition of TFA to the DMS solutions of 2-oxo-1,2-dihydroquinoxalines, the enaminic NH exchanges very slowly with respect to amidic NH. This seems to suggest that the former are intramolecularly hydrogen bonded. However, many compounds are stable in the enaminic form (D) in solution of TFA, which is a known strong hydrogen-bonding medium; moreover, the same tautomeric form is present also in some lepidine derivatives (VI) (Table IV) where chelation is not possible. The latter arguments

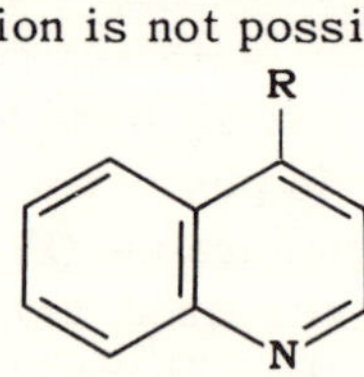

(VI)

suggest that the chelated ring is not necessary to stabilize the enaminic structure.

TABLE IV

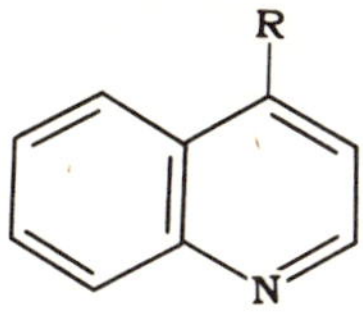

R	TFA			DMS			$CDCl_3$		
	CH_2	=CH	NH^+	CH_2	=CH	NH	CH_2	=CH	NH
CH_3	3.17	–	13.7	2.68[a]	–	–	2.68[a]	–	–
CH_2CN	4.86[b]	–	14.2	4.65[a]	4.97	11.0	4.13[a]	–	–
CH_2COOEt	4.67	–	14.1	4.25	–	–	4.06	–	–
$CH_2COCOOEt$	5.22	7.58	13.5	–	6.76	Not det.		Insol.	

[a]Doublet, $J \sim 1$ cps.
[b]Doublet, $J \sim 0.7$ cps.

Note added in proof

We have observed later an allylic coupling between $—CH_3$ and =CH groups in 2-pyridylacetone, which exists in CCl_4 solution as an equilibrium mixture between the keto form (~ 90%) and the enol form (~ 10%). See R. Mondelli and L. Merlini, *Tetrahedron* (in press).

References

1. A. R. Katritzky and J. M. Lagowski, *Advan. Heterocyclic Chem.*, **1**, p. 311 (1963).
2. W. von Philipsborn, H. Stierlin, and W. Traber, *Helv. Chim. Acta* **46**, 2592 (1963); L. Merlini, W. von Philipsborn, and M. Viscontini, *ibid.* **46**, 2597 (1963).
3. G. W. H. Cheeseman, *Advan. Heterocyclic Chem.* **2**, 229 (1963).
4. J. L. Burdett and M. T. Rogers, *J. Am. Chem. Soc.* **86**, 2105 (1964). R. W. Hay and P. P. Williams, *J. Chem. Soc.*, p. 2270 (1964); also references quoted therein.

Study of Hydrogen Bonding of Pyrazole in Carbon Tetrachloride by Proton Magnetic Resonance

ANGELO PEROTTI and MARIO COLA

Istituto di Chimica Generale dell' Università di Pavia,

In a previous paper (to which refer for a full bibliography) (*1*), we have undertaken a systematic interpretation of proton magnetic resonance data of pyrazole and some of its derivatives, which were substituted in various positions.

In this manner we have confirmed, also as far as the chemical shielding is concerned, the equivalence of positions 3 and 5 in the pyrazole ring in all those derivatives in which the nitrogen-bound hydrogen was not removed by substitution.

Among the factors concurring to give a greater degree of symmetry in the pyrazole ring we have considered the exchange reactions (*2*) and the considerable tendency of pyrazole to form hydrogen-bonded associations, with particular reference to the cyclic dimer and trimer as pointed out by Anderson *et al.* (*3*) in the interpretation of IR data for solutions of pyrazole in carbon tetrachloride.

Since we have found a marked dependence of the NH proton chemical shift on the concentration, it seemed very suitable to apply PMR to investigate the association of pyrazole by hydrogen bonding. We have hence carried out here a systematic study of the chemical shifts at different concentrations and temperatures of pyrazole in CCl_4, with particular reference to the very low-concentration region, to obtain the value of δ_m and to study the dimerization equilibrium for the purpose of elucidating the nature of the various species involved in the association equilibria.

Experimental and Results

Reagents. The purification and handling of the reagents have been reported previously (*1*). In the pyrazole we have left trace amounts of moisture to produce a sharp NH proton resonance by catalyzing proton exchange; checks showed that NH chemical shift was practically unaffected by this means. (A similar expedient—addition of traces of HCl—has already been used to study hydrogen bonding in hydroxylic compounds (*4*)).

All the solutions have been prepared with pyrazole so obtained by dilution of gravimetrically prepared stock solutions; their concentration is given in mole fraction units, x, which avoids temperature dependence.

Measurement of Chemical Shifts. All the spectra were recorded on a Varian DP operating at 60 Mcps, equipped with superstabilizer, field homogeneity control system kit, and variable temperature conversion kit. The temperature was continuously controlled and monitored with a thermocouple fitted in the probe insert; it was checked that the sample temperature could be determined to within $\pm 1\,^{\circ}C$.

The NH proton chemical shifts have been measured by the sideband technique for more concentrated solutions, using as reference the central position of the doublet of the protons 3 and 5. For the more diluted solutions δ was determined from an average of at least five measurements of the shift between NH signal and the central position of the doublet of H (*3, 5*), using as internal calibration the known chemical shift between the 3, 5, and 4 protons and the relative splittings.

The values of δ are expressed in parts per million and are reported in Table I with the concentrations x, at the different temperatures; their accuracy is ca. $\pm$ 0.007 ppm. In the same table are also reported the values of $(\partial\delta/\partial x)_{x \longrightarrow 0}$, i.e., the limiting slopes of the plots δ *versus* x, and the values of δ_m obtained as discussed below.

Discussion

The results of Table I show a marked dependence of the NH proton chemical shift on concentration and temperature, while the relative positions of the 3, 5, and 4 proton signals are practically not affected.

Figure 1 reports a typical set of data in the low-concentration region and at $75\,^{\circ}C$; measurements at different temperatures give similar plots.

Curves of this general shape, including reversal of curvature at $x \simeq 0.004$ (this value depends on the temperature) are consistent for systems involving monomers, dimers, and higher polymers. All these curves have been extrapolated to $x = 0$, with the assumption (confirmed by the experimental data) that δ is a linear function of x for $x < 0.003$. The values of δ obtained by these extrapolations are practically temperature independent within the limits of the experimental errors and represent the true δ for the monomer, labeled as δ_m. Its value is -0.77 ± 0.05 in the given reference.

We do not discuss here the high-concentration region of the plots. The curves in these regions fit equations of higher order and their interpretation is very hard owing to the complexity of the system which possibly contains a number of polymers with different degrees of association. Such

TABLE I
CHEMICAL SHIFT OF THE NH PROTON RESONANCE OF PYRAZOLE IN CCl_4

x	29°	40°	54°	60°	75°
0.3109	−6.250	−6.078	−6.000	−5.651	−5.616
0.2179	−6.233	−5.921	−5.810	−5.500	−5.487
0.1420	−6.166	−5.978	−5.835	−5.320	−5.480
0.1035	−6.166	−6.070	−5.350	−5.320	−5.315
0.0537	−5.666	−5.585	−5.030	−5.120	−4.957
0.0433	−5.416	−4.960	−4.612	−4.552	−4.615
0.0361	−5.283	−4.756	−4.570	−4.312	−4.321
0.0271	−5.002	−4.687	−4.365	−4.188	−4.105
0.0217	−4.916	−4.450	−4.360	−4.077	−3.911
0.0180	−4.633	−4.266	−4.205	−3.808	−3.744
0.0135	−4.534	−4.105	−4.080	−3.813	−3.620
0.0108	−4.313	−4.041	−3.697	−3.675	−3.392
0.0090	−4.276	−3.989	−3.753	−3.642	−3.227
0.0077	−4.034	−3.860	−3.740	−3.491	−3.134
0.0068	−3.792	−3.760	−3.638	−3.465	−3.036
0.0060	−3.630	−3.546	−3.421	−3.017	−2.803
0.0049	−3.327	−3.144	−3.098	−2.717	−2.251
0.0042	−3.103	−3.097	−2.930	−2.365	−2.050
0.0036	−2.837	−2.733	−2.459	−2.146	−1.755
0.0032	−2.504	−2.413	−2.175	−1.810	−1.523
0.0029	−2.339	−2.223	−2.060	−1.697	−1.394
0.0026	−2.150	−2.069	−1.962	−1.630	−1.317
0.0024	−2.060	−1.970	−1.843	−1.563	−1.291
0.0022	−1.960	−1.880	−1.762	−1.493	−1.233
δ_m	−0.76	−0.82	−0.80	−0.80	−0.72
$(\partial\delta/\partial x)_{x\to 0}$	−550 ± 10	−480 ± 10	−400 ± 8	−320 ± 7	−250 ± 5

a system contains so many parameters that they cannot be determined uniquely from any curve, and PMR alone does not suffice to show the various species present: a proper interpretation must combine data from additional sources.

We will consider here the situation at very low concentration: it is reasonable in these conditions to assume that monomer-dimer equilibrium is by far dominant (in the same way as accepted for other hydrogen-bonded systems) (*5*).

Huggins *et al.* (*6*) have shown that for such a system the equilibrium constant for dimer formation is related to the limiting slope of the δ *versus* x plot by the equation

$$\left(\frac{\partial\delta}{\partial x}\right)_{x\to 0} = 2K_d\Delta_d, \tag{1}$$

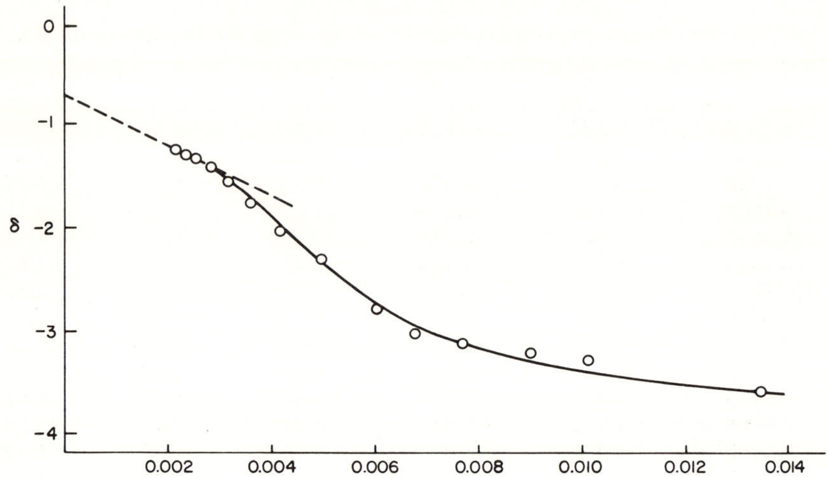

FIG. 1. Chemical shift of the NH proton resonance, δ, *versus* mole fraction of pyrazole at 75 °C.

where

$K_d = x_d/x_d^2$ is the dimerization equilibrium constant;

$\Delta_d = \delta_d - \delta_m$; δ_d is the NH proton chemical shift of the dimer (see below);

x_d, and x_m are the mole fractions of dimer and monomer respectively.

From our measurements we can deduce the values of the limiting slope and δ_m, but unfortunately, owing to the presence of higher polymers, it is impossible from PMR data alone to obtain the value of δ_d and thereby K_d.

However, we can assume that δ_d is temperature and concentration independent like δ_m, as accepted for certain carboxylic acids and alcohols (*7, 8*) in analogous studies of hydrogen bonding.

With this assumption and remembering that Δ_d is a negative quantity because, as a general rule, the proton resonance of hydrogen-bonded species is at lower magnetic field than the resonance of the monomer, we can conclude from the van't Hoff equation and Eq. (1) that

$$-\Delta H = R \frac{\partial \ln (\partial\delta/\partial x)_{x \longrightarrow 0}}{\partial\,(1/T)},$$

and thus calculate the ΔH of dimerization.

Figure 2 shows a plot of log $(\partial\delta/\partial x)_{x \longrightarrow 0}$ *versus* $1/T$. The experimental points fit a straight line, from which we can obtain $\Delta H = -3.4 \pm$

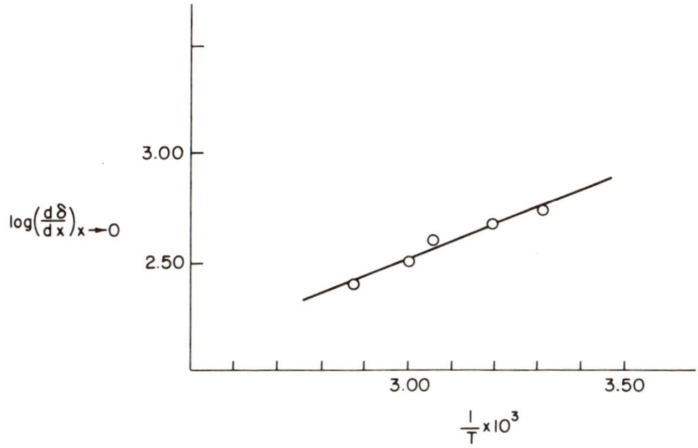

FIG. 2. Dependence of the limiting slope on the temperature.

0.3 kcal/mole. This value is of the same order of magnitude as those reported in the literature for the hydrogen bond NH $\cdots$ N (*5*).

The known values of ΔH for hydrogen bonding of this kind are too few for drawing final conclusions by comparing them with ours; moreover, we must not forget the simplifications made in our discussion. However, we can guess that with our experimental conditions and at very low concentrations only one hydrogen bond is statistically involved between two molecules in the dimerization equilibrium of pyrazole. It would seem, that is, that the dimer has an open structure.

This conclusion does not exclude the possibility that in regions of moderately greater concentration the cyclic dimer and trimer are important.

References

1. M. Cola and A. Perotti, *Gazz. Chim. Ital.* **94,** 1268 (1964).
2. N. Joop and H. Zimmermann, *Z. Elektrochem.* **66,** 140 (1962).
3. D. W. M. Anderson, J. L. Duncan, and F. J. C. Rossotti, *J. Chem. Soc.*, p. 140 (1961).
4. M. Saunders and J. B. Hyne, *J. Chem. Phys.* **29,** 1319 (1958).
5. G. C. Pimentel and A. L. McClellan, "The Hydrogen Bond." Freeman, San Francisco, 1960.
6. C. M. Huggins, G. C. Pimentel and N. J. Shoolery, *J. Phys. Chem.* **60,** 1311 (1956).
7. J. C. Davis, Jr. and K. S. Pitzer, *J. Phys. Chem.* **64,** 386 (1960).
8. J. C. Davis, Jr., K. S. Pitzer, and C. N. R. Rao, *J. Phys. Chem.* **64,** 1744 (1960).

Proton Magnetic Resonance Spectra of a Series of Cyclic Imides

GIOVANNI CASINI

Istituto di Chimica Farmaceutica e Tossicologica dell'Università di Roma, Roma, Italy

and

MARIA LETIZIA SALVI

Istituto di Chimica dell'Università Cattolica del Sacro Cuore, Roma, Italy

We have determined the proton magnetic resonance (PMR) spectra of a series of five- and six-membered cyclic imides, including succinimides, glutarimides and glutaconimides, as shown in Fig. 1.

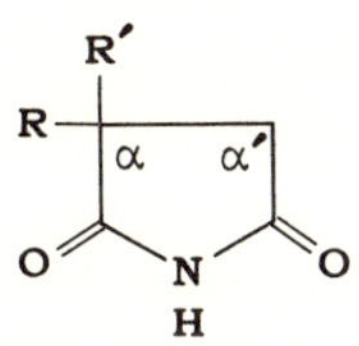

(I) $R = R' = H$
(II) $R = C_6H_5$ $R' = H$
(III) $R = C_6H_5$ $R' = CH_3$
(IV) $R = C_6H_5$ $R' = C_2H_5$

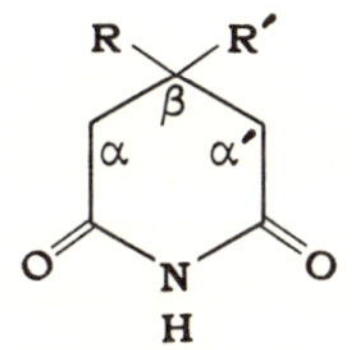

(V) $R = R' = H$
(VI) $R = R' = CH_3$
(VII) $R = CH_3$ $R' = C_2H_5$
(VIII) $R = R' = C_2H_5$
(IX) $R = C_6H_5$ $R' = H$
(X) $R = C_6H_5$ $R' = CH_3$
(XI) $R = C_6H_5$ $R' = C_2H_5$

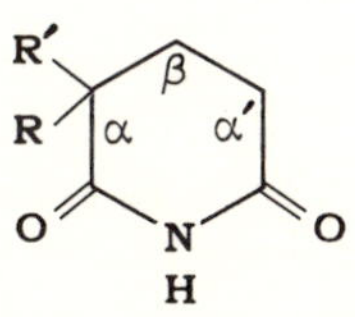

(XII) $R = C_6H_5$ $R' = H$
(XIII) $R = C_6H_5$ $R' = CH_3$
(XIV) $R = C_6H_5$ $R' = C_2H_5$

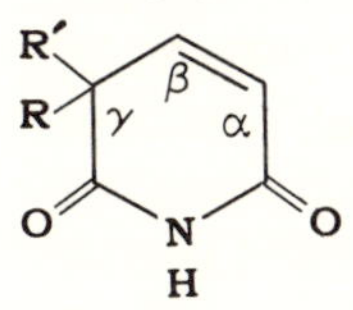

(XV) $R = C_6H_5$ $R' = CH_3$
(XVI) $R = C_6H_5$ $R' = C_2H_5$

FIG. 1.

Some of the compounds studied are of pharmacological interest as CNS stimulants or depressants. We have examined the PMR characteristics of these substances to provide information relevant to the conformation of these molecules, in order to have a better understanding of structure-activity relationships.

All the spectra were taken in $CDCl_3$ solution, at the same molar concentration (0.4 *M*), with a Varian A-60 instrument. Tetramethylsilane was used as an internal reference.

Only a few of the spectra showed overlapping signals or complicated spin-spin coupling patterns which could not be analyzed. Most of the spectra were easily interpreted on the basis of the chemical structure of the compounds. We have collected in Table I all the chemical shifts and coupling constants which were easily measured.

Some of the most interesting features which were observed in this series of spectra are the following.

First of all, the imidic proton gives a rather broad signal at very low fields, as expected. Here we observed a consistent difference between the chemical shift in five- and six-membered imides, i.e., the five-membered imides (succinimides) give signals at lower fields than the six-membered imides:

five membered, $\delta = 9.1$ ppm,
six membered, $\delta = 8.6$ ppm.

This observation is in agreement with the known difference in acidity of the two kinds of imides: Succinimides are more acidic than glutarimide (*1*, *2*, *3*), the ratio of ionization constants being about 100:1. There is at present no satisfactory explanation for this behavior. Hydrogen bonding, which is known to occur through dimerization in imides (*4*), might play an important role in determining the differences both in acidity and in shift value.

Some of the signals arising from the ring protons were interpreted from the standpoint of ring conformation, as follows.

In those molecules possessing isolated methylene groups in the ring (III, IV, VI, VII, VIII, X, XI), we observed sharp singlets when the ring substituents were exclusively alkyl groups (VI, VII, VIII), and signals deriving from an AB system when a phenyl substituent was present. In the latter case we could measure the corresponding geminal coupling constants. Here again we found a constant difference in the absolute magnitude of coupling constants between five- and six-membered rings:

five membered $|J_{AB}| = 18$ cps,
six membered $|J_{AB}| = 16$ cps.

This fact can be at least qualitatively explained by supposing a small difference in the H–C–H bond angles in the two series (*5*) arising from a difference in conformation and ring strain. An inspection of models of the two kinds of rings shows indeed that the succinimide ring should be planar with some bond angle compression across the ring. The glutarimide ring instead should be nonplanar (*6*) with one of the six atoms (the β atom) out of the plane of the other five. Some x-ray analysis data are also in agreement with the latter conclusion (*4*). In the glutarimide ring there is no angle strain.

In one case, α-phenylsuccinimide (II), we also measured the two *cis*- and *trans*- vicinal coupling constants of the ring protons whose signals appear to be generated from an ABX system (see Fig. 2). The greater

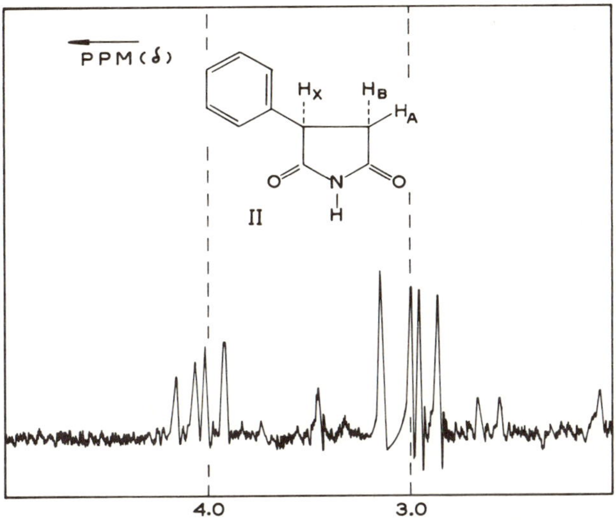

FIG. 2. ABX part of the PMR spectrum of α-phenylsuccinimide (II).

absolute value has been assumed for the coupling between *cis* (eclipsed) protons:

$$|J_{\text{cis}}| = 8 \text{ cps},$$
$$|J_{\text{trans}}| = 6 \text{ cps}.$$

It is interesting to note that if we assume a perfectly planar structure for the succinimide ring (dihedral angles of about 0° and 120°) these J values are not quantitatively in agreement with Karplus' curve (*7*).

In those rings where only olefinic protons are present (the two glutaconimides XV and XVI) the spectra are of course quite different: they show a clean AB pattern for the two ring protons at relatively low fields (coupling constant $J = 10$ cps), with a further splitting of the higher-field signal ($J = 1.5$ cps) (see Fig. 3).

The most logical explanation of this second-order splitting is a long-range coupling between the α proton of the ring and the imidic proton. In

TABLE I

CHEMICAL SHIFTS δ (PPM), AND COUPLING CONSTANTS $|J|$ (CPS)

Compound	Imidic NH	Ring protons: Aliphatic α—CH_2		Aliphatic α—CH	Aliphatic β—CH_2	Aliphatic β—CH	Olefinic α—CH	Olefinic β—CH	Substituent protons: —C_6H_5	—CH_3	—CH_2—	—CH_3
I	...[a]	2.74		–	–	–	–	–	–	–	–	–
II	9.2	2.8 3.2	$J_{AB} = 18.5$	$J_{cis} = 8$ 4.04 $J_{trans} = 6$	–	–	–	–	7.25	–	–	–
III	9.2	2.87 3.12	$J_{AB} = 18.5$	–	–	–	–	–	7.32	1.71	–	–
IV	9.0	2.91 3.11	$J_{AB} = 18$	–	–	–	–	–	7.33	–	2.07	0.90
V	8.6	2.58		–	1.95	–	–	–	–	–	–	–
VI	8.6	2.41		–	–	–	–	–	–	1.11	–	–
VII	8.6	2.39		–	–	–	–	–	–	1.03	1.41	0.88
VIII	8.6	2.40		–	–	–	–	–	–	–	1.41	0.83
IX	8.7	...		–	–	...	–	–	7.24	–	–	–
X	8.6	2.72 3.04	$J_{AB} = 16.3$	–	–	–	–	–	7.23	1.39	–	–
XI	8.6	2.69 3.07	$J_{AB} = 16.5$	–	–	–	–	–	7.21	–	1.71	0.65
XII	8.6	2.62		3.75	2.24	–	–	–	7.24	–	–	–
XIII	8.6	...		–	...	–	–	–	7.24	1.57	–	–
XIV	8.5	...		–	...	–	–	–	7.24	–	...	0.86

XV	8.6	–	–	–	–	6.19	6.70 $J = 10$	7.27	1.78	–	–
XVI	8.6	–	–	–	–	6.28	6.68 $J = 10$	7.27	–	...	0.90

[a]The three points indicate that the corresponding shift value has not been determined because of superimposed signals, or complicated spin-spin coupling, or not visible signal.

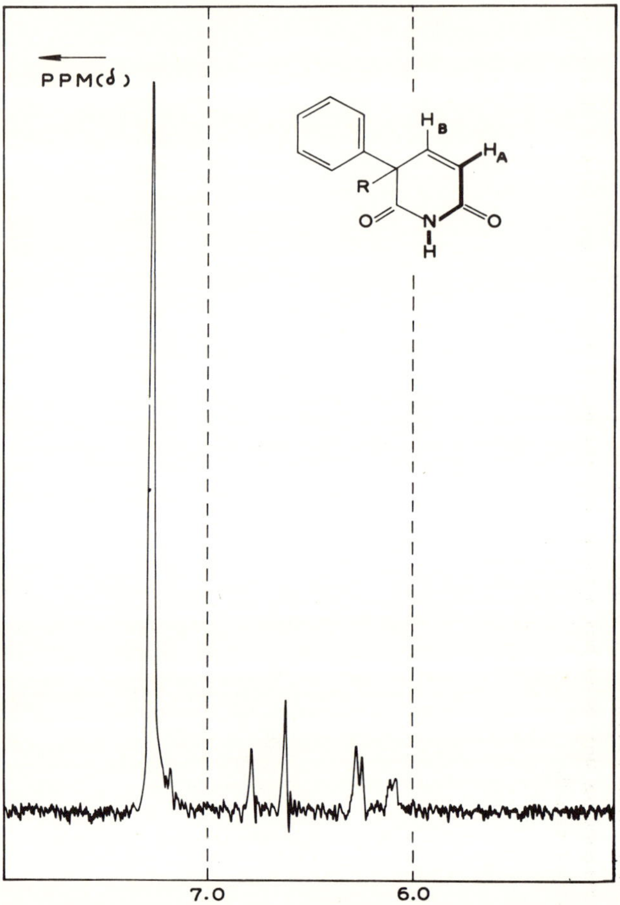

FIG. 3. Aromatic-olefinic portion of the PMR spectrum of one of the disubstituted glutaconimides (XV).

this H–C–C–N–H four-bond system we find indeed the characteristic W planar disposition that has been observed in other cases of similar long-range coupling (*8*). That the glutaconimide ring is planar can be easily seen from an inspection of the model (five trigonal atoms in a six-membered ring) and has been confirmed by x-ray structural determination (*9*).

Thus, the higher-field signal in the AB portion of the two last mentioned spectra is to be assigned to the α ring proton. The signal from the β proton is probably shifted to a lower field by the influence of the phenyl group's magnetic anisotropy. These assignments are confirmed by deuteration of the NH proton, which eliminates the second-order splitting in the higher-field signal.

In the spectrum of one of the two last-mentioned compounds, γ-phenyl-γ-ethylglutaconimide (XVI), we observed another very interesting feature: The signal arising from the methylene group of the ethyl substituent does not appear as a simple quartet, as in the other compounds of the series, but shows a more complex splitting pattern (see Fig. 4). This behavior

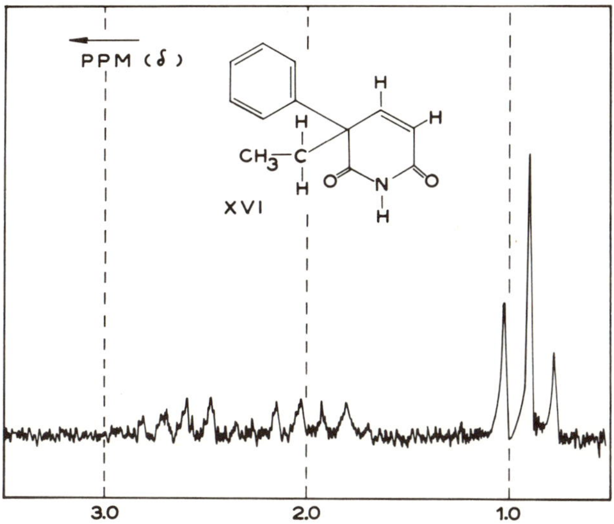

FIG. 4. PMR spectrum of the ethyl group of γ-phenyl-γ-ethylglutaconimide (XVI).

is probably due to an inequivalence of the two protons of the methylene group. The asymmetry of the neighboring carbon atom of the ring might in principle justify such an inequivalence (*10*), but since we did not find a similar splitting pattern in the other compounds of the series with the same substituents (IV, XIV), it is necessary to consider another explanation. We have interpreted this situation as resulting from a restricted rotation about the CH_2–C_{ring} bond, probably due to the particular conformation of this molecule. The lack of rigidity in the analogous glutarimide compound (XIV), and the lesser steric hindrance in the succinimide analog (IV), would explain the normal appearance of the methylene group signal in these compounds.

References

1. H. von Euler and A. Oelander, *Z. Physik. Chem.* (*Frankfurt*) (Abt. A) **137**, 393 (1928).

2. A. Foucaud and M. Duclos, *Compt. Rend.* **256**, 4033 (1963).

3. G. Schwarzenbach and K. Lutz, *Helv. Chim. Acta* **23**, 1162 (1940).

4. M. Bonamico, F. Coppola, and G. Giacomello, *Gazz. Chim. Ital.* **91**, 193 (1961); M. Bonamico, F. Coppola, G. Giacomello, and L. Zambonelli, *ibid.* **92**, 587 (1962).
5. M. Karplus and H. S. Gutowsky, Abstract, Am. Chem. Soc. Meeting, San Francisco, 1958.
6. G. Casini, A. Schimberni, and V. Zevi, *Ric. Scient.* **29**, 761 (1959).
7. M. Karplus, *J. Chem. Phys.* **30**, 11 (1959).
8. A. Rassat, C. W. Jefford, J. M. Lehn, and B. Waegell, *Tetrahedron Letters*, p. 233 (1964).
9. M. Bonamico, F. Coppola, G. Giacomello, A. Vaciago, and L. Zambonelli, *Gazz. Chim. Ital.* **92**, 1319 (1962).
10. J. D. Roberts, "Nuclear Magnetic Resonance; Applications to Organic Chemistry," p. 58. McGraw-Hill, New York, 1959.

A Proton Magnetic Resonance Study of the Structure of the Adducts Formed by the Addition of Dimethyl Acetylene Dicarboxylate to Thiazoles

W. BONTHRONE and F. S. SKELTON

Milstead Laboratory of Chemical Enzymology, Sittingbourne, Kent, England

and

D. H. REID

Department of Chemistry, St. Salvator's College,
University of St. Andrews, St. Andrews, Fife, Scotland

We studied the addition of dimethyl acetylene dicarboxylate to various thiazoles in the hope of finding a satisfactory preparation of the pyrrolo [2,1-b] thiazole ring system. This seemed reasonable since pyridine reacts with acetylene dicarboxylic ester to give indolizine 1,2,3-triboxylic ester (Fig. 1). We allowed the ester to react with the thiazoles in 2 : 1

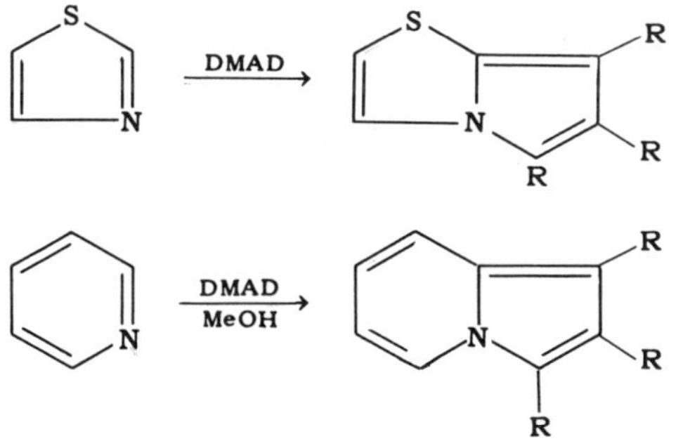

FIG. 1. DMAD ≡ Dimethyl acetylene dicarboxylate. R ≡ COOMe in all cases.

molecular proportions at room temperature for 3 to 4 days and found that the nature of the products was markedly solvent dependent.

2,4-Dimethyl thiazole in DMF gave as major product a 2 : 1 adduct for which we propose the [3,4,0]-bicyclic structure (I) in Fig. 2. The chemical shift values shown in the figure are parts per million downfield from the tetramethyl silane signal and the numbers in brackets refer to the integral intensities of the various signals. The NMR spectrum of the ad-

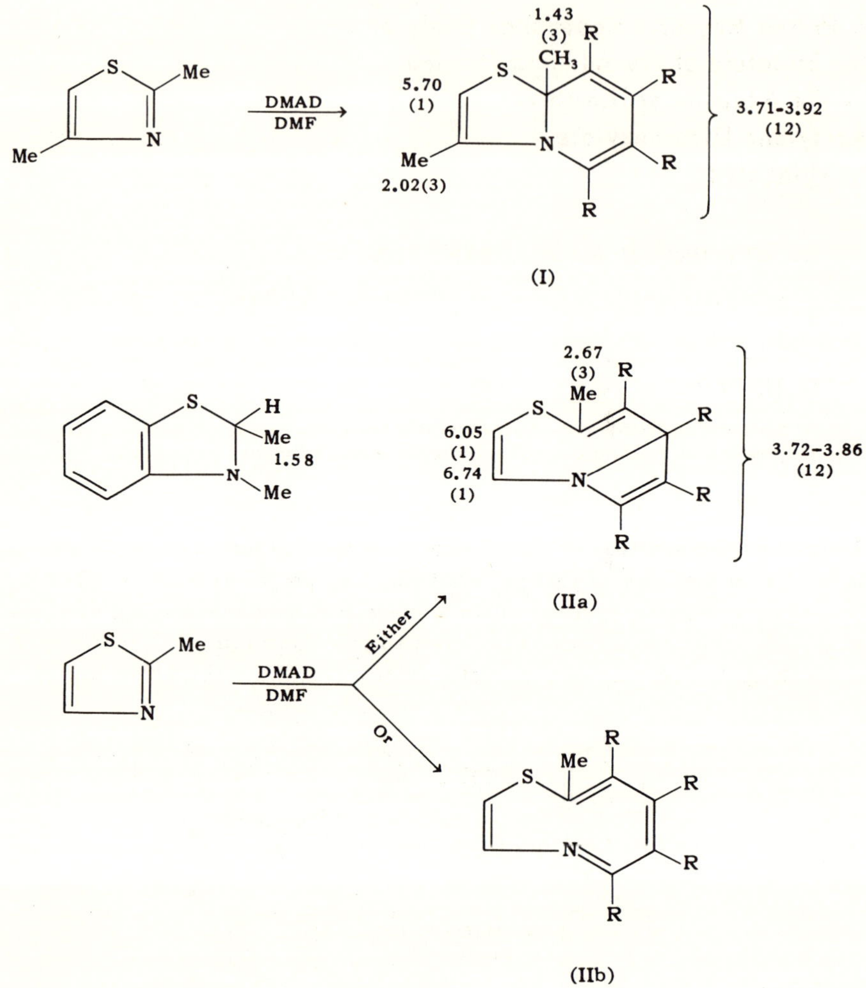

FIG. 2.

duct shows a group of signals corresponding to twelve protons in the range δ 3.71–3.92, a singlet at δ 1.43, a weakly resolved doublet at δ 2.02, and a broad unresolved band at δ 5.70 in the vinylic region. The methyl group at δ 2.02 is weakly coupled to the vinyl proton with a coupling constant of 1.1 cps, and it corresponds to the 4-methyl group of the reactant. The other methyl group must be located on an Sp^3 carbon atom to account for its relatively shielded environment; its chemical shift compares favorably with the chemical shift (1.58) of the methyl group in 2,3-dimethyl 2,3-dihydrobenzothiazole.

In contrast, we have assigned either of the structures (IIa) or (IIb) to the major product from 2-methyl thiazole in DMF solution. The NMR spectrum of the adduct shows a group of signals at δ 3.72–3.86 corresponding to four carbomethoxy groups, a singlet at δ 2.67, and two doublets of an AB system centered at δ 6.05 and 6.74 with a coupling constant of 5 cps. The methyl resonance is incompatible with an angular methyl group as present in an adduct of type (I) and it requires the methyl group to be vinylic. The spectrum then allows only two possible structures for this adduct—the [5,2,0] bicyclic structure (IIa) and nine-membered monocyclic structure (IIb). The monocyclic structure appears less likely on stereochemical grounds but it cannot be excluded with certainty on the available evidence.

We obtained similar adducts (IIIa) or (IIIb) and (IVa) or (IVb) from thiazole and 4-methyl-thiazole, respectively (Fig. 3). Besides the methoxy

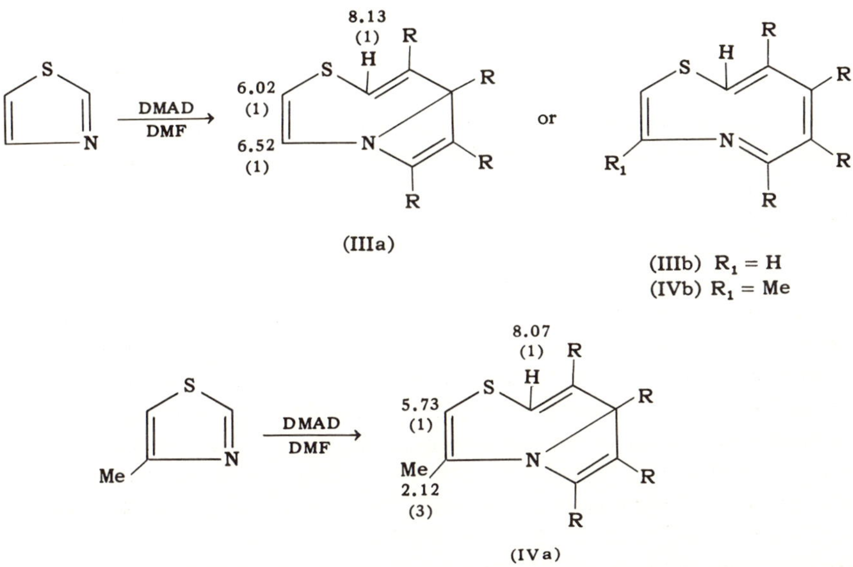

FIG. 3.

signals between δ 3.72 and 3.92, the NMR spectrum of the product from thiazole contains two doublets of an AB system centered at δ 6.02 and δ 6.52 with a coupling of 4.75 cycles, and a singlet at δ 8.13. The low-field position of the singlet requires the proton to be vinylic and therefore to be on what was C-2 of the thiazole. This limits the structure of the adduct to either (IIIa) or (IIIb). On the basis of similar reasoning we assign structure (IVa) or (IVb) to the adduct from 4-methyl thiazole.

Supporting evidence for structures of this type comes from the NMR spectra of some model compounds containing a proton in the same general environment as the vinylic proton in these two adducts.

The general environment is as shown in Fig. 4, and the NMR spectra of open-chain and cyclic model compounds containing this structure all

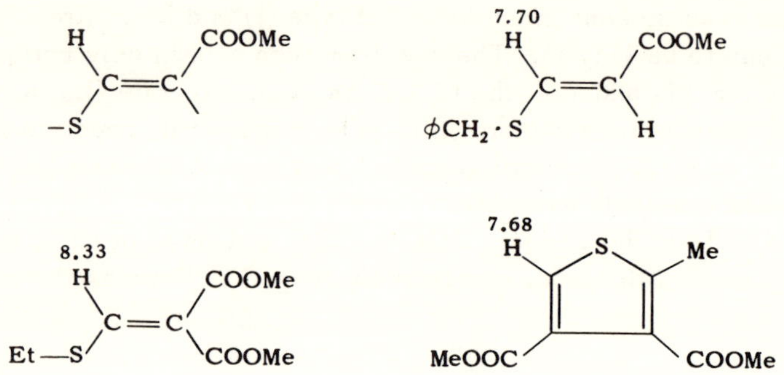

FIG. 4.

showed strong deshielding of the β proton. The chemical shifts of the β protons in methyl *trans*-β-benzyl thioacrylate (δ 7.70), dimethyl ethylthiomethylene malonate (δ 8.33), and dimethyl-2-methylthiophene-3,4-dicarboxylate (δ 7.68) are all of the same order as the chemical shifts of the vinylic protons in the adducts just discussed.

A radically different course was followed when the reactions were carried out in methanol or acetonitrile. 2-Alkyl-substituted thiazoles formed 2 : 1 adducts containing a seven-membered ring by addition involving nitrogen and the 2-alkyl group. The evidence favors structure (V) in Fig. 5 for the adduct from 2-ethylthiazole. The NMR spectrum showed, in addition to the methoxy signals, two doublets of an AB system centered at δ 6.29 and 6.72 with a coupling of 4.5 cps, two doublets of a second AB system at δ 5.36 and 5.66 with a coupling of 5.5 cps, and a sharp singlet at δ 1.82. The methylene group of the 2-ethyl substituent must have reacted because methylene and methine signals are absent and the methyl signal at δ 1.82 is unsplit. These facts strongly suggest that a seven-membered ring has been formed. The adjacent pair of thiazole ring protons in structure (V) accounts for one of the AB systems. The other must arise from an adjacent pair in the newly formed ring. The absence of splitting from the methyl signal indicates that the methyl group is vinylic. The seven-membered ring must accommodate a vinylic methyl group, an AB pair of protons, and two double bonds. Only two structures, (V) and (VI), satisfy these requirements. Evidence obtained from the adduct from 2-methyl thiazole favors structure (V).

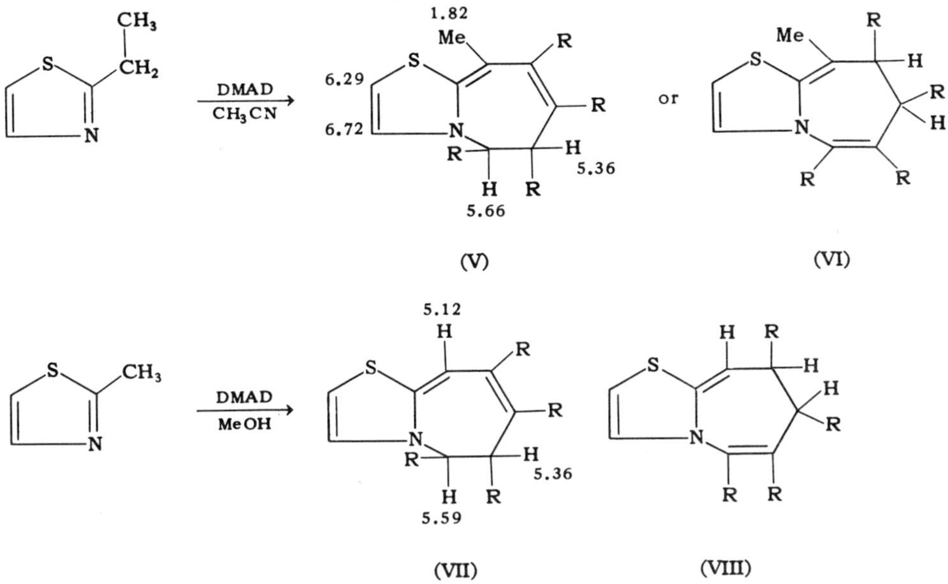

FIG. 5.

2-Methylthiazole in methanol gave an analogous product, whose ultraviolet and visible spectrum was very similar to that of the adduct from 2-ethylthiazole; this indicates that the double-bond distribution must be the same in both adducts. The NMR spectrum of the 2-methylthiazole adduct showed the same pattern of signals as the spectrum of the 2-ethylthiazole adduct with the only difference that the ring methyl signal in the latter is replaced by a singlet integrating for one proton at δ 5.12. The fact that the proton signal is a sharp singlet rules out a structure with the double bond distribution in (VIII) and establishes (VII) as the structure of the adduct. By analogy we favor (V) for the structure of the adduct from 2-ethylthiazole.

Thiazoles unsubstituted at position 2 reacted in methanol to give the desired pyrrolo [2,1-b] thiazole derivatives. Thiazole yielded the tricarboxylic ester in Fig. 6. The NMR spectrum showed a singlet at

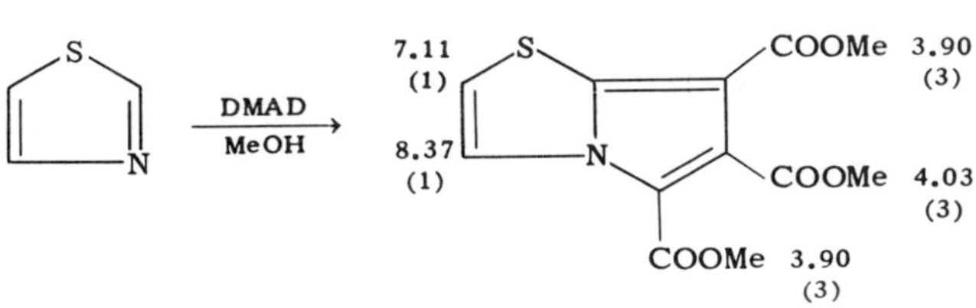

FIG. 6.

δ 3.90 corresponding to two carbomethoxy groups, a singlet at δ 4.03, and an AB pair of doublets at δ 7.11 and 8.37 with a coupling constant of 4.5 cps.

Acetylene dicarboxylic ester also forms adducts with thiazoles saturated at the 4 and 5 positions and we are now investigating the structures of these compounds.

Nitrogen-15 Magnetic Resonance Spectroscopy. III[1]

JOSEPH B. LAMBERT,[2] BRYAN W. ROBERTS,[3] GERHARD BINSCH,[4] and JOHN D. ROBERTS

Gates and Crellin Laboratories of Chemistry, California Institute of Technology, Pasadena, California

The relatively lightly explored field of nitrogen magnetic resonance spectroscopy shows considerable promise as the result of studies employing the isotope of mass number 15. Under conditions of high resolution, possible to achieve with ^{14}N only in highly symmetrical molecular environments because of the quadrupole interaction, relatively small spin couplings may be routinely observed. Saturation effects which arise from unusually long relaxation times may sometimes be alleviated by addition of traces of paramagnetic substances. We have used reasonably high-resolution nitrogen-15 NMR spectroscopy (*1, 2*) to obtain information concerning the molecular structure of nitrogen compounds, the electronic structure and hybridization of nitrogen, the mechanism of spin-spin coupling, the site of protonation of nitrogen compounds, the effect of protonation on chemical shifts, and the rates of proton exchange at nitrogen. Almost all studies have thus far been performed on isotopically enriched samples. However, ^{15}N resonances may be observed in natural abundance (*3*), and computer-averaging techniques are being applied to enhance the signal-to-noise ratios.

Table I lists the ^{15}N chemical shifts of exemplary compounds. As is the case with fluorine (*4*) the paramagnetic term (*5*), which arises from electronic asymmetries at the nucleus, provides the dominant contribution to ^{15}N chemical shifts (*1, 6*). Thus the range of chemical shifts in parts per million is large, and functional groups are readily distinguished (see Fig. 1). Only aliphatic amines resonate at higher field than the spherically symmetrical ammonium ion, because of the shielding effect of the lone pair. As the anisotropy of the screening constant increases in going progressively from amines to amides, nitriles, imines, and azo compounds, the resonances occur at increasingly lower field.

Protonation of amines, for which the paramagnetic shift is near a minimum, removes the shielding effect of the nonbonded electron pair and

[1]Contribution No. 3163. Supported in part by the Public Health Service Research Grant 11072–01 from the Division of General Medical Sciences, the Office of Naval Research, and the National Science Foundation.

[2]Now at Dept. of Chem., Northwestern U., Evanston, Ill.

[3]Now at Dept. of Chem., U. of S. California, Los Angeles, Calif.

[4]Now at Inst. für org. Chemie, U. of Munich, Germany.

TABLE I
NITROGEN-15 CHEMICAL SHIFTS

Compound	Solvent	Chemical shift (ppm downfield)[a]
Ammonia (anhydrous)	None	0
Methylamine	None	2
Ammonium chloride	Water	24
Methylammonium chloride	Water	28
Glycine, pH 6	Water	31
Anilinium chloride	Water	55
Aniline	None	59
Methyl isothiocyanate	None	93
Hydrazobenzene	DMF	96
Benzamide	Acetone	100
Acetamide	Water	113
Acetanilide	Acetone	135
Diphenylketimine hydrochloride	SO_2	168
Acetonitrile	None	245
Benzonitrile	None	259
Potassium cyanide	Water	279
Diphenylketimine	None	308
Benzalmethylamine	None	325
trans-Azoxybenzene	Ether	324
		328
Conjugate acid of *trans*-azobenzene	Acid[b]	360
Nitric acid, 8.57 *M*	Water	367
Nitrobenzene	None	372
trans-Azobenzene	Ether	510
Sodium nitrite	Water	608

[a]The precision is about 1%.
[b]65% Concentrated sulfuric acid, 20% ethanol, and 15% water.

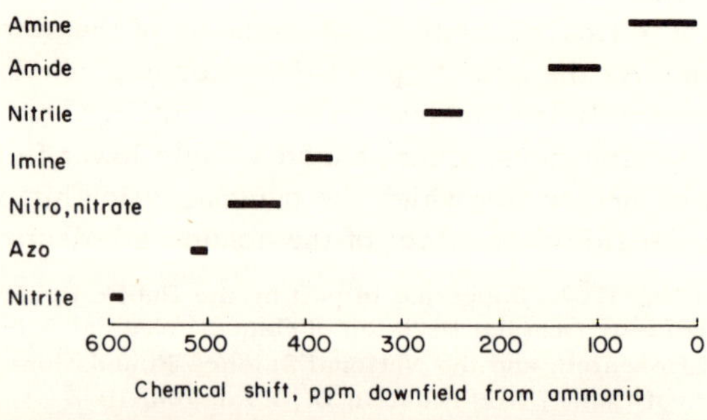

FIG. 1. Nitrogen-15 chemical shift of functional groups.

causes a downfield shift. If the lone pair is involved in partial double bonding, the shielding effect will be reduced or eliminated in the neutral species, and protonation causes only a small chemical shift. Thus, there is little difference between the shifts of aniline and anilinium ion. The situation would be similar for amides if they protonated on nitrogen.

The large paramagnetic shift of multiply bonded systems, such as imines, azo compounds, and azoxy compounds, has been associated with a low mean triplet electronic excitation energy (*1*). Protonation of diphenylketimine or *trans*-azobenzene increases the excitation energy, most likely through a raising of the energy of the $n \longrightarrow \pi_1{}^*$ transition (*7*). As a result, the protonated compounds should have ^{15}N resonances at higher fields, and, in fact, shifts of 140 ppm and of 150 ppm are associated with protonation of diphenylketimine and *trans*-azobenzene, respectively. Thus, as a general rule, the farther downfield the resonance of the neutral species, the smaller the downfield shift or the larger the upfield shift with protonation. This is a corollary of the dominance of the paramagnetic term in determining the chemical shift. A second corollary will be described in the context of coupling constants.

The quadrupole interaction of ^{14}N not only broadens signals and militates against accurate chemical-shift determinations, but it also very largely obscures spin-spin coupling. Only in relatively few cases, even at high temperatures (*8*), can coupling with ^{14}N be observed. A study of coupling could help elucidate the problem of hybridization and that of the mechanism of coupling itself. Couplings between directly bonded atoms, e.g., $J_{^{13}CH}$, have proved to be particularly useful (*9*) in studies of hybridization in bonds to carbon. A rather good linear relationship [Eq. (1)] between percentage of *s* character of the orbitals furnished by carbon in bonding to hydrogen and $J_{^{13}CH}$ has been established empirically (*9*) using

$$s = 0.20\, J_{^{13}CH} \tag{1}$$

conventional hybridization assignments and has since been widely used to determine *s* from measured values of $J_{^{13}CH}$. Examination of $J_{^{15}NH}$ and $J_{^{15}N^{13}C}$ values shown in Tables II and III indicates that somewhat similar relationships appear to exist between *s* and $J_{^{15}NH}$ or $J_{^{15}N^{13}C}$. Thus, ammonium ion and other nearly sp^3 hybridized systems have coupling constants near 75 cps, whereas sp^2 systems such as diphenylketimine hydrochloride, or nearly sp^2 systems such as amides, have couplings near 90 cps. The two points from ammonium ion ($s = 25\%$) and the imine hydrochloride ($s = 33.3\%$) may be taken to determine the relationship

$$s = 0.43\, J_{^{15}NH} - 6. \tag{2}$$

TABLE II

Compound	Solvent	$\lvert J_{^{15}NH} \rvert$ (cps)
$(C_6H_5)_2C{=}NH$	Pentane	51.2 ± 0.4
NH_3 [a]	None	61.2 ± 0.9
NH_4Cl	Water	73.2 ± 0.4
CH_3NH_3Cl	Water	75.6 ± 0.2
$H_2N(CO)NH_2$	DMSO	89 ± 1
CH_3CONH_2	Water	89 ± 2
H_2NCN	DMSO	89.4 ± 1.0
$C_6H_5NHNHC_6H_5$	DMF	90.5 ± 0.4
$C_6H_5NHCOCH_3$	Acetone	90.9 ± 0.2
$(C_6H_5)_2C{=}NH_2Cl$	SO_2	92.6 ± 0.4
Phthalimide	DMSO	93.0 ± 0.8

[a] R. A. Bernheim and H. Batiz-Hernandez, *J. Chem. Phys.* **40**, 3446 (1964).

TABLE III

	Compound	Solvent	$\lvert J_{^{15}N^{13}C} \rvert$ (cps)
1.	$C_6H_5CH{=}^{15}N{-}^{13}CH_3$	None	<3
2.	$^{13}CH_3{-}^{15}NH_2$	None	7 ± 1
3.	$C_6H_5{}^{13}CH{=}^{15}N{-}CH_3$	None	7.1 ± 0.3
4.	$^{13}CH_3{-}^{15}\overset{\oplus}{N}H_3\overset{\ominus}{Cl}$	Water	<8
5.	$C_6H_5{}^{15}NH(^{13}CO)CH_3$	Methanol	13.1 ± 0.5
6.	$^{13}CH_3{}^{15}N{=}C{=}S$	None	13.4 ± 0.2
7.	$CH_3(^{13}CO)^{15}NH_2$	Water	<15
8.	$CH_3{}^{13}C{\equiv}^{15}N$	None	17.5 ± 0.2

If reasonable values are assigned to the carbon and nitrogen hybridizations (Table III), compounds 2 and 4–7 determine the following relationship:

$$s_N s_C = 80\, J_{^{15}N^{13}C}. \tag{3}$$

Equation (2) suggests that s is about 20% for ammonia. This does not agree with the suggestion of Pauling (*10*) that nitrogen in ammonia is nearly p hybridized (which would correspond to a coupling of almost zero) but accords moderately well with the hybridization predicted (*2, 11*) from orthogonality and normalization relations, $s = 22.9\%$, wherein s is taken to be determined entirely from interbond angles.

Relationships, such as Eqs. (1–3), between hybridization and coupling constants require that the Fermi contact term give rise to the predominant portion of the coupling (*9*). This postulate has received good experimental support for the case of $J_{^{13}CH}$ and for most other examples of directly bonded couplings. Certain nitrogen couplings, however, fail to correlate according to Eqs. (2) and (3). Thus the observed couplings for diphenylketimine ($J_{^{15}NH}$) and for benzalmethylamine and acetonitrile ($J_{^{15}N^{13}C}$) predict unreasonably low *s* values, which deviate widely from correlations (2) and (3). The ketimine, which should contain a nearly sp^2 hybridized nitrogen and thus have one of the larger couplings, exhibits the smallest observed coupling. Apparently there is more than one factor contributing to the magnitude of the coupling constant. Pople (*12*) has suggested that contributions from orbital motion become significant in fluorine couplings when the screening constant of one of the nuclei is highly anisotropic. Since the three compounds in question all have large downfield shifts (Table I), associated with anisotropic screening constants, it is possible that orbital as well as contact contributions are important in such cases. Thus, a second corollary states that coupling to nuclei with large paramagnetic shifts is not necessarily determined exclusively by the Fermi contact interaction. This corollary is substantiated by the properties of the protonated form of diphenylketimine. The "normal" coupling constant (92.6 cps) suggests that the anisotropy of the screening constant and hence the magnitude of the paramagnetic shift are small. The ketimine hydrochloride resonates 140 ppm to higher field than the imine itself. Thus the absolute values of the chemical shift, their changes with protonation, and the apparent deviations from the relationships expressed by Eqs. (2) and (3) appear to be interrelated in a predictable manner. The extremely low-field resonance of the azo group (Fig. 1) suggests that $J_{^{15}NH}$ for a system such as phenyldiimide (I) should not be determined by the contact term alone.

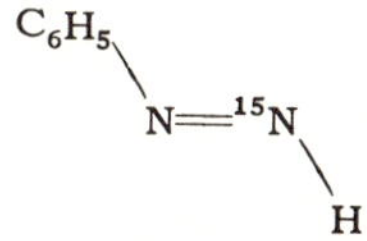

(I)

Nitrogen-15 magnetic resonance spectroscopy is also a potentially powerful method for investigating certain physicochemical properties of nitrogen-containing systems of biological interest. The more clearly defined proton spectra to be expected for ^{15}N-labeled substances, together with the available fund of empirical chemical shifts and coupling constants of ^{15}N in various bonding states, should permit the unambiguous solution of problems such as the detection of tautomeric forms and the location of sites of protonation. A case in point is the nucleotide analog 1-methylcytosine (II) and its hydrochloride (III). In a recent PMR study

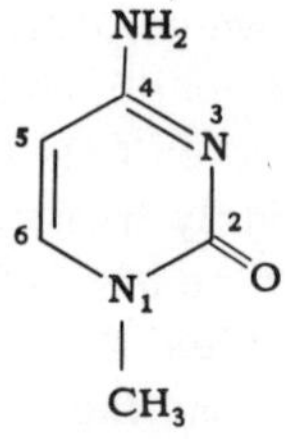

(II)

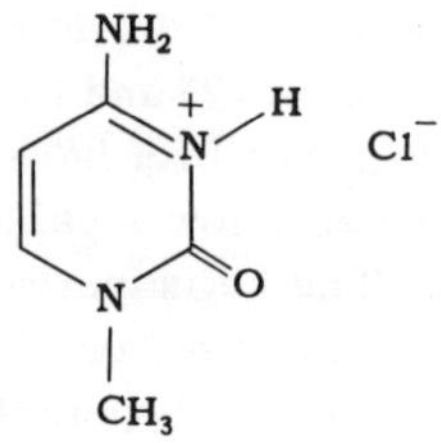

(III)

of material labeled with ^{15}N in the amino group, Miles *et al.* (*13*) demonstrated that the neutral base exists in dimethyl sulfoxide solution predominantly in the amino form and that it protonates at N-3. We are now extending the scope of this method in a study of 1-methylcytosine fully labeled with ^{15}N. The chemical shifts and coupling constants extracted from the ^{15}N spectrum should complement data from the ^{1}H spectrum and allow an unambiguous assignment of structure. Thus, appearance of a characteristic ^{15}N-H doublet in the ^{15}N and ^{1}H spectra of the nonexchanging hydrochloride would confirm N-3 as the site of protonation.

The use of ^{15}N NMR in the study of biological systems on a practical basis depends upon the development of routine methods for detecting the isotope in low isotopic concentration and in natural abundance. As one means toward this end, we have found the computer-averaging repetitive-scan technique useful in detecting ^{15}N-H satellites in the ^{1}H spectra of a number of N-H compounds (*2*). This approach is suited for detecting nonexchanging N-H groups, since the satellite peaks are usually well separated from the central ^{14}N-H peak (compare the large $J_{^{15}NH}$ values in Table II).

We have recently observed a large ^{15}N-H coupling over two bonds in the heteroaromatic system, 2,4-dichloropyrimidine-$^{15}N_2$ (IV). First-order

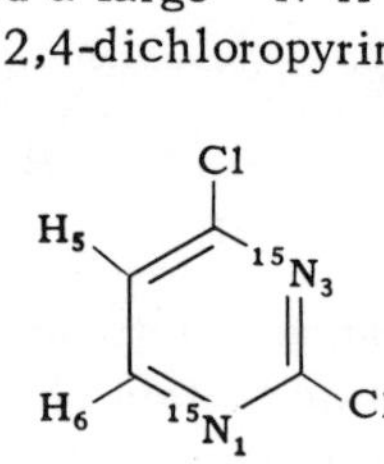

(IV)

analysis of the proton spectrum indicates apparent ^{15}N-H couplings of 12.5 and 0.9 cps for H-6 and 1.4 and 0.8 cps for H-5. The couplings of H-6 in particular are larger than in similarly constituted systems. The complexities of the ^{15}N spectra produced by the proton couplings are in this case so great as to require proton spin decoupling before it will be possible to obtain $J_{^{15}N^{15}N}$.

References

1. J. B. Lambert, G. Binsch, and J. D. Roberts, *Proc. Nat. Acad. Sci. US* **51**, 735 (1964).
2. G. Binsch, J. B. Lambert, B. W. Roberts, and J. D. Roberts, *J. Am. Chem. Soc.* **86**, 5564 (1964).
3. J. D. Ray, *J. Chem. Phys.* **40**, 3440 (1964).
4. A. Saika and C. P. Slichter, *J. Chem. Phys.* **22**, 26 (1954).
5. J. A. Pople, W. G. Schneider, and H. J. Bernstein, "High-resolution Nuclear Magnetic Resonance," pp. 165 f. McGraw-Hill, New York, 1959.
6. B. E. Holder and M. P. Klein, *J. Chem. Phys.* **23**, 1956 (1955); B. M. Schmidt, L. C. Brown, and D. Williams, *J. Mol. Spectr.* **2**, 539, 551 (1958); *ibid.* **3**, 30 (1959).
7. H. H. Jaffé, S.-J. Yeh, and R. W. Gardner, *J. Mol. Spectr.* **2**, 120 (1958).
8. J. D. Roberts, *J. Am. Chem. Soc.* **78**, 4495 (1956).
9. N. Muller and D. E. Pritchard, *J. Chem. Phys.* **31**, 768 (1959). J. N. Shoolery, *ibid.* **31**, 1427 (1959). C. Juan and H. S. Gutowsky, *ibid.* **37**, 2198 (1962).
10. L. Pauling, "The Nature of the Chemical Bond," pp. 120f. Cornell Univ. Press, Ithaca, New York, 1960.
11. C. A. Coulson, "Valence," pp. 165 f., 203 f. Oxford Univ. Press, London, 1961.
12. J. A. Pople, *Mol. Phys.* **1**, 216 (1958).
13. H. T. Miles, R. B. Bradley, and E. D. Becker, *Science* **142**, 1569 (1963).

NMR Studies of Some Formamides Containing ^{15}N

A. J. R. BOURN,[1] D. G. GILLIES,[2] and E. W. RANDALL

Queen Mary College, London, England

Introduction

The positions in the amides are numbered as before (*1*):

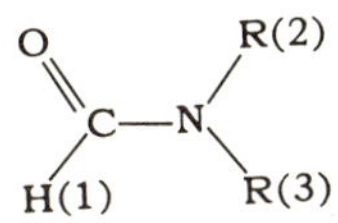

The signs of coupling constants in N-substituted formamides have been studied by a variety of means. Two proton-proton signs in *N*-methyl formamide [R(2) ≡ Me; R(3) ≡ H] were determined from spectral analysis of single (*2*) and H{^{14}N} double (*1*) resonance spectra. Both proton-proton and proton-nitrogen-15 signs have been obtained from H{H} double resonance experiments on formamide ^{15}N and *N*-methyl formamide-^{15}N (*3*). Similar experiments for *N,N*-dimethyl formamide-^{15}N (*4*), and proton single resonance work employing dilution shifts for the cases of *trans*-formanilide [R(2) ≡ H; R(3) ≡ C_6H_5] and *cis*-formanilide [R(2) ≡ C_6H_5; R(3) ≡ H] gave the proton-nitrogen-15 signs in these cases (*5*).

The above results are summarized in Fig. 1. Couplings related by H{H} decoupling are enclosed by single lines. Couplings enclosed by broken lines were related by inspection of the single resonance spectrum.

One ambiguity remaining after all these studies was the sign relation between the set of proton-proton couplings on the one hand, and the proton-nitrogen-15 couplings on the other. This ambiguity is now removed by a heteronuclear, H{^{15}N}, decoupling experiment on *N,N*-dimethyl formamide-^{15}N.

Experimental

The basic proton magnetic resonance spectrometer operated at 40 Mc/s and utilized a Mullard permanent magnet at approximately 9400 gauss which incorporates an improved thermostatic control. This modification

[1]Now postdoctoral Fellow at the Department of Chemistry, University of California, Los Angeles, California.

[2]*Present address*: Division of Pure Chemistry, National Research Council, Ottawa, Canada.

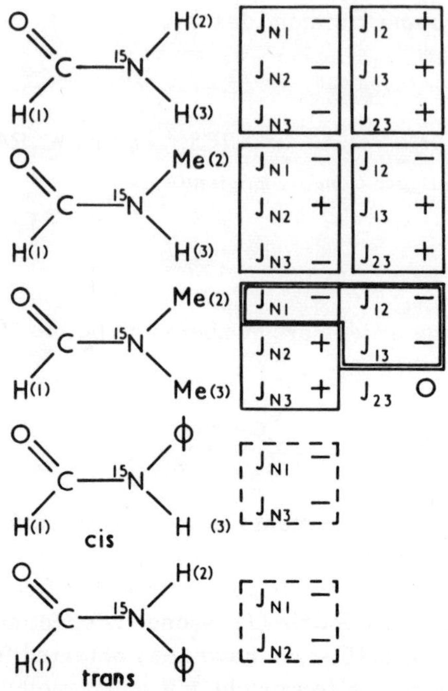

FIG. 1. The absolute signs of coupling constants in N-substituted formamides containing ^{15}N.

was introduced to minimize magnetic field drift and change in homogeneity with time.

The operating frequency for protons was derived by doubling three times the stable 5 Mc/s output from a crystal-controlled oscillator (Standard Telephones and Cables, Type 90 LQE 811). The probe was of the single-coil type, and was part of a twin T bridge. The bridge was followed by a tuned preamplifier with a low-noise, cascade-input stage (*6*) and then by a tuned amplifier.

Either straight diode or phase-sensitive detection was available. The latter was preferable on grounds of baseline stability. Audio field modulation was employed. The phase-sensitive detector and associated amplifier were supplied by Brookdeal Electronics Ltd., London.

The field was homogenized by conventional, printed-circuit, Golay coils kindly donated by Perkin Elmer Ltd. Linewidths of 0.3 c/s have been obtained.

The sample tubes were of the conventional Varian A-60 type and were spun by means of an A-60 spinner which was driven by preheated air to minimize any temperature effect on the permanent magnet.

The spectrometer used a field sweep from a modified circuit originally due to Keuning (*7*).

The requirement of a stable but variable known radio frequency for precise irradiation in the nitrogen-15 region was met by single sideband, suppressed carrier, modulation (*8*) of a stable radio frequency derived from a crystal-controlled oscillator at 4.0538 Mc/s (Cathodeon Type SFG1). The modulation was performed by means of a Heathkit single sideband adaptor, SB-10U.

Normally, it was sufficient to monitor the audio frequency with a counter (Venner 1 Mc/s Type TSA 3336). The actual applied radio frequency could, however, be checked by using a frequency divider (Venner TSA 851) in conjunction with the 1 Mc/s counter. An attenuator placed between the single sideband output and the probe enabled the power to be set to the desired level.

Irradiation of the sample at this second frequency was achieved by means of a second coil mounted orthogonally to the 40-Mc/s detecting coil. The coil was tuned, using an externally mounted variable capacitor.

Results

The nitrogen-15 region in *N,N*-dimethyl formamide-^{15}N is basically a doublet arising from a coupling, $J_{N1} = 15.6$ c/s, to the CH proton. Each methyl feature has four peaks (*4*) which will be numbered here 1, 2, 3, 4 with increasing field. Irradiation in the nitrogen-15 region of the low-field feature (high ^{15}N frequency) produced collapse of high-field features of each methyl group (lines 2 and 4). Similarly, irradiation of the high-field feature (low ^{15}N frequency) produced collapse of low-field features of each methyl group (lines 1 and 3).

The most widely used definition of signs of coupling constants, due to Karplus (*9*), will be used: a positive coupling constant signifies that an antiparallel nuclear spin orientation is a more stable state than the parallel orientation. This convention is adopted irrespective of the sign of γ, the magnetogyric ratio.

The decoupling experiment shows that, since γ is negative for nitrogen-15, J_{N1} has the same sign as both J_{12} (H-Me) and J_{13} (H-Me). This result is indicated in Fig. 1 by the couplings enclosed by the double lines.

Discussion

It will be assumed that analogous couplings in different molecules retain the same absolute sign.

All signs can now be related to the directly bonded coupling $J_{^{15}NH}$.

Directly bonded $J_{^{13}CH}$ couplings were predicted to be absolutely positive by Karplus and Grant (*10*). This has been confirmed experimentally by Buckingham and McLauchlan (*11*). The calculation of the signs and magnitudes of directly bonded couplings has now been improved by Pople and Santry (*12*). They predict that the reduced coupling $(2\pi/R\gamma_A\gamma_B)\,J_{AB}$ (when J_{AB} is the usual coupling constant) will be positive unless one of the atoms is flourine. Hence, directly bonded $J_{^{14}NH}$ should be positive (positive γ) and $J_{^{15}NH}$ negative (negative γ).

Using this result, then, Fig. 1 shows the absolute signs of all the coupling constants. The vicinal coupling constant, J_{13}, in formamides is confirmed to have an absolute positive sign as does the analogous coupling in ethylene.

Recently Manatt (*13*) and co-workers have determined the relative signs in ethyl isonitrile, $CH_3CH_2{}^{14}N \equiv C$. This compound is remarkable in that the proton lines are not broadened by quadrupole effects of nitrogen-14. They find that the coupling $^{14}N—C—H$ is opposite in sign (negative) to the $^{14}N—C—C—H$ and H—C—C—H couplings, the latter being assumed to have an absolute positive sign.

We suggest that this $^{14}N—C—H$ coupling is analogous to $J_{N2(N\text{-}Me)}$ and $J_{N3(N\text{-}Me)}$ in methyl substituted nitrogen-15 amides. The couplings in each type of molecule are between nitrogen and a freely rotating sp^3 hybridized alkyl group. The couplings are of the same order of magnitude. The sign of $^{15}N—C—H$ is found to be positive and hence in accord with the negative sign for $^{14}N—C—H$ when the negative γ of nitrogen-15 is allowed for.

In methyl acetylene, $HC \equiv C—CH_3$, the coupling $^{13}C—C—H$ is found to be negative relative to $^{13}C—H$ (*14*). In ethane (*15*) and ethyl iodide (*16*) the same result is found. The situation is summarized in Table I.

TABLE I

ABSOLUTE SIGNS OF MH COUPLING CONSTANTS FOR M—H, M—CH_3, AND M—CH_2—CH_3 GROUPS

M	M—H	M—C—H	M—C—C—H
^{14}N	+	−	+
^{15}N	−	+	
^{13}C	+	−	

Nitrogen appears similar to carbon with regard to the alternation of sign with the number of bonds involved.

Further H{^{15}N} irradiation experiments are being carried out on the remaining amides with a view to testing the above results.

Acknowledgments

We are grateful for research studentships to A. J. R. Bourn and D. G. Gillies from D. S. I. R., and to the University of London and I. C. I. Ltd., for equipment grants.

References

1. E. W. Randall and J. D. Baldeschwieler, *J. Mol. Spectr.* **8**, 365 (1962).
2. V. J. Kowalewski and D. G. de Kowalewski, *J. Chem. Phys.* **32**, 1272 (1960).
3. A. J. R. Bourn and E. W. Randall, *Mol. Phys.* **8**, 567 (1964).
4. A. J. R. Bourn and E. W. Randall, *J. Mol. Spectr.* **13**, 29 (1964).
5. A. J. R. Bourn, D. G. Gillies, and E. W. Randall, *Tetrahedron* **20**, 1811 (1964).
6. H. Wallman, A. B. MacNee, and C. P. Gadsden, *Proc. Inst. Radio Eng.* **36**, 700 (1948).
7. R. Keuning, *Rev. Sci. Instr.* **31**, 1839 (1960).
8. D. W. Turner, *J. Chem. Soc.*, p. 847 (1962).
9. M. Karplus, *J. Chem. Phys.* **84**, 2458 (1962).
10. M. Karplus and D. Grant, *Proc. Nat. Acad. Sci. US* **45**, 1269 (1959).
11. A. D. Buckingham and M. McLauchlan, *Proc. Chem. Soc.*, p. 144 (1963).
12. J. A. Pople and D. P. Santry, *Mol. Phys.* **8**, 1 (1964).
13. S. Manatt, personal communication.
14. H. Dreeskamp, E. Sackmann, and G. Stegmeier, *Ber. Bursenges.* **67**, 860 (1963).
15. R. Lynden-Bell and N. Sheppard, *Proc. Roy. Soc. (London)* **A269**, 385 (1962).
16. H. Dreeskamp and E. Sackmann, *Z. Physik. Chem. (Frankfurt)* **34**, 261 (1962).

Interpretation of NMR Spectra of Complicated Mixtures in Terms of Large Molecular Species

J. R. VAN WAZER and L. C. D. GROENWEGHE

Monsanto Company, Central Research Department, St. Louis, Missouri

During the 10-year period that our laboratory has been engaged in high-resolution NMR, most of the effort has been devoted to the use of ^{31}P spectra in solving problems of phosphorus chemistry (*1–5*). However, during the past few years, we have become active in the use of high-resolution NMR for studying chemical systems which reach equilibrium with respect to exchange of parts between the constituent molecules (*6, 7*). The theory (*8*) of such scrambling equilibria has been treated and in a number of recent publications we have given examples of equilibrated families of compounds (*9–16*) as well as simple systems (*17–21*) undergoing scrambling reactions leading to equilibrium.

The purpose of this paper is to show the type of NMR spectra obtained in these studies and to review briefly the methods whereby such spectra are interpreted. For quantitative data and detailed interpretation, the reader is referred to the specific publication concerned with each individual system.

The Simple System CH_3GeCl_3 *versus* CH_3SiBr_3 (22)

When mixtures of trichloromethylgermane with triboromethylsilane are allowed to stand for several days or more at room temperature or above, scrambling of chlorine and bromine atoms is expected, with the methyl groups remaining firmly affixed to their respective germanium or silicon atoms since their removal would require high temperatures, > ca. 300°, for long periods of time. Eight compounds should result from scrambling of chlorine and bromine atoms between the methyl germanium and methyl silicon moieties. These are CH_3GeBr_3, CH_3GeBr_2Cl, $CH_3GeBrCl_2$, and CH_3GeCl_3, as well as CH_3SiBr_3, CH_3SiBr_2Cl, $CH_3SiBrCl_2$, and CH_3SiCl_3.

Spectra of such equilibrated mixtures were taken at 60.0 Mc with a Varian A-60 spectrometer, thus giving the proton NMR resonances of all the methyl groups. As shown in Fig. 1, up to eight different resonance peaks could be seen (in addition to some satellites), depending on the over-all composition of the mixture. This suggests that each of these peaks corresponds to one out of the eight possible compounds. The sim-

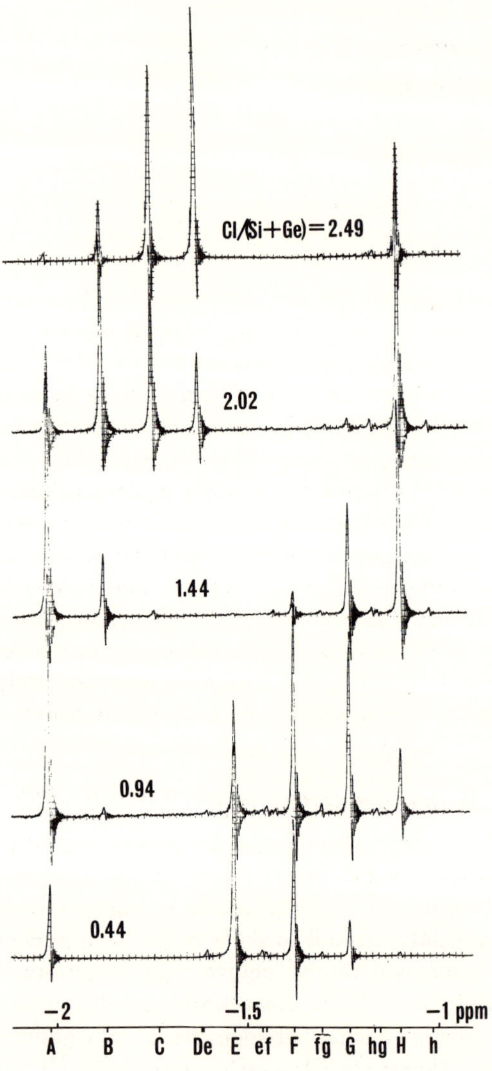

FIG. 1. ^{1}H NMR spectra of randomly reorganized mixtures of CH_3GeCl_3 and CH_3SiBr_3 at various Cl/(Si + Ge) ratios. Sweep rate, 0.2 cps/sec.

ilarity of the spectra indicates that a peak with the same gross chemical shift in different spectra represents the same compound, although there are minor variations in these shifts and even the distances between the peaks in each spectrum may vary slightly.

The problem of assigning these various resonances to their respective chemical structures may be solved either by preparing all eight compounds in reasonably pure form and measuring their shifts or, more easily, from the following rationale: First, since we started with CH_3GeCl_3 and CH_3SiBr_3, the shifts of these two compounds are known and correspond to resonances D and E, respectively, of Fig. 1. Since successive substitution of bromine for chlorine will probably have the same effect on the chemical shifts of the protons of $CH_3Ge\lessgtr$ or $CH_3Si\lessgtr$ and since peaks D and E lie toward the middle of the spectrum, it seems reasonable to ascribe peaks A through D to the methyl germanium compounds ranging from the tribromide to the trichloride, and groups E through G to the methyl silicon compounds also ranging from the tribromide to the trichloride. Furthermore, the presence of a pair of satellites (e through h on Fig. 1) encompassing the more intense peaks in the series E through H gives strong indication that these are silicon compounds—an indication which is borne out by the fact that the areas of the satellites relative to those of their main peaks, as well as the satellite–main peak spacings in cps., are equal for both satellites and correspond to the expected spin-spin splitting of the hydrogen of methyl groups bonded to the ^{29}Si atoms.

An independent check of these assignments is obtained from the material balance. Thus, on the assignments established in the preceding paragraph for the eight observed resonances, we find that the over-all ratio Cl/(Ge + Si) varies from the top to the bottom spectrum of Fig. 1 in the sequence 2.49 (2.44), 2.02 (1.98), 1.44 (1.41), 0.94 (0.90), and 0.44 (0.44), where the first numbers listed result from the stoichiometry of mixing (i.e., from the reagents) and the following numbers in parentheses are calculated from the relative areas of the NMR peaks. Good agreement between the Ge/(Ge + Si) ratios from the reagents and from the NMR analysis are also obtained. This strongly supports the peak assignments and the differences can be used to estimate statistically the reproducibility of the quantitative NMR results (*28*). We have found that the application of material-balance requirements to the interpretation of a stoichiometrically related series of NMR patterns is a powerful tool for making or verifying NMR assignments.

The reader is referred to the specific paper (*22*) on this system for interpretation of the curves of Fig. 1 in terms of equilibrium constants and other chemical parameters.

The Family of Polyphosphoryl Dimethylamides (5)

As evidenced by the spectra of Fig. 1 and, indeed, the entire body of NMR chemical-shift data, successive substitution of one substituent by another on a polyfunctional moiety causes a usually more-or-less regular stepwise change in the chemical shift of a magnetically active nucleus embodied in this moiety. Work published several years ago (*6*, *23*) on ^{31}P spectra measured at 16.2 Mc demonstrated this effect in complicated molecules in which the magnetically active phosphorus nuclei appear in the molecular backbone. A good example is found in the reaction products (*5*) obtained by heating various mixtures of hexamethylphosphorylamide, $OP[N(CH_3)_2]_3$, with phosphorus pentoxide, $[P_2O_5]$, so as to achieve equilibrium with respect to the scrambling of bridging oxygens (—O—) and monofunctional dimethylamine $[(CH_3)_2N—]$ groups around the phosphoryl $\left(\mathrm{OP}\leqslant\right)$moiety.

Four resonance peaks were observed in the spectra of these reaction products, only one of which, that of the hexamethylphosphorylamide at −23 ppm (referenced to 85% H_3PO_4, using the usual convention of assigning the negative sign to downfield shifts) was directly assignable. The other starting material, phosphorus pentoxide, is difficult to handle in liquid form and hence a directly measured chemical shift has not even been reported for it yet. Primarily on the basis of material balance, the other three resonances at −11.5, +12.0, and +40 ppm were assigned to end, middle, and branch structure-building units, respectively. These various structure-building units in the phosphoryl dimethylamides, including the *neso*[1] molecule—hexamethylphosphorylamide—are shown here. The $O_{1/2}$ represents half of an oxygen—an oxygen which is shared

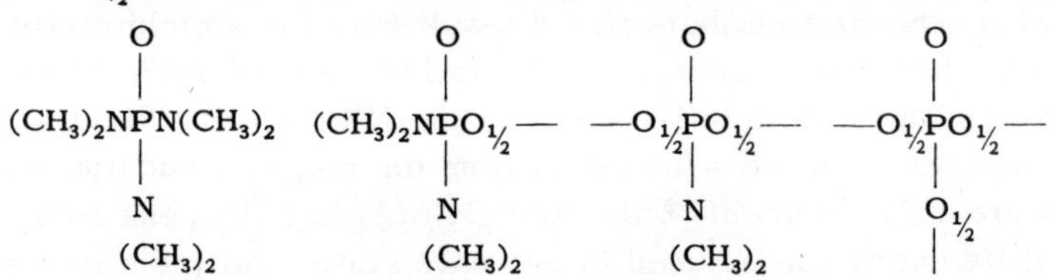

(*Neso*, n) (End, e) (Middle, m) (Branch, b)

between two neighboring phosphorus atoms. A typical straight chain molecule in this system, composed of two end groups and three middle groups, has the following structure.

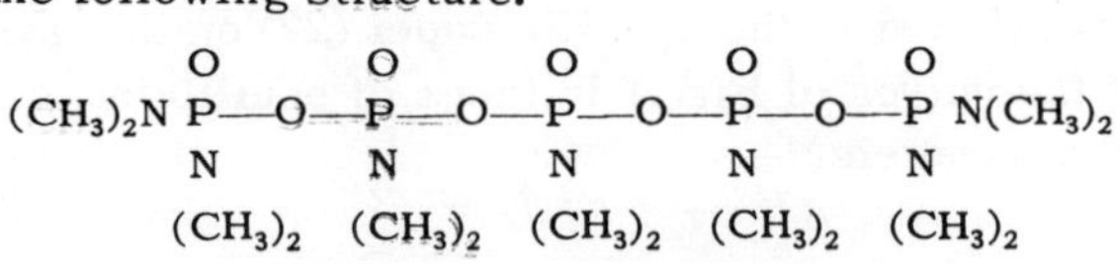

Abbreviation: emmme

[1]The *neso* molecule is the smallest member of a family of compounds, being a central atom or moiety surrounded by monofunctional substituents. Methane is the *neso* compound of the aliphatic hydrocarbons.

The Family of Chloro-Terminated Polydimethyl-*N*-Methyl-silazanes (*14*)

When dichlorodimethylsilane, $(CH_3)_2SiCl_2$, is mixed in various proportions with any of the cyclic dimethyl-*N*-methylsilazanes, $[(CH_3)_2SiNCH_3]_x$, there is scrambling on the dimethylsilicon moiety, $(CH_3)_2Si{<}$, of the chlorine atoms with the methylimino bridges, $CH_3N{<}$, since Si–Si, N–N, and N–Cl bonds do not form under these conditions and are probably thermodynamically unstable. In the proton NMR spectrum measured at 60.0 Mc with a Varian A-60 spectrometer, two distinct regions of peaks are observed: one in the range of –2.3 to –2.7 ppm and the other in the range of 0 to –0.9 ppm, as referenced to tetramethylsilane. Comparison with spectra of a number of known compounds as well as with the spectra of the two starting materials shows that the downfield groups of resonances correspond to the hydrogen of the $CH_3N{<}$ moiety, whereas, the upfield group of peaks is attributable to the $(CH_3)_2Si{<}$ moiety. A series of spectra corresponding to equilibrated systems obtained by combining various proportions of $(CH_3)_2SiCl_2$ with $[(CH_3)_2SiNCH_3]_x$ are presented in Fig. 2.

The scrambling reaction can be written in terms of structure-building units, as follows:

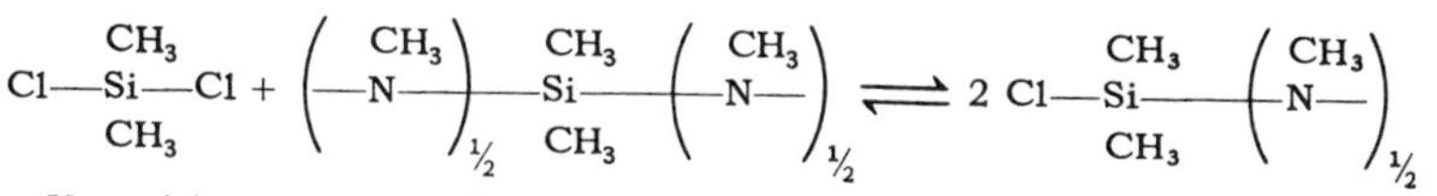

Neso (n) Middle (m) End (e)

Obviously when a small amount of dimethyl-*N*-methylsilazane is added to dichlorodimethylsilane, the product will be all or nearly all the end-end molecule:

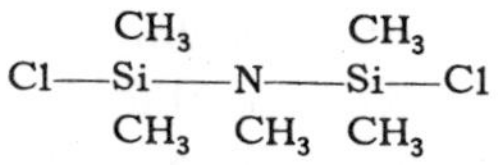

which from the vantage point of the methyl group on the methylimino bridge may be abbreviated as e—e but from the vantage point of the methyl groups on a silicon moiety is best depicted as **ee**. These e—e and **ee** NMR peaks exhibited relative areas of 1 : 4, respectively, as expected, since there are one CH_3 group on the nitrogen and four CH_3 groups corresponding to the two silicon atoms of this molecule.

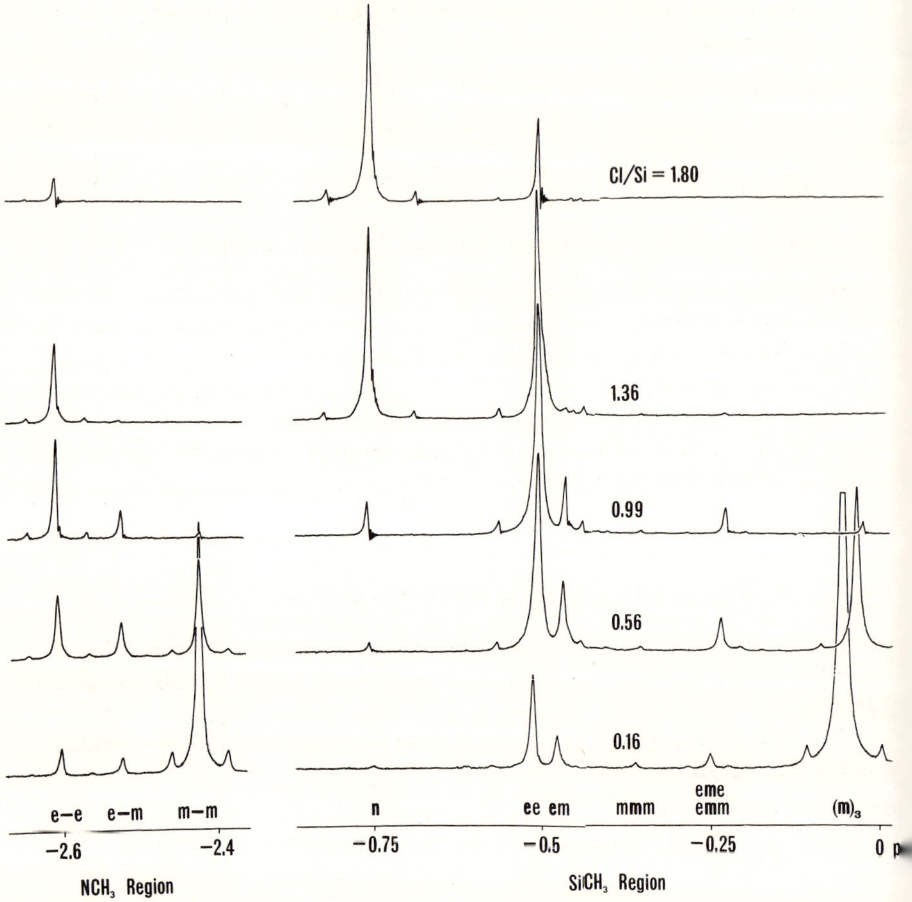

FIG. 2. ^{1}H NMR spectra of randomly reorganized mixtures of $(CH_3)_2SiCl_2$ and $[(CH_3)_2SiNCH_3]_x$ at various Cl/Si ratios. Sweep rate, 0.1 cps/sec.

Under the conditions of resolution employed in this study, it was found that the chemical shift of the hydrogens of a dimethylsilicon moiety was not only affected by the substitution of a methylimino bridge by a chlorine atom on that particular silicon but that such substitution on a neighboring silicon atom bonded to the given dimethylsilicon moiety through a methylimino bridge also has a secondary effect on the chemical shift. This means that in the region where end group resonances occur (−0.48 to −0.51 ppm), two kinds of end groups were seen: An end bonded to another end (−0.51 ppm), maximizing at the over-all composition of $R \equiv [Cl]/[Si] \cong 1$, and an end bonded to a middle (−0.48 ppm), maximizing at $R \cong 0.5$ where the abundance of middle groups is more plentiful. Likewise, to the same degree of structure resolution, one should see at least three resonances for the hydrogens of the dimethylsilicon moieties bearing a pair of methylimino bridges in the middle-group region (– 0.06 to – 0.37 ppm). The corresponding three atomic arrangements are e**m**e, e**m**m, and m**m**m, with the *a priori* possibility of several m**m**m resonances corresponding to middles in chains and different-size rings.

Similarly one should also be able to "see in depth" into the molecule from the hydrogens of each methylimino bridge. If the chemical shift of these hydrogen atoms is sensitive to the substitution of a methylimino bridge by a chlorine atom on each of the silicon atoms bridged by this group, three kinds of resonances should be seen. These are e—e, e—m, and m—m, where the bold-faced bond refers to the methylimino bridge.

In the spectra shown in Fig. 2, there are a number of quantitative relationships between resonance areas in addition to the 1 : 4 ratio of the e—e peak to the **e**e peak. For example, there must always be a 1 : 2 relationship between the areas of the e—m and **e**m peaks. A relationship which often is applicable (but not useful in this case because of peak overlap) is [**e**m] = 2[e**m**e] + [e**m**m]. In addition, from quantitative measurements of the areas of the $CH_3N<$ resonances and the stoichiometry of the starting ingredients, one can calculate the amount of *neso* compound by difference since the *neso* compound has no $CH_3N<$ bridge. This calculated amount of *neso* should and does agree with the directly measured amount of $(CH_3)_2SiCl_2$ obtained from the $(CH_3)_2Si<$ region of the spectrum. Furthermore, material-balance calculations must again agree with the peak assignments, as must the relative areas and coupling constants of the satellites due to spin-spin splitting of the hydrogens by ^{29}Si and ^{15}N. These various requirements generally lead to unambiguous structural assignments.

In addition, there is another set of requirements applicable to equilibrated systems which may be used to further verify the correctness of an assignment. These generally appertain to the form of the mathematical function describing the variation of the amount of a given atomic arrangement with over-all composition. Thus, under equilibrium conditions, the relative areas of the total resonances corresponding to a given structure-building unit must maximize at the over-all composition corresponding to this unit. One can mathematically derive a number of relationships between the rate of change with over-all composition of the area of a given resonance peak to the set of equilibrium constants controlling the system. Detailed information on these relationships will be published shortly (*24*).

A further check on the validity of the peak assignments is obtained from kinetic data. When one is observing reactions by NMR in well-filled sealed tubes where no phase separation occurs, the over-all composition must remain fixed so that the material balance retains a given value from start to finish. In following the kinetics of many scrambling reactions, it is found that certain peaks appear, maximize, and then disappear at a given over-all composition, whilst at another over-all composition, one of these peaks may be seen under equilibrium conditions. Obviously, the correct structural assignment will lead to acceptable material balances during and at the end of a set of rate measurements.

Although we do not claim that all of our published assignments are unequivocal, we firmly believe that this general technique of studying the arrangement of atoms in mixtures of molecules (as developed in our laboratories and reviewed here) can be used with great reliability. NMR as usually employed on a single pure sample has been dubbed "a sporting method of structural determination." By varying widely the over-all composition in studying systems of molecules and taking advantage of the various stoichiometric and, where pertinent, thermodynamic relationships, much of the "sporting" aspect is removed.

The Family of Polyarsenous Oxyfluorides (*10*)

When various proportions of arsenous oxide, As_2O_3, were dissolved in arsenic trifluoride, AsF_3, the spectra shown in Fig. 3 were obtained on a Varian HR-60 spectrometer run at 56.4 Mc. By interchanging fluorine atoms with bridging oxygens on arsenic, the following four structure-building units result:

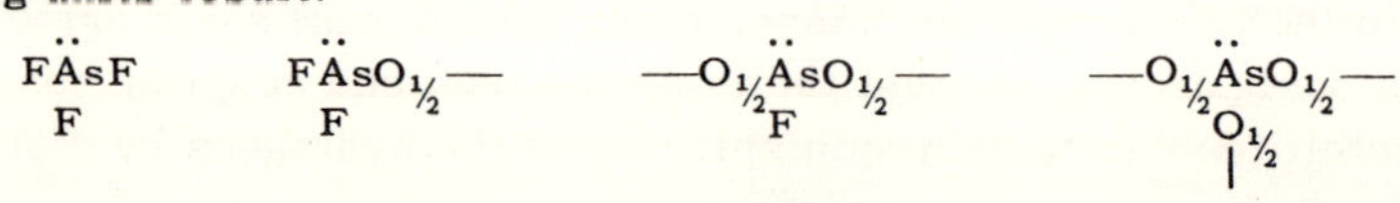

Neso End Middle Branch

The first three of these units show up in ^{19}F spectra since they contain fluorine, but the amounts of branch groups must be calculated by difference. Since there appear to be about nine peaks in the NMR spectrum, we must assume that the resolution is such that we can distinguish the effect of various arrangements of oxygen and fluorine atoms on an arsenic one removed from the arsenic to which the resonating fluorine is attached. Under such circumstances, we would, of course, expect only one *neso* peak, three end-group peaks (**e**†e, **e**†m, and **e**†b), and six middle-group peaks (**em**†e, **em**†m, **em**†b, **mm**†m, **mm**†b, and **bm**†b). The dagger, †, employed in describing these atomic arrangements means that the NMR data are obtained on an *exchangeable* monofunctional substituent (*8*).

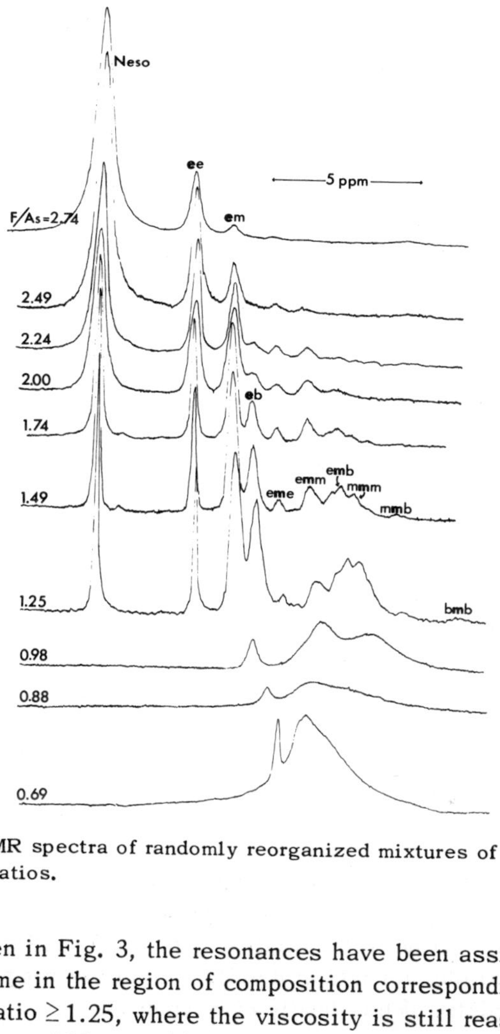

FIG. 3. ^{19}F NMR spectra of randomly reorganized mixtures of AsF_3 and As_2O_3 at various F/As ratios.

As can be seen in Fig. 3, the resonances have been assigned according to this scheme in the region of composition corresponding to an overall F/As mole ratio ≥ 1.25, where the viscosity is still reasonably small. This assignment could be established by following a reasoning similar to that outlined above for the chloro-terminated polydimethyl-*N*-methylsilazanes. Here, however, although the *neso* compound stands out by itself, the regions where end and middle groups resonate are close together. Taking into account the sharpness of the peaks and the separation gaps between peaks, in addition to the criteria of maximization at the expected F/As ratios and relative chemical shift, a reasonable assignment was possible. This assignment was verified by material balances and the equations giving quantitative relations between certain e and m resonance peaks, as explained earlier.

For values of the F/As mole ratio below 1.25, the viscosity becomes large, owing to the predominance of branch groups which bind the structure-building units into infinite (wall-to-wall) molecules.[2] This leads to very high viscosities, with the rate of flow apparently being controlled by making and breaking of As–O–As linkages. As expected under these conditions, the NMR spectra are very broad (the three spectra at the bottom of Fig. 3) so that we have not tried to interpret them. (It should also be noted that stochastic theory (*8*) has not been extended in an adequate manner beyond the gel point.)

There is an interesting parallel between the type of NMR spectroscopy described here and the early observations of optical spectroscopists. Because it is a single molecule (and also usually a molecule without different configurational possibilities which might cause line broadening), the resonance attributed to the *neso* compound is *sharp*. On the other hand, since most NMR spectroscopy of the type described in this paper is confined to the composition region lying between the pure *neso* compound and the gel point at which three-dimensional infinite network polymers are formed, the end-group peaks might be called the *principal* ones in the spectrum. Moreover, the resonances ascribed to middle groups and branch groups are generally quite *diffuse* because of the large number of possibilities corresponding to the arrangement of atoms in the molecules at a distance where structural variations do not result in clearly separated peaks. The diffuseness of the latter resonances may also be enhanced by possibilities of spatial configurational changes (due to restricted rotation about bonds). Thus, we see that on a molecular level we can divide the lines in these NMR spectra into sharp, principal, and diffuse and that the reasons for this division are basically similar to those on an electronic level underlying the *s*, *p*, and *d* notation of optical spectra—a notation now used in quantum mechanics.

The reader is referred to the original paper (*10*) dealing with the polyarsenous oxyfluorides for further details and a more comprehensive picture of the chemistry of this family of compounds.

Other Families of Arsenic Compounds

As evidenced by the work on polyarsenous oxyfluorides, the analytical technique described in this paper has proven especially valuable in the study of arsenic compounds, since the rate of exchange of substituents (other than those exhibiting a carbon–arsenic bond) on arsenic is too fast for the use of classical procedures in separating compounds but is relatively slow with respect to the coalescence of NMR lines due to exchange

[2]That is, the gel point of the system is exceeded. See a text on polymer chemistry for definition of "gel point."

processes. This means that the NMR application described in this manuscript is particularly useful for elucidating the chemistry of arsenic so that in a short time we have been able to obtain a rather complete picture of hitherto unknown or obscure areas of arsenic chemistry. Much more remains to be done along these lines in order to enlarge the knowledge of arsenic and its compounds to the same state as that of phosphorus chemistry (*1*).

When the bird-cage structure having the formula $As_4(NCH_3)_6$ is dissolved in an arsenous halide, exchange of halogen with methylimino bridges occurs rapidly. When the halide is fluoride (*11*), ^{19}F NMR spectra prove to be rather uninteresting because of rapid exchange of the fluorines between chemically different sites. However, the hydrogen spectra of the methylimino bridges give a wealth of information at 60.0 Mc when the measurements are carried out on a Varian A-60 spectrometer. In this case, however, the *neso* compound, AsF_3, cannot be seen. Typical spectra are shown in Fig. 4. Assignments in this figure have been made following procedures as outlined above. It is clear that differences in the arrangement of fluorines and methylimino bridges on the second arsenic atom in the molecular structure on either side of the resonating methylimino bridge cause resolvable different peaks. For clarification of the assignments, the structures corresponding to two of these atomic arrangements are given below:

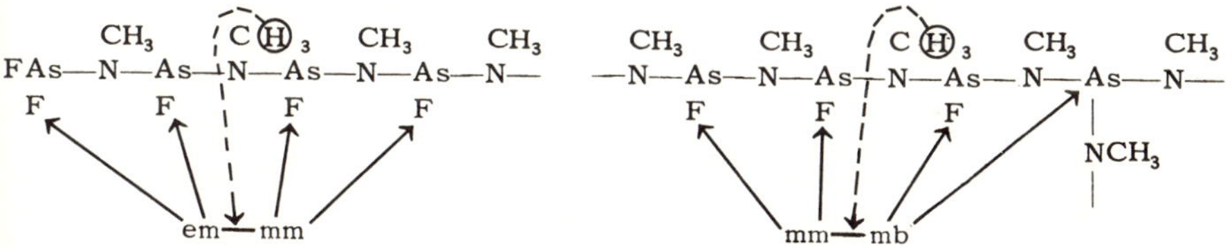

The Σ appearing at the lower middle of Fig. 4 is used as a label for a resonance corresponding to the sum of several atomic arrangements. Spin-spin splitting due to coupling with the fluorine atoms could not be observed, a fact which can be attributed to the rapid exchange of the fluorine atoms with each other, as has been concluded from the ^{19}F spectrum. However, the various spectra show that the relative chemical shifts of the resonating nuclei vary in a pronounced manner as a function of the over-all composition ratio, F/As, although we concluded that they appear in the same sequence in each spectrum. This is not necessarily the case and sometimes, when going through spectra obtained with increasing over-all composition ratios, one notices a peak disappearing under another one and then reappearing on the other side of that other peak (examples not given herein).

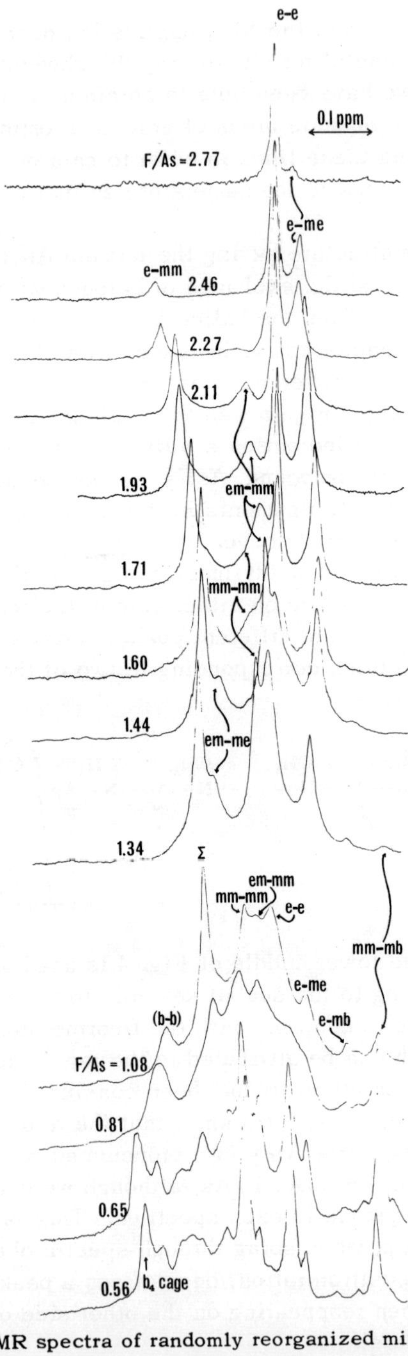

FIG. 4. ^{1}H NMR spectra of randomly reorganized mixtures of AsF_3 and $As_4(NCH_3)_6$ at various F/As ratios. Sweep rate, 0.2 cps/sec.

Further details of the interpretation may be obtained from the paper dealing with this system (*11*).

The hydrogen NMR spectra shown in Fig. 5 for the related equilibrated system where the halide is the chloride (*25*) are extremely simple, show-

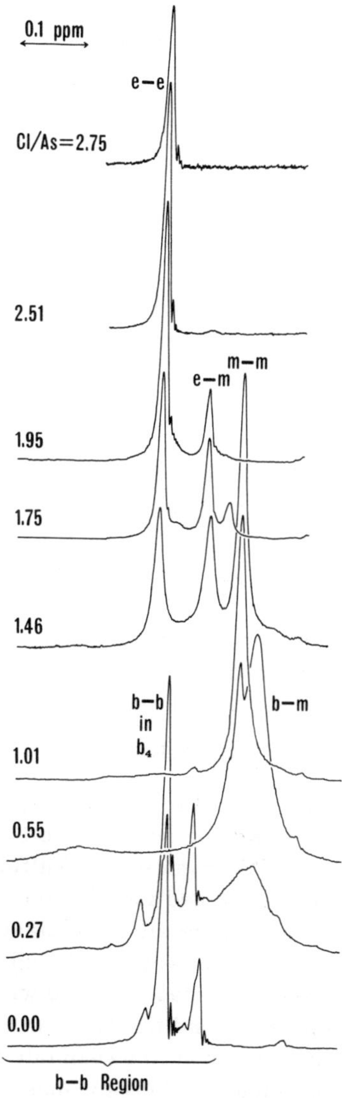

FIG. 5. ^{1}H NMR spectra of randomly reorganized mixtures of $AsCl_3$ and $As_4(NCH_3)_6$ at various Cl/As ratios measured as 1 : 4 solutions in toluene. Sweep rate, 0.2 cps/sec.

ing a total of only three resonances in the composition region which can be studied adequately by high-resolution NMR ($R \equiv Cl/As \geq 1$). These resonances have been attributed to e—e, e—m, and m—m. Because of the equilibrium constants controlling the molecules appearing in this system, branch groups do not appear to an appreciable amount until infinite network polymers are formed. The startlingly great differences for

$3 > R \equiv X/As \geq 1$ between the highly complex spectra of the system

AsX_3–$As_2(NCH_3)_3$ for X = F and the extremely simple spectra of the system for X = Cl are not surprising when the spectra have been fully interpreted, as may be judged by reading the detailed papers (*11*, *25*) describing each system.

Seeing in Depth within Molecules

In the chloro-terminated polydimethyl-*N*-methylsilazanes, the chemical shift of the encircled hydrogen is sensitive to exchange of a chlorine for a nitrogen atom at the end of either of the following chains of atoms:

Ⓗ CSiNSi— or Ⓗ NSi—

Similarly, in the polyarsenous oxyfluorides, the chemical shift of the encircled fluorine is sensitive to interchanging an oxygen for a fluorine at the end of a four-atom chain:

Ⓕ AsOAs—

However, in the polyfluoroarsenousmethylimides, the chemical shift of the encircled hydrogen is sensitive to the interchange of a fluorine for a nitrogen atom at the end of a chain of six atoms:

Ⓗ CNAsNAs—

NMR chemical shifts have served to plumb to even greater depths certain straight-chain molecules having sulfur or selenium in their molecular backbone. Thus, for the alkyl-terminated polysulfides, a reasonably good quantitative analysis may be obtained for the molecule $(CH_3)_3CS_nC(CH_3)_3$, where $n = 10$ in mixtures where there are appreciable amounts of similar chains for which $n = 7$, 8, and 9, as well as 11 and larger (*13*). This is done by measuring ^1H resonance areas obtained with a Varian A-60 spectrometer, running under optimum conditions, using a sweep width of 25 or 50 cycles for the entire scale and a slow sweep rate of 0.1 or 0.05 cps/sec. Likewise, reproducibly distinguishable resonance peaks can be observed (*26*) for n as great as 9 in the dimethyl polysulfate family of molecules having the general formula

$$CH_3O\left[\begin{matrix} O \\ SO \\ O \end{matrix}\right]_n CH_3$$

For the polysulfides, this means that the substitution of a carbon for a sulfur atom noticeably affects the chemical shift of the encircled hydrogen in the following sequence of 13 atoms:

Ⓗ CCSSSSSSSSSS—

For the polysulfates, the sequence is even greater, being twenty-one atoms:

Ⓗ COSOSOSOSOSOSOSOSO—

On the assumption of a random-flight orientation of the molecules, using appropriate bond angles but no rotational barrier, we have been able (*27*) to fit the observed sequences of chemical shifts to the assumption that the magnetic contribution of each group in the molecule to the magnetically active nucleus is proportional to the reciprocal of the cube of the distance of that group from the nucleus in question, as measured through space (i.e., the "neighbor-anisotropy effect"). Of course, such curve fitting does no more than show that the data are not inconsistent with this concept. We believe that a more complete theory should include electronic effects propagated through the molecular chain itself.

Conclusions

The interpretation of NMR data obtained on complicated reaction mixtures can be interpreted with a high degree of veracity in terms of the arrangements of neighboring atoms within molecules. This technique has proven to be of great value in elucidating the structure and thermodynamic properties of entire families of compounds which reach equilibrium with respect to exchange of parts between the molecules making up a mixture. Since much of inorganic chemistry is equilibrium controlled with equilibration being achieved rather rapidly, classical methods for separating and studying molecules have often proved inapplicable. Generally this more-or-less pronounced structural lability (ready exchange of parts between molecules) throughout most of the Periodic Table appears to be the main reason that only carbon (and, to a lesser extent, silicon, phosphorus, and sulfur) out of the hundred-odd elements has a replete chemistry. We believe that full chemistries involving many molecules ranging from small to large structures can be developed for about three-quarters of the elements. The NMR approach described herein is a very useful tool for the systematic development of such chemistries.

References

1. J. R. Van Wazer, "Phosphorus and Its Compounds," Vol. 1, "Chemistry." Wiley (Interscience), New York, 1958.
2. J. R. Van Wazer, C. F. Callis, and J. N. Shoolery, *J. Am. Chem. Soc.* **77**, 4945 (1945).
3. C. F. Callis, J. R. Van Wazer, J. N. Shoolery, and W. A. Anderson, *J. Am. Chem. Soc.* **79**, 2719 (1957).
4. D. P. Ames, S. Ohashi, C. F. Callis, and J. R. Van Wazer, *J. Am. Chem. Soc.* **81**, 6350 (1959).
5. E. Schwarzmann and J. R. Van Wazer, *J. Am. Chem. Soc.* **82**, 6009 (1960).
6. J. R. Van Wazer, *Am. Scientist* **50**, 450 (1962).
7. J. R. Van Wazer and K. Moedritzer, *Angew. Chem.* (in press).
8. D. W. Matula, L. C. D. Groenweghe, and J. R. Van Wazer, *J. Chem. Phys.* **41**, 3105 (1964).
9. K. Moedritzer and J. R. Van Wazer, *J. Am. Chem. Soc.* **86**, 802 (1964).
10. J. R. Van Wazer, K. Moedritzer, and D. W. Matula, *J. Am. Chem. Soc.* **86**, 807 (1964).
11. M. D. Rausch, J. R. Van Wazer, and K. Moedritzer, *J. Am. Chem. Soc.* **86**, 814 (1964).
12. H. K. Hofmeister and J. R. Van Wazer, *J. Inorg. Nucl. Chem.* **26**, 1201 (1964).
13. D. Grant and J. R. Van Wazer, *J. Am. Chem. Soc.* **86**, 3102 (1964).
14. J. R. Van Wazer and K. Moedritzer, *J. Chem. Phys.* **41**, 3122 (1964).
15. K. Moedritzer, J. R. Van Wazer, and C. H. Dungan, *J. Chem. Phys.* (in press).
16. K. Moedritzer and J. R. Van Wazer, *Inorg. Chem.* **3**, 943 (1964).
17. J. R. Van Wazer and L. Maier, *J. Am. Chem. Soc.* **86**, 811 (1964).
18. K. Moedritzer and J. R. Van Wazer, *Inorg. Chem.* **3**, 268 (1964).
19. K. Moedritzer and J. R. Van Wazer, *Inorg. Chem.* **3**, 139 (1964).
20. M. D. Rausch and J. R. Van Wazer, *Inorg. Chem.* **3**, 761 (1964).
21. H. K. Hofmeister and J. R. Van Wazer, *J. Inorg. Nucl. Chem.* **26**, 1209 (1964).
22. K. Moedritzer and J. R. Van Wazer, *Inorg. Chem.* (to be submitted for publication).
23. J. R. Van Wazer, "Inorganic Polymers" (F. G. A. Stone and W. A. G. Graham, eds.), Chapter 2. Academic Press, New York, 1962.
24. D. W. Matula, L. C. D. Groenweghe, and J. R. Van Wazer (manuscript in preparation).
25. J. R. Van Wazer, K. Moedritzer, and M. D. Rausch, *J. Chem. Phys.* **42**, 3302 (1965).
26. J. R. Van Wazer, D. Grant, and C. H. Dungan, *J. Am. Chem. Soc.* (in press).
27. J. R. Van Wazer and D. Grant, *J. Am. Chem. Soc.* **86**, 1450 (1964).
28. L. C. D. Groenweghe, J. R. Van Wazer, and A. W. Dickinson, *Anal. Chem.* **36**, 303 (1964).

Résonance magnétique du fluor dans les monofluoroalcools

G. ARANDA, J. JULLIEN, J. A. MARTIN et R. RAMANADIN[1]

Institut d'Electronique et Laboratoire de Chimie Structurale Organique, Faculté des Sciences, Orsay, France

Nous avons étudié la RMN du fluor dans une série homogène de monofluoroalcools du type $CFR_1R_2—C(OH)R_3R_4$ dans lesquels les substituants R sont des protons, des méthyles, ou des noyaux benzéniques. Il s'agissait d'abord de vérifier la structure des produits obtenus par une synthèse originale, ensuite d'examiner les corrélations entre le déplacement chimique, les constantes de couplage et la structure.

Les spectres ont été réalisés à 56.4 MHz. Ils sont pour la plupart du premier ordre et d'interprétation immédiate. Des difficultés apparaissent pour certains par suite de la non équivalence des protons méthyléniques. Pour $CH_2F—CHOH(C_6H_5)$ cette non équivalence ne se manifeste pas dans le spectre et pour $CHF(CH_3)—CH_2OH$ elle est peu gênante ; pour $CH_2F—CHOH(CH_3)$ et $CHF(C_6H_5)—CH_2OH$ elle rend l'interprétation immédiate impossible.

Le déplacement chimique ϕ du fluor a été mesuré par rapport à CCl_3F utilisé également comme solvant. Pour certains corps, on a effectué une extrapolation à dilution infinie comme le recommandent Filipovich et Tiers (*1*). Pour les autres, on n'a effectué de mesure qu'à une seule concentration voisine de 10% en volume et la valeur trouvée est inférieure de 0.1 à 1.0 ppm à celle à dilution infinie. Cet écart est assez faible pour que l'on puisse faire des comparaisons valables. Les mesures faites par d'autres expérimentateurs et citées ici ont été effectuées dans le même solvant servant aussi de référence. Il existe d'autre part une liaison hydrogène intramoléculaire entre F et OH qui peut être très forte, comme dans le cas du fluoroéthanol (*2*). Nous supposerons que son influence sur ϕ est sensiblement constante dans la série étudiée.

Nous avons cherché à interpréter nos résultats d'après la théorie de Saika et Slichter (*3*), selon laquelle le déplacement du fluor dépend essentiellement du terme paramagnétique local, terme qui augmente avec le caractère covalent de la liaison dans laquelle le fluor est engagé.

[1]*Adresse actuelle* : Laboratoire des Recherches, Péchiney Saint-Gobain, Salindres (Gard), France.

TABLEAU I

DÉPLACEMENT CHIMIQUE DU FLUOR

	ϕ ppm		ϕ ppm
$CH_2F—CH_2OH$	226.2		
$CH_2F—CHOH(CH_3)$	226.0	$CH_2F—CHOH(C_6H_5)$ M_1	220.6 E
$CH_2F—COH(CH_3)_2$	224.4		
$CHF(CH_3)—CH_2OH$	182.5	$CHF(C_6H_5)—CH_2OH$ M_1	186.5 E
$CHF(CH_3)—CHOH(CH_3)$,		$CHF(C_6H_5)—CHOH(CH_3)$,	
thréo	183.5	*thréo* $M_{2,3}$	182.9 E
érythro	182.7	*érythro* M_2	191.0 E
		$CHF(CH_3)—CHOH(C_6H_5)$,	
		thréo M_3	180.7 E
		erythro M_2	179.4 E
		$CHF(C_6H_5)—CHOH(C_6H_5)$	
		thréo *	181.3
		érythro *	183.9
$CHF(CH_3)—COH(CH_3)_2$	183.4	$CHF(C_6H_5)—COH(CH_3)_2$ M_4	188.4 E
$CF(CH_3)_2—CH_2OH$	148.8		
$CF(CH_3)_2—CHOH(CH_3)$	147.3	$CF(CH_3)_2—CHOH(C_6H_5)$ M_4	146.6 E
$CF(CH_3)_2—COH(CH_3)_2$	151.7		

M : mélange ; E : valeur extrapolée à dilution infinie ; * : corps cristallisés, insolubles dans $CFCl_3$ et CCl_4, mesuré dans $CHCl_3 + CFCl_3$.

Effet des substitutions des méthyles

On constate que les déplacements correspondant à un même degré de substitution du carbone porteur du fluor se montrent remarquablement groupés. Nos résultats sont contenus dans les limites suivantes :

	ϕ ppm
Fluor sur carbone:	
primaire	225.3 ± 0.9
secondaire	183.0 ± 0.5
tertiaire	149.5 ± 2.2

Ce fait confirme l'utilité de la valeur de ϕ pour la détermination de la structure des composés fluorés. La variation du déplacement avec le degré de substitution est d'autre part très régulière, et cette régularité se conserve si nous considérons la valeur 271.9 ± 1.7 ppm mesurée pour FCH_3 (*4*) dont le degré de substitution est égal à zéro. La substitution sur le carbone porteur de F d'un H par un radical hydrocarboné provoque un "apport d'électrons" vers ce carbone et rend la liaison C—F plus "covalente": le déplacement de F diminue conformément à la théorie.

Pour préciser cette corrélation, nous avons calculé la charge ε du carbone porteur du fluor (5) en utilisant une méthode semi-empirique dont le point de départ est la "série de paramètres" qui reproduisent en termes de charges effectives les moments dipolaires des molécules (6,7).

TABLEAU II
CHARGE ε DU CARBONE FLUORÉ[a]

	−0.3 D	0	+0.3 D
CH_3F	−0.19	0.88	1.94
CH_2F—CH_2OH	0.029	0.91	1.78
CH_2F—$CHOH(CH_3)$	0.05	0.89	1.74
CH_2F—$COH(CH_3)_2$	0.06	0.88	1.70
$CHF(CH_3)$—CH_2OH	0.08	0.82	1.55
$CHF(CH_3)$—$CHOH(CH_3)$	0.10	0.81	1.51
$CHF(CH_3)$—$COH(CH_3)_2$	0.11	0.80	1.47
$CF(CH_3)_2$—CH_2OH	0.13	0.75	1.36
$CF(CH_3)_2$—$CHOH(CH_3)$	0.14	0.74	1.32
$CF(CH_3)_2$—$COH(CH_3)_2$	0.15	0.73	1.29

[a]En 10^{-10} ues cgs.

Cette méthode permet de calculer ces moments dipolaires de façon satisfaisante pour trois valeurs du moment de la liaison C—H: −0.3 D; 0; +0.3 D (+0.3 D correspondant à C^+H^-). Les Fig. 1a,b et c, representant la variation de ϕ avec la valeur calculée de ε dans ces trois hypothèses, montrent que le meilleur accord entre nos mesures et celle sur CH_3F est obtenu pour le cas +0.3 D. Dans ce cas, le sens de variation de ϕ avec ε est en accord avec la théorie de Saika et Slichter; la très bonne corrélation obtenue suggère que ε est le facteur principal dont dépend le terme paramagnétique local dans la série étudiée et que l'hypothèse + 0.3 D est la plus satisfaisante dans notre méthode de calcul de charge.

L'alignement des points expérimentaux est encore meilleur si l'on compare les corps dont le carbone hydroxylé est substitué de façon identique. On obtient alors trois droites parallèles nettement distinctes, ce qui tient vraisemblablement à l'approximation du calcul de ε et à la simplicité de la façon dont on fait intervenir ε.

Le déplacement du fluor dans n-$C_6H_{13}F$ est 219.0 ppm (*1*); dans nos fluoroalcools correspondants, où le fluor est sur le carbone primaire, il est supérieur de 6.3 ppm à cette valeur. On a constaté que ϕ était peu dépendant des substituants méthyles sur le carbone hydroxylé. En sup-

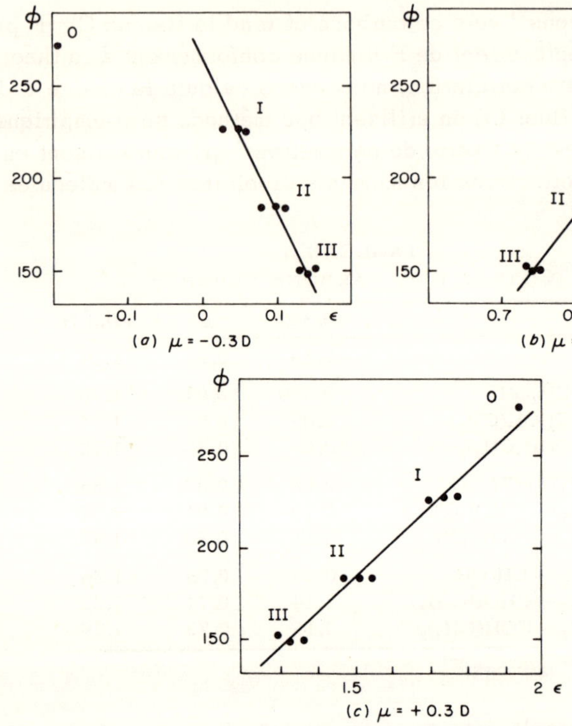

FIG. 1. Déplacement du fluor en fonction de la charge $\mathcal{E}$ du carbone fluoré : ϕ en ppm; $\mathcal{E}$ en 10^{-10} ues cgs ; 0, I, II, III : degré de substitution du carbone fluoré; μ : valeur du moment de la liaison C—H.

posant que l'effet de la substitution par OH reste faible également, on déduit que la liaison H intramoléculaire augmente ϕ de quelques parties par million.

Dans des composés cycliques nous avons mesuré 180.8 ppm pour le fluorocyclohexanol *trans*, 179.8 ppm pour le fluorocyclopentanol *trans*, et 143.6 ppm pour le méthyl-2-fluoro-2-cyclohexanol *trans*. Ces valeurs sont sensiblement inférieures à celle des composés linéaires.

Effet des substitutions des noyaux benzéniques

Le Tableau I montre que les déplacements observés, là aussi bien groupés selon le degré de substitution du carbone, s'écartent cependant sensiblement des valeurs précédentes.

La méthode simple de calcul de charge ne s'applique pas ici. Nous ne ferons que comparer l'effet de ces substituants à celui des méthyles :

— substitution sur le carbone fluoré: ϕ est plus grand de 2.8 à 5.0 ppm que dans la substitution de CH_3. L'anisotropie du noyau benzénique seule donnerait une diminution. Le remplacement du méthyle donneur d'électrons par le noyau benzénique attracteur a donc augmenté la charge du carbone porteur du fluor, et par suite ϕ.

— substitution sur le carbone hydroxylé: ϕ est ici plus petit de 0.6 à 5.4 ppm. L'effet d'anisotropie parait prépondérant.

— substitution sur les deux carbones: la différence de solvant ne permet pas de considérer la différence de 0.5 ppm comme significative.

Références

1. G. Filipovich et G. V. D. Tiers, *J. Phys. Chem.* **63,** 761 (1959).
2. P. J. Krueger et H. D. Mettee, *Can. J. Chem.* **42,** 326 (1964).
3. A. Saika et C. P. Slichter, *J. Chem. Phys.* **22,** 26 (1954).
4. S. G. Frankiss, *J. Phys. Chem.* **67,** 752 (1963).
5. R. Ramanadin, Thèse ing. doc. (1964).
6. H. Eyring, J. Magee, T. Ree et R. P. Smith, *J. Am. Chem. Soc.* **73,** 2263 (1951).
7. P. Smith et E. M. Mortensen, *J. Am. Chem. Soc.* **78,** 3932 (1956).

Magnetic Resonance in Some Deuterated Compounds

S. S. DHARMATTI, V. SARASWATI, and R. VIJAYARAGHAVAN

Tata Institute of Fundamental Research, Bombay, India

Introduction

Proton magnetic resonance is usually employed to gain important information about a crystal, molecular structure, and the nature of hydrogen bonding. In this process, the magnetic effects of the surroundings on the resonance line shape of a nucleus are analyzed to get relevant information. However, if a deuteron replaces a proton, the quadrupole moment of the deuteron interacts with the field gradient and the resulting quadrupolar interaction is very sensitive even to small changes in the environment. Such data, besides supplementing the proton results, are sometimes unique in unambiguously deciding the structure.

Our studies were originally restricted to hydrates of sodium tetraborates, where certain important conclusions were reached about the crystal structures using proton resonance (*1*). We have now extended our investigations to some deuterated compounds to determine the arrangement of hydrogen atoms and molecular motions in the crystal.

Experimental

All deuterated compounds were prepared by repeated exchange of compounds with heavy water and the proton content after deuteration was found negligible by high-resolution NMR. The measurements were made using a crossed coil spectrometer. Some compounds were irradiated to a dose of 60 megaroentgens of γ-rays from ^{60}CO to provide better relaxation times. Samples investigated are in the polycrystalline form.

Results

The compounds studied were borax, its lower hydrates, oxalic acid dihydrate, glycine, biphenyl deuterate, formaldehyde, and boric acid. The results are shown in Table I. ^{2}D has spin I = 1 and a small quadrupolar moment; hence only a first-order perturbation effect is expected. This has been confirmed by studying the resonance at two fields in all the compounds. The resonance line consists of two humps symmetrically situated about the center for vanishing value of the asymmetry parameter. Line shapes for finite values of η have been given elsewhere (*2*).

TABLE I

Compounds	Molecular formula	Coupling constants (kc/s)	Remarks
1. Borax	$Na_2B_4O_5(OD)_48D_2O$	124	A pair of doublets
2. Tincalconite	$Na_2B_4O_6(OD)_24D_2O$	130	Asymmetric line irradiated
3. $Na_2B_4O_7 \cdot 3D_2O$	—	—	Very weak central line, irradiated
4. Boric acid	D_3BO_3	136	Line very weak, irradiated
5. Oxalic acid dihydrate	$(COOD)_2 \cdot 2D_2O$	125 and 121	Three doublets
6. Glycine-d_5	ND_2CD_2COOD	36.8, 61.2, and 97.2	Three doublets
7. Glycine-d_3	ND_2CH_2COOD		
8. Glycine-d_2	NH_2CD_2COOH	No line	—
9. Biphenyl deuterate	$C_{12}D_{10}$	172	—
10. Formaldehyde	HCDO	157.2	

Borax ($Na_2B_4O_7 \cdot 10D_2O$)

The ^2D resonance consists of a pair of doublets (Fig. 1). Probably one doublet arises from OD and the other from the $8D_2O$ molecules. The situation is very well reconcilable with the crystal structure (*3*) $Na_2B_4O_5(OD)_48D_2O$. However the low temperature measurements lead us to the conclusion that the room temperature doublets correspond to just one group of deuterons, probably those of D_2O, with a finite $\eta \sim 0.8$ and $e^2qQ/h = 124 \pm 2$ Kc/s.

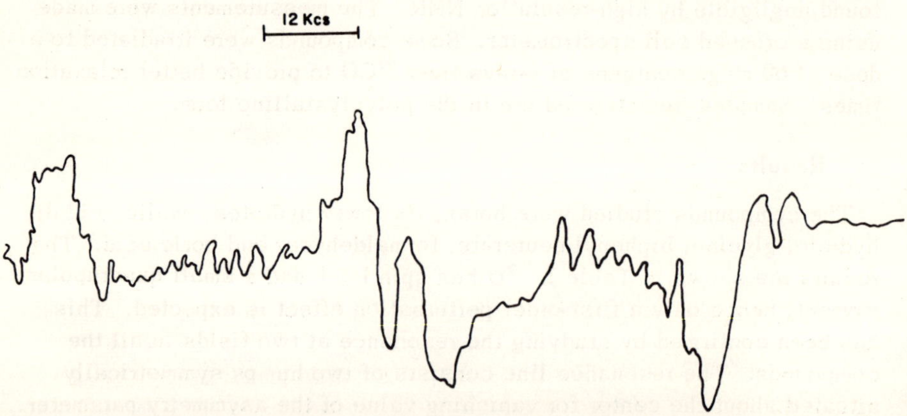

FIG. 1. ^2D resonance in borax ($Na_2B_4O_5(OD)_2 \cdot 8D_2O$).

Tincalconite ($Na_2B_4O_7 \cdot 5D_2O$)

A central line of width 30 Kc/s accompanied by humps separated by 180 ± 5 Kc/s is observed. The central line is broadened due to the high sweep utilised to observe the signal, which is weak in spite of γ-irradiation. The lines could be attributed to a coupling constant of 130 Kc/s and $\eta \sim 0.9$.

$Na_2B_4O_7 \cdot 3D_2O$

A very weak line was obtained in spite of irradiation. The coupling constant could not be measured with any certainty. ^{23}Na and proton resonances (*1*) show that this compound is amorphous and protons exist not as water molecules but as OH groups. It could be that the amorphous nature produces a random distribution in field gradients at deuterium nuclei and broadens the line. This view is consistent with the observation on protons (*1*).

Boric Acid (D_3BO_3)

Here again, the line is very weak in spite of γ-irradiation. The coupling constant turns out to be 136 kc/s, and from the line shape η does not seem to be zero.

Oxalic Acid, Deuterate ($(COOD)_2 \cdot 2D_2O$)

Richards and Smith (*4*) in their PMR investigation, observed a single line, smooth at the top with small kinks at the center. They ruled out oxonium structure and instead proposed that a simple line due to the OH group is superposed on a doublet from water molecules. Subsequently this was examined by a Japanese group (*5*) in a single crystal. They concluded that the line was due to a three-proton system, two protons of H_2O and one from OH forming a triangle of different sides.

Bersohn (*6*) has shown that the coupling constants of ^2D in OD, D_2O, and D_3O should be different. Also the resonance line is a doublet for each of them. We obtain three doublets for deuteron resonance in oxalic acid dideuterate similar to those for borax. Evidently two groups of deuterons are present. This confirms the conclusion of Richards and Smith that there exist two types of hydrogen groups in oxalic acid dihydrate. In our case, η seems to be close to zero for one group of deuterons and for the other $\eta \simeq 0.8$ with e^2qQ/h as 121 Kc/s.

Glycine

The x-ray analysis of Albrecht and Corey (*7*) indicated a Zwitter ion structure $NH_3^+CH_2COO^-$ for this compound. Proton resonance studies of

Shaw *et al.* (*8*) supported this. Kromheut and Moulton (*9*) calculated the second moment for various molecular models of glycine and came to the conclusion that the Zwitter ion structure was the most probable though the others were not impossible. Recently the presence of a Raman line at 870 cm^{-1} in triglycine sulfate (*10*) and other additional compounds of glycine has been attributed to COOH. We have studied the proton and deuteron resonances in partially and completely deuterated glycine with the aim of elucidating the possible structure (Figs. 2 and 3).

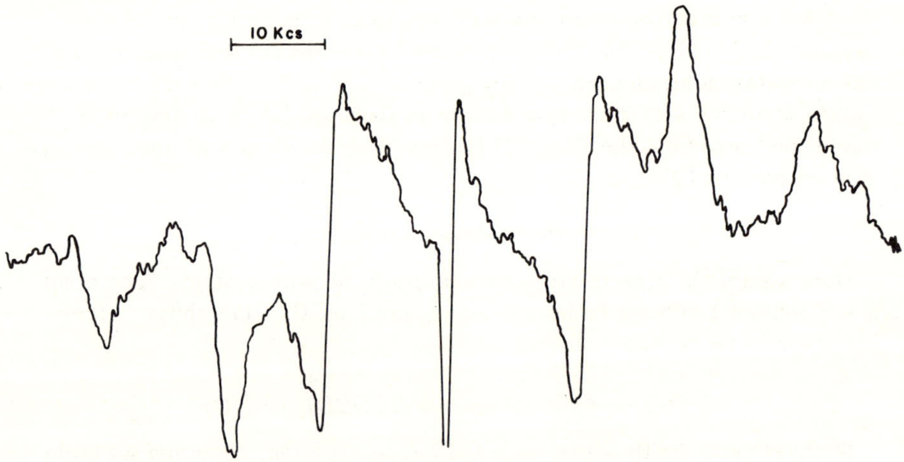

FIG. 2. ^{2}D resonance in glycine-d_5 (ND_2CD_2COOD) at 8.0 Mc/s.

The proton resonance in normal glycine, NH_2CH_2COOH, had a width equal to 9.5 gauss. The line had a smooth top and kinks near the center. At about 180°K (Fig. 4) this line broadened to 20 gauss and the shape was simple. The room temperature line of 9.5 gauss could be resolved into two, one of width 10 gauss arising from CH_2 and temperature independent, and the other 4 gauss in width and changing to ~20 gauss at 180°K. This was confirmed by measuring the temperature dependence of the proton resonance line in two partially deuterated compounds, NH_2CD_2COOH (glycine-d_2) and ND_2CH_2COOD (glycine-d_3). In the former a 4 gauss proton line broadened to ~20 gauss at nearly 180°K (Fig. 3). In the latter the linewidth remained unchanged on lowering the temperature (line shape was a doublet as expected). The low-temperature line in glycine-d_2 does not clearly indicate the presence or absence of NH_3^+.

The deuteron resonance in ND_2CD_2COOD (glycine-d_5) consists of three pairs of lines. For ND_2CH_2COOD a similar spectrum was observed and for NH_2CD_2COOH no line was detected for a scan of 300 kc/s on

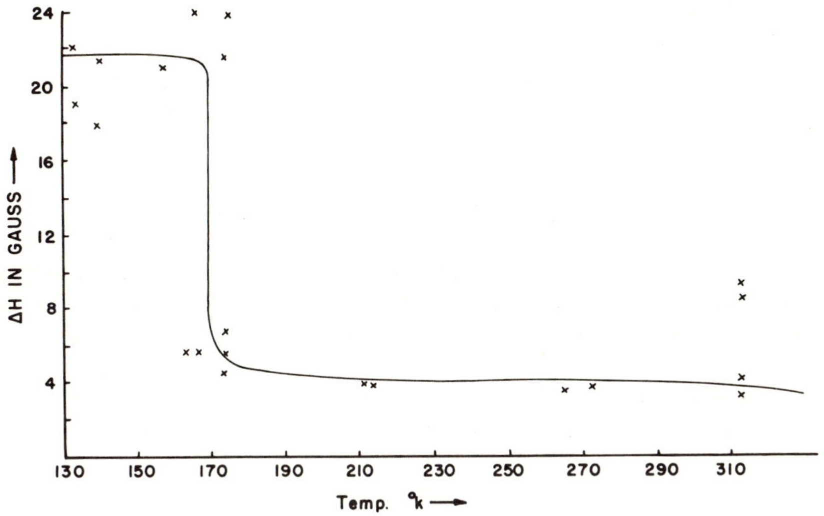

FIG. 3. Temperature variation of linewidth in glycine-d_2 (NH_2CD_2COOH).

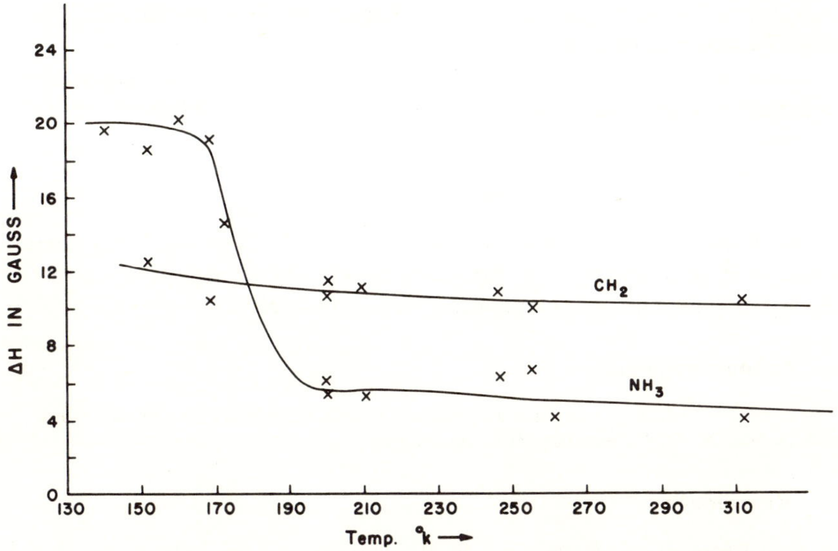

FIG. 4. Temperature variation of linewidth in glycine (NH_2CH_2COOH).

either side of the center. This indicates that three pairs of lines arise from the ND_2 and COOD groups. If a Zwitter ion is assumed we have only ND_3^+; three pairs could arise only if (i) the three deuterons of ND_3 are in nonequivalent sites or (ii) there is dipolar splitting of the quadrupolar split line. Calculations were carried out for the latter possibility and it was found that the splittings are too large to be explained by dipolar interaction. The three pairs correspond to coupling constants 36.8, 61.2, and 97.2 kc/s. The values are small compared with expectation and preliminary work on temperature variation showed broadening at about 200°K, close to the transition for the proton case. Hence nonequivalence of all the deuterons in ND_3^+ is also difficult to presume.

The three pairs can also be explained by postulating the simultaneous existence of Zwitter and nonzwitter ion structures in glycine. The quadrupole coupling constants could then be ascribed to ND_3^+, ND_2, and COOD. Infrared and Raman spectra in several glycine compounds have been seen to contain lines corresponding to un-ionized COOH, apart from those of NH_3^+ and COO . Hence the simultaneous existance of Zwitter ion and un-ionized molecules in glycine cannot be ruled out.[1]

Biphenyl Deuterate ($C_{12}D_{10}$)

This gives only one doublet with a coupling constant of 172 kc/s. This seems to be small for a strong C–D bond where the e^2Qq/h valve expected is 350 kc/s but is close to that of benzene (169 kc/s). The observed value can be understood by assuming existence of a torsional oscillation of one ring about the planar axis of the other. A temperature variation investigation which is being undertaken should verify this assumption.

Formaldehyde (HCDO)

The deuteron resonance is from the C–D bond. Here again the coupling constant is only 157.2 kc/s, far less than for the rigid system, which needs verification from temperature variation studies.

Acknowledgments

Our thanks are due to Mr. Jagannathan for help in preparation of the compounds and Dr. C. R. Kanekar for discussion.

References

1. S. S. Dharmatti, S. A. Iyer, and R. Vijayaraghavan, *Proc. Phys. Soc. Japan* **17**, 1736 (1962).
2. T. Chiba, *J. Chem. Phys.* **36**, 1122 (1962).
3. N. Morimoto, *Mineral. J.* (*Japan*) **2**, 1 (1956).
4. R. E. Richards, and J. A. S. Smith, *Trans. Farad. Soc.* **47**, 1261 (1951).

[1]*Note added in proof*: Detailed observations on the temperature variation of deuteron and proton resonance spectrum in glycine and its addition compounds confirm the Zwitter ion nature of glycine.

5. J. Itoh, R. Kusaka, R. Kiriyama, and S. Yabumoto, *J. Chem. Phys.* **21**, 1895 (1953).
6. R. Bersohn, *J. Chem. Phys.* **32**, 85 (1960).
7. G. Albrecht and R. B. Corey, *J. Am. Chem. Soc.* **61**, 1087 (1939).
8. T. M. Shaw, R. H. Elsken, and K. J. Palmer, *Phys. Rev.* **85**, 762 (1952).
9. R. A. Kromhout and W. G. Moulton, *J. Chem. Phys.* **23**, 1673 (1955).
10. R. S. Krishnan and P. S. Narayanan, in "Crystallography and Crystal Perfection" (G. N. Ramachandran, ed.), p. 329. Academic Press, New York, 1963.

High-Resolution Nuclear Magnetic Resonance Spectra of Molecules in Inclusion Compounds

K. A. McLAUCHLAN

Basic Physics Division, National Physical Laboratory, Teddington, England

Introduction

In 1961 it became apparent that serious discrepancies existed between the signs of spin-spin coupling constants observed in saturated molecules and those predicted by the current valence-bond theory. It became essential, therefore, to determine the absolute sign of at least one coupling constant. However, it is an inherent property of a NMR spectrum that its appearance, whilst depending upon the relative signs of the coupling constants, is unaffected by a complete sign reversal. Consequently no method utilizing the normal energy level system can yield the absolute sign of the spin-spin coupling constants. Buckingham (*1*) suggested, therefore, that the nuclear magnetic dipole-dipole interaction, which normally averages to zero by molecular tumbling, should be caused to have a nonzero value by partial orientation of the molecules and that this quantity, whose sign is fully calculable if the orientation direction of the molecule is known, should be compared in sign with the spin-spin coupling constant. This experiment was successfully performed in 1963 (*2*) by causing partial orientation of polar molecules in an applied electric field. However, the theory had shown that with reasonable degrees of orientation, much greater than can be obtained in an electric field, accurate measurements of chemical shift anisotropies and internuclear distances should be possible. Furthermore it was shown that magnetically equivalent protons should exhibit nondegenerate transitions under these conditions. The higher degrees of orientation required were obtained by Phillips (*3*) and Englert and Saupe (*4, 5*) in the nematic phase of a liquid crystal in which orientation results from the magnetic polarization induced between molecules and by the applied magnetic field. It was shown that comparatively small aromatic molecules added to this phase themselves became aligned and displayed well-resolved NMR spectra. This paper describes attempts to produce similar alignment in small aliphatic molecules by limiting their free motion inside a crystal lattice which contains molecular-size diffusion paths. This technique was shown to be possible by Ducros *et al.* (*6, 7*), who discovered that partial orientation of water and heavy water occurred in zeolites. In this communciation these results are extended to HDO and methyl cyanide included in zeolite crystals and a further study of some organic inclusion compounds is described (*8*).

Principle of the Method

The normal high-resolution NMR spectrum is characterized by two sets of parameters, the chemical shifts and the spin-spin coupling constants. If the normal free tumbling of the molecule in solution is restricted, an additional interaction between the magnetic moments of the nuclear spins is noted. For two equivalent nuclei this interaction is given, in frequency units, by $\frac{3}{2}(\mu^2/hr^3)(3\cos^2\theta - 1)$, where μ is the nuclear magnetic moment, h is Planck's constant, r is the internuclear distance, and θ is the angle between the direction of orientation of the spins and the vector joining the nuclei. This interaction is very large compared with the electron-coupled spin-spin interaction which, therefore, often appears in the spectra as a small perturbation. As an example, an isolated water molecule held rigid with the H–H vector perpendicular to the applied magnetic field would exhibit a doublet with a dipolar splitting of 25·43 kc/s. The actual splitting observed is less than this and depends upon the mean value of $(3\cos^2\theta - 1)$. One may define a "percentage degree of orientation" as the percentage of the optimum value which the observed value is. In the case of an array of protons with C_{3v} symmetry, such as in a CH_3X molecule, it may easily be shown that a 1 : 2 : 1 triplet should be observed.

Experimental

Spectra were obtained at 100 Mc/s on a Varian HR 100 spectrometer and were calibrated by the normal sideband technique. In order to obtain an optimum signal-to-noise ratio and also to obviate confusion from 2-kc/s sidebands, the spectrometer was operated with rf detection only, without the aid of a baseline stabilization device.

The requirement for a suitable crystal lattice is that it should contain diffusion channels or large cages in which the included molecule can move quite freely so that any dipolar interaction with the lattice should be averaged to zero. Furthermore, these channels should have a preferred axis. These conditions are met by a number of naturally occurring zeolites (*9*) and by numerous inclusion compounds of organic materials. The zeolites were kindly donated by various museums and private collectors. The organic crystals were obtained simply by crystallizing the host lattice from a solution containing the molecule to be included, usually as the solvent.

Results

It is convenient to discuss the inorganic and organic lattices separately.

Molecules Included in Zeolites

Zeolites are alumino-silicates in which AlO_4 and SiO_4 tetrahedra join together to form four-, six-, and eight-membered rings of various conformations (*9*). These rings associate to form continuous three-dimensional structures which are characterized by containing large diameter channels in specific crystal directions. In these channels small molecules may be included and undergo reasonably free diffusion and furthermore very small molecules may also diffuse through the eight-membered rings. Since the lattice components are anionic, the electrical neutrality of the crystal is preserved by cations absorbed on the inside of these channels and these tend, therefore, to obstruct the diffusion path.

Figure 1 shows the spectra of water at various orientations inside a gmelinite lattice. This zeolite has a nonintersecting unidirectional sys-

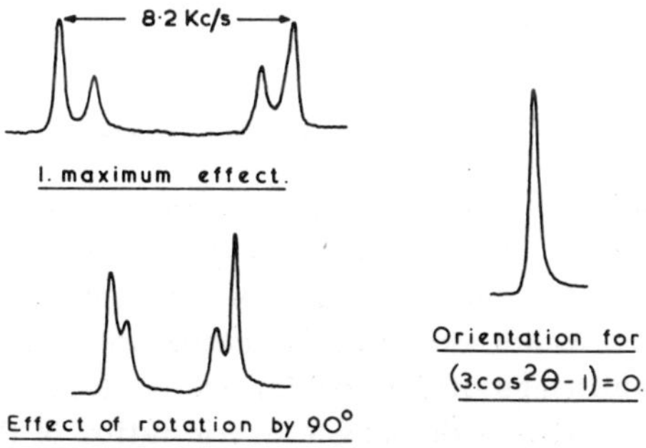

FIG. 1. H_2O oriented in gmelinite. The two doublets result from two distinct diffusion paths.

tem of channels but the observation of two doublets is clear evidence of diffusion occurring along two separate directions. The linewidth is about 300 c/s and may be due to exchange or to non-averaged-out intermolecular dipolar interactions. With a similar zeolite, chabazite, an optimum peak separation of 8.5 kc/s was observed corresponding to a degree of orientation of about 33%. This contrasts with a degree of 0.01% obtained in the electric field experiments (*2*). It has been shown that powdered zeolites will, after outgassing to a critical degree above which the pore structure disintegrates, include many organic molecules (*10*), including *n*-paraffins. Attempts to perform similar experiments on single crystals after degassing under the conditions suggested in this work, however, produced no detectable inclusion of the guest molecule. This is presumably due to the much smaller surface area.

It is, however, possible to react with the water *in situ*, e.g., exposure to ammonia gas leads to the observation of a single line, completely independent of the crystal orientation. This is due to the spherically symmetric ion NH_4^+. Similarly heating the crystal with heavy water produces a certain amount of HDO as shown in Fig. 2. This crystal contained ca.

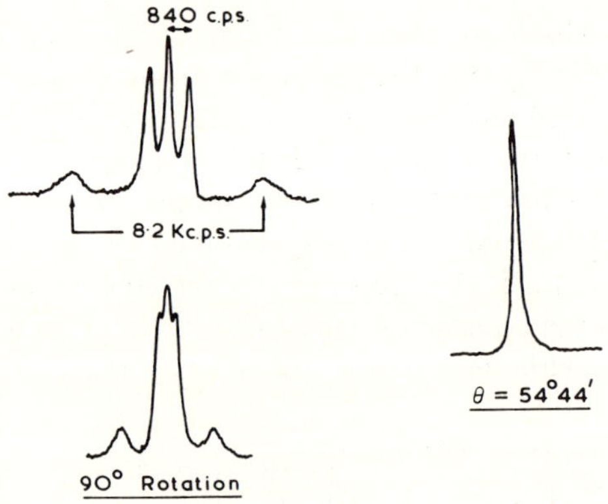

FIG. 2. HDO oriented in chabazite.

40% HDO and exhibits a triplet proton spectrum due to coupling to the spin one deuterium nucleus besides a low intensity doublet due to residual H_2O. The 1 : 1 : 1 intensity ratio of the triplet is perturbed by overlap of the lines which are actually sharper than the ones observed in H_2O. The total splittings of the doublet and triplet are, within experimental error, in the ratio of the magnetogyric ratios of the proton and the deuteron times a numerical factor of 3/2.

In only one case has it so far proved possible to introduce an organic molecule into the zeolite lattice. This was accomplished by baking the crystal in an oven for 3 hours at 220 °C, which removed ca. 80% of the water from the channels, and then subjecting it immersed in methyl cyanide to a pressure of 3000 atmospheres for half an hour. The result is shown in Fig. 3, which shows a broad triplet, which was orientation dependent, of intensity ratio 1 : 2 : 1. The lines are very broad and the maximum total splitting observed between the outer lines of the triplet was 2550 c/s corresponding to 29% orientation.

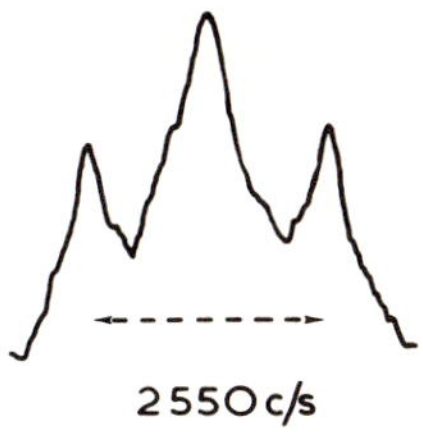

FIG. 3. Methyl cyanide in chabazite.

Molecules Included in Organic Lattices

Many systems are known in which lattices crystallized from organic solvents include an appreciable amount of the solvent. Preliminary investigations of a number of such systems have shown that the included molecules exhibit quite sharp (ca. 60 c/s broad) NMR lines but that the over-all degrees of orientation are small. The systems investigated include urea and tri-*o*-thymotide inclusion compounds, in which the lattice structure depends upon the solvent from which crystallization occurs, and the hydroquinone clathrates. The former compounds may crystallize with unidirectional channels, but from solvents of low molecular weight cavity structures or structures including no solvent molecule may result. Furthermore, in order to observe genuine spectra great care must be taken to avoid twinned crystals, whilst stacking faults and general crystal imperfections appear to inhibit molecular diffusion. The spectra normally observed, therefore, tend to be superpositions of numerous overlapping spectra. This problem can only be resolved to some degree by utilizing very small single crystals in which signal-to-noise ratio problems become acute. Occasionally suitable crystals are obtained. Figure 4 shows the triplet of methyl in methyl alcohol in which the hydroxyl proton is undergoing exchange. The spectrum was obtained from two crystals of urea in about one thousand grown from methanol solution. In the hexagonal form of urea diffusion channels exist parallel to the c axis

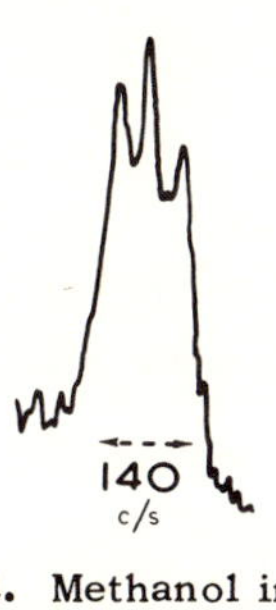

FIG. 4. Methanol in urea.

so that if the crystal is rotated about this axis no change in the spectrum should result. This was verified experimentally. The degree of orientation is only about 1.6%. It has not yet proved possible to obtain any simple spectra from hydroquinone clathrates although quite sharp lines can be observed.

Conclusions

It has been demonstrated that crystal lattices may be employed to orientate small molecules other than water. However, the lines are in general broad and spinning the crystal cannot be performed since it is usually anisotropic. Except in systems in which chemical shift anisotropies may be expected to be large, therefore, no great accuracy can be expected from the method. However, a detailed study of specific inclusion compounds may give considerable information on the physical state of the included molecule.

Acknowledgments

This work forms part of the research program of the National Physical Laboratory and is published by permission of the Director.

References

1. A. D. Buckingham and E. G. Lovering, *Trans. Faraday Soc.* **58**, 2077 (1962).
2. A. D. Buckingham and K. A. McLauchlan, *Proc. Chem. Soc.*, p. 144 (1963).
3. W. D. Phillips, 5th Exp. NMR Conf. Meeting, Pittsburgh, Pennsylvania, 1964.
4. G. Englert and A. Saupe, *Phys. Rev. Letters* **11**, 462 (1963).
5. G. Englert and A. Saupe, *Z. Naturforsch.* **19a**, 172 (1964).
6. P. Ducros and X. Pare, *Arch. Sci. (Geneva)* **13**, 383 (1960).
7. Y. Ayant, P. Ducros, and M. Sortif, *Compt. Rend.* **252**, 550 (1961).
8. K. A. McLauchlan, 5th Exp. NMR Conf. Meeting, Pittsburgh, Pennsylvania, 1964.
9. R. M. Barrer and I. S. Kerr, *Trans. Faraday Soc.* **55**, 1915 (1959).
10. R. M. Barrer, *Quart. Rev. (London)* **3**, 293 (1949).

High-Resolution Nuclear Resonance Studies of Polymer Chain Conformation in Solution

T. M. CONNOR and K. A. McLAUCHLAN

Basic Physics Division, The National Physical Laboratory, Teddington, England

We have begun a series of experiments which we hope will give us some information about the conformation of polymer chains in solution, using high-resolution nuclear resonance techniques. Previous studies of polymers by this method have been largely concerned with the question of the stereoregularity of polymer chains (*1*), that is, of determining the relative amounts of isotactic, syndiotactic, and heterotactic groupings present in the polymer chain. We are interested in other aspects of chain micro-structure and in particular in the energy differences between the various rotationally isomeric forms of the monomer unit in the polymer chain. Such quantities are of great importance in theories of the configurational statistics of polymer chains (*2*).

For our experiments we have chosen a simple oxide polymer, polyethylene oxide, $-\!\!\left[CH_2-CH_2-O\right]\!\!-_n$. Since this polymer has no side-chain groups there are no complicating effects due to stereoregularity. The molecular weight is 6000, which corresponds to $\sim$136 monomer units, the chain being sufficiently long for end-group effects to be negligible, but short enough to ensure adequate solubility. We have measured the temperature dependence of the splittings in the $^{13}C-H$ proton satellite bands derived from monomer units of the type

$$-\!\!\left[{}^{13}CH_2-CH_2-O\right]\!\!-_n$$

Although the ordinary proton spectrum from this polymer is a single line, the proton satellite spectra caused by coupling to the ^{13}C nuclei approximate to half an AA′XX′ spectrum. Figure 1 shows the spectrum for a saturated aqueous solution at 31.5°C and shows a satellite band similar to that seen in dioxane, and 1,2-dibromoethane, etc. (*3*). The splitting between the outermost lines is $J + J'$ where J and J' are the two coupling constants between the A and X groups. Owing to the rather large line-width of $\sim$1.25 cps, the coupling $J - J'$ in the central region of the satellite is not resolved.

The samples studied were polyethylene oxide in H_2O and $CHCl_3$ solutions and in the melt. Three small molecules which are more or less

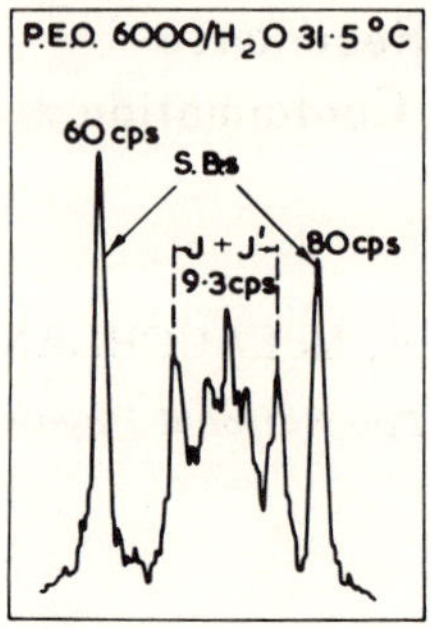

FIG. 1. ^{13}C sideband of proton spectrum of polyethylene oxide in water at 31.5°C. (Measuring sidebands also shown at 60 and 80 cps.)

satisfactory as model compounds have also been studied, i.e. 1,2-dimethoxyethane, 1,2-dichloroethane, and ethylene glycol. Figure 2 shows $J + J'$ in cycles per second plotted as a function of temperature for these systems. The polymer solutions show a variation of ~1 cps in $J + J'$

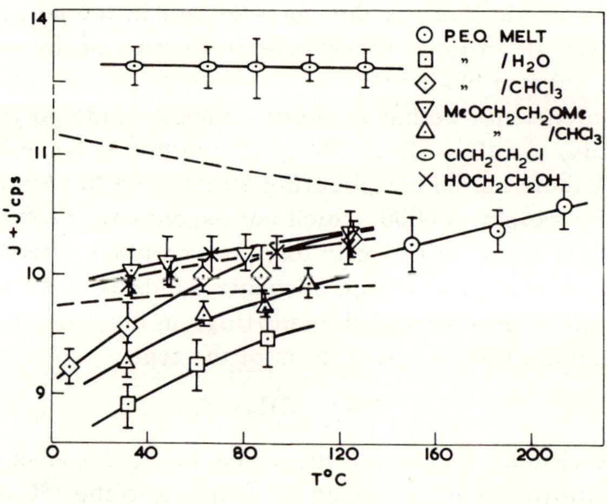

FIG. 2. $J + J'$ as a function of temperature for polyethylene oxide and related compounds.

over a 100°C temperature range. Pure 1,2-dimethoxyethane and ethylene glycol show much smaller variations, whilst 1,2-dimethoxyethane in $CHCl_3$ shows a similar variation to the polymer solutions. The polymer melt could not be studied until a temperature of 150°C was reached, i.e., 80°C above its melting point.

We have used the current theory for a preliminary analysis of the results (*3,4*). In general J and J' may be expressed in terms of the coupling constants in the stable isomers, and in terms of the fractional populations of these isomers. In the case of polyethylene oxide monomer units there are three stable conformations, one *trans* and two *gauche*, which are shown in Fig. 3, with appropriate coupling constants. It is

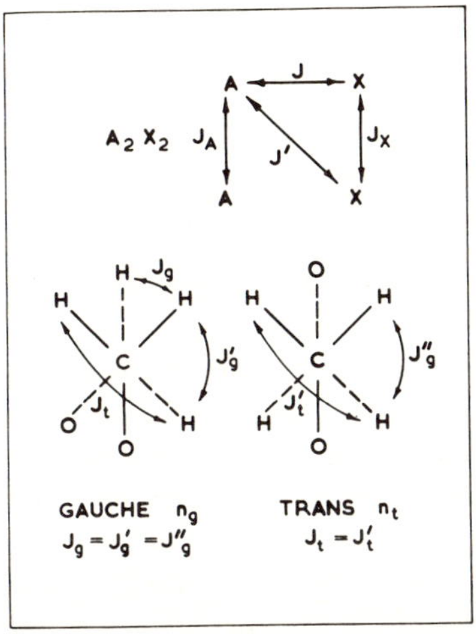

FIG. 3. *Gauche* and *trans* forms of monomer unit of polyethylene oxide.

assumed that $J_g = J_g' = J_g''$ and that $J_t = J_t'$. The fractional populations n_g and n_t are related by the expressions

$$n_g/n_t = 2\exp[\Delta E/RT],$$
$$n_g + n_t = 1,$$

where ΔE is the energy difference between *gauche* and *trans* positions. It can be shown that

$$J + J' = \tfrac{1}{2}[J_t + 3J_g + n_t(J_t - J_g)].$$

Since $J + J'$ increases with increasing temperature in all cases (apart from 1,2-dichlorethane) the above expression shows that the *gauche* form is the more stable isomer in all these systems. To find values of n_t and hence ΔE we must assume values of J_t and J_g. It has been shown that $J_t + 2J_g$ should be a constant (*4*) and in the case of 2-methoxyethanol, which is a reasonable model compound, is equal to 15.8 cps. This defines ranges of J_t and J_g values (assuming both are positive) so that values of n_t can be calculated. If the theory outlined is applicable and correct values of J_t and J_g have been chosen, by plotting $\ln[(1 - n_t)/n_t]$ against $1/T \times 10^3$, a line of slope $\Delta E/R$ should be obtained, cutting the y axis at ln 2 at infinite temperature. Figure 4 shows the results for the

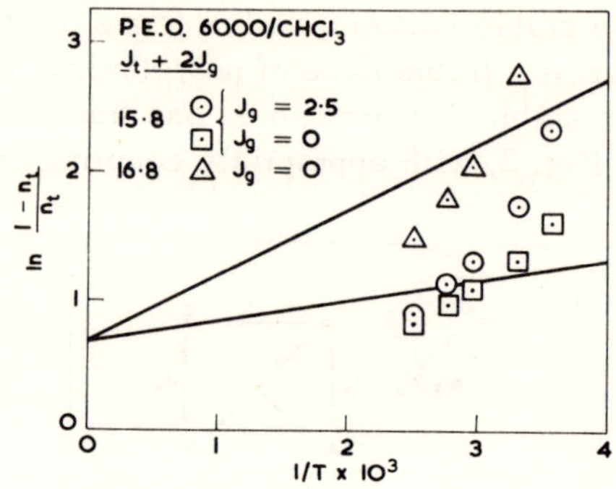

FIG. 4. ln $[(1-n_t)/n_t]$ *versus* $1/T \times 10^3$ for polyethylene oxide in chloroform.

polymer, compared with theoretical curves for ΔE values of 300 and 1000 calories. Figure 2 also contains theoretical curves, the upper one corresponding to $\Delta E = -1000$ cal and the lower one to $\Delta E = +1000$ cals. Values of $J_g = 2.5$ and 0 cps have been chosen, these approximating to limiting values defined by the conditions that $J_g > 0$ and that $1 > n_t > 0$. Figure 5 shows similar results for ethylene glycol and in both cases it is seen that the results do not fit the simple theory used, the variations in $J + J'$ being larger than predicted by the theory.

Various explanations of this discrepancy are possible:

(1) That J_t and J_g are themselves temperature dependent. Experiments on acetaldehyde have shown that small variations of coupling constants with temperature are possible (*5*), although it seems unlikely that these can explain all the present discrepancy.

(2) Variations of ΔE with temperature. The energy difference possibly depends on the dielectric constant of the medium and this is itself temperature dependent (*6*).

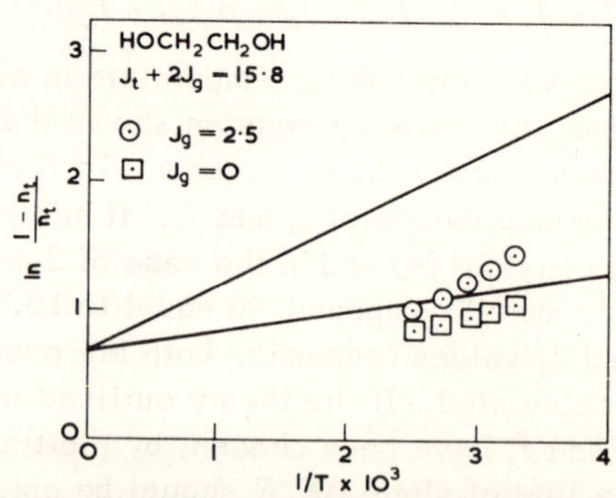

FIG. 5. ln $[(1-n_t)/n_t]$ versus $1/T \times 10^3$ for ethylene glycol.

(3) It is an approximation to consider one value of ΔE. A more sophisticated treatment would consider a distribution of ΔE values. Other properties of polymer chains have frequently to be described by distributions of a particular parameter, e.g., activation energies for molecular motion (7).

It is hoped to continue these investigations on this and other polymers and to attempt a more sophisticated interpretation in terms of the above suggestions. Since the high-resolution spectra of polymers are often complicated owing to the spins being highly coupled, the study of ^{13}C–H sidebands seems to offer some hope of reducing the complexity, since an effective chemical shift equal to the ^{13}C–H coupling (~ 140 cps) is introduced. On the debit side, we have to cope with the much reduced intensity of the sidebands as compared with the parent peak. However, the use of signal averaging techniques and possibly ^{13}C-enriched samples may offer solutions of this difficulty. Also irradiation in the ^{13}C region whilst observing the proton region may reduce the sideband complexity and facilitate interpretation.

References

1. F. A. Bovey and G. V. D. Tiers, *Fortschr. Hochpolymer. Forsch.* **3**, 139 (1963).
2. M. V. Volkenstein, "Configurational Statistics of Polymer Chains," Vol. XVII of *High Polymers*. Wiley (Interscience), New York, 1963.
3. N. Sheppard and J. J. Turner, *Proc. Roy. Soc. (London)*, **A252**, 506 (1959).
4. R. J. Abraham and K. G. R. Pachler, *Mol. Phys.* **7**, 165 (1963).
5. J. G. Powles and J. H. Strange, *Mol. Phys.* **5**, 329 (1962).
6. R. J. Abraham, L. Cavalli, and K. G. R. Pachler, paper in this volume.
7. T. M. Connor, *Trans. Farad. Soc.* **60**, 1574 (1964).

Quelques résultats obtenus dans l'étude des polymères par la RMN

G. LANZI

Istituto di Fisica dell'Università di Pavia, Pavia, Italy

On peut dire qu'après la découverte de la RMN on a entrevu immédiatement les possibilités de cette nouvelle technique pour rechercher les caractéristiques des polymères.

On a, en effet, en 1949 un premier travail par Alpert (*1*), qui concerne l'étude du mécanisme des transitions de phase, dans lequel on examinait aussi le comportement de la gomme.

Tout de suite ce premier travail a été suivi par des autres, dans lesquels on rapportait des expériences avec la gomme aussi bien qu'avec le polythène et le polyméthylmétacrylate (*2, 3*). Actuellement les travaux publiés sur ce sujet sont quelques centaines.

Avant d'examiner ce que l'on a pu faire dans le champ des polymères avec la RMN, et de discuter les informations qu'on peut en obtenir, il est convenable de mettre en évidence le fait que pendant les premiers dix ans de travail la recherche a été fondée uniquement sur des études de relaxation et de largeur de la raie. Dans ce cas le polymère est considéré plus souvent du point de vue de ses propriétés macroscopiques que du point de vue de ses propriétés moléculaires, et les informations qu'on peut obtenir sont souvent comparables avec celles trouvées par des autres moyens de recherche, comme les mesures de relaxation diélectrique ou mécanique.

On peut obtenir des informations de nature moléculaire plus détaillées soit en augmentant la température à laquelle on effectue la mesure, soit en diluant le polymère dans des dissolvants convenables. La réduction des interactions dipolaires qu'on obtient dans ce cas permet en effet de mettre en évidence les causes les plus faibles qui provoquent un élargissement de la raie de résonance.

Il est possible, par conséquent, d'observer une structure dans le signal de résonance, structure due à l'existence des noyaux placés en différentes positions chimiques. Ces considérations ont porté à l'emploi de la technique de la haute résolution pour l'étude des polymères en solution. Cette technique a été introduite assez récemment. Les premiers travaux, publiés en 1959 (*4–6*), ont été suivi immédiatement de plusieurs autres et ils ont mis en évidence l'importance de ce moyen de recherche qui s'est révélé en effet très efficace pour l'étude des structures des chaînes polymériques.

Les mesures des temps de relaxation ont habituellement comme but l'étude de la nature des mouvements moléculaires. En ce qui concerne les techniques de mesurer T_1 et T_2, on utilise presque toujours la méthode de la saturation pour T_1 et celle de la mesure de la largeur de raie pour T_2 dans le cas des polymères à l'état vitreux, la méthode des impulsions étant mal appropriée à cause des temps très courts (qui correspondent à des raies très larges) caractéristiques de ces types de polymères. La technique des impulsions a été cependant appliquee avec des résultats excellents dans le cas des polymères à l'état gommeux et des polymères gonflés par des dissolvants convenables.

Dans le cas particulier de T_1 pas très courts, c'est-à-dire de polymères partiellement polymérisés ou gonflés, soit avec le monomère respectif, soit avec des dissolvants organiques, on a effectué récemment à Pavia des mesures même en utilisant une modification de la méthode des passages rapides adiabatiques (*7*).

Il a été possible de réaliser une mesure distincte du T_1 relatif au monomère ou au dissolvant (lorsque celui-ci contient des protons) et du T_1 relatif au polymère, dans le cas de deux temps assez différentes (*8*).

Suivant la polymérisation de quelques polymères vinyliques, on a mis en évidence deux différents processus de relaxation en constatant que le T_1 (de l'ordre de quelques seconds) du monomère décroît lentement pendant la polymérisation, tandis que la viscosité augmente rapidement. Cela démontre que les mouvements responsables de la relaxation sont surtout de type rotationnel. Au contraire le T_1 des protons des chaînes polymériques est de l'ordre de 10^{-2} seconds et à son tour décroît lentement pendant le processus de polymérisation.

La dépendance en fonction de la température du T_1 relatif au monomère pour des échantillons différemment polymérisés a permis d'obtenir des informations à l'égard de l'énergie d'activation des mouvements et en particulier de mettre en evidence sa variation relativement au pourcentage de polymère présent. Au cours du processus de polymérisation, la valeur de l'énergie d'activation paraît augmenter et l'on met en évidence la formation d'une distribution d'énergies d'activation dont la largeur croît aussi.

Toujours avec la méthode des passages rapides adiabatiques on a obtenu des informations intéressantes à propos des mouvements moléculaires dans les polymères gonflés (*9*).

Même l'étude de la dépendance en fonction de la température de la largeur de la raie δH peut donner beaucoup d'indications concernant l'établissement des mouvements moléculaires dans un polymère aussi bien qu'au sujet de quelques caractéristiques de ces mouvements, en particulier l'énergie d'activation et la fraction de substance intéressée à ces mouvements.

L'effet des mouvements est de rétrécir la raie de résonance; mais seulement les mouvements d'amplitude suffisamment grands et ayant une fréquence supérieure à la fréquence de résonance sont efficaces pour le rétrécissement. Cela élimine comme cause de rétrécissement les mouvements vibratoires liés aux énergies thermiques, puisque ces derniers, tout en ayant une fréquence convenable, sont de trop petite amplitude.

De la variation de δH avec la température on peut déduire l'énergie d'activation des mouvements. Les énergies d'activation trouvées par cette méthode sont en général plus petites que celles déduites par des autres méthodes. Ce désaccord peut être due à la présence d'une distribution de fréquences de corrélation. Même les différences possibles de forme de la raie de résonance, dues elles-mêmes à la présence des mouvements, ne sont pas prises en considération dans la mesure de δH. Pourtant il est convenable d'utiliser le second moment de la raie de résonance. Gutowsky et Pake (*10*) ont développé l'expression de Van Vleck du second moment pour montrer comment les mouvements réduisent la valeur prévue pour le réseau rigide.

Plusieurs expérimentateurs ont suivi cette voie et ont interprété les divers types des mouvements moléculaires par l'étude du second moment de la raie en fonction de la température.

L'existence de deux phases, une phase cristalline et une phase amorphe, donne lieu à des raies qui sont visiblement la superposition d'une raie large et d'une raie plus étroite. La raie large est caractéristique de la phase cristalline, la raie étroite de la phase amorphe. Cela permet, bien que grossièrement, la détermination du pourcentage de cristallinité d'un échantillon.

La liberté des mouvements, influencée par la cristallinité, est modifiée soit par les traitements thermiques auxquels l'échantillon a été soumis, soit par la tacticité du même.

Contrairement à ce que l'on croyait d'abord, la RMN s'est demontrée très sensible aux petites différences dans les échantillons; cela pourrait expliquer les divers résultats obtenus par différents expérimentateurs, dus à des échantillons qui, malgré leur similarité, ne sont pas rigoureusement identiques.

On donne maintenant les résultats d'une recherche sur la polymérisation du styrène effectuée au moyen de la mesure de la largeur de la raie à différentes températures (*11*).

Les mesures ont été effectuées sur des mélanges monomère–polymère avec des pourcentages de polymère comprises entre 10% et 90%, obtenues par une polymérisation thermique du monomère à 90 °C et sur des échantillons de monomère pur et de polymère.

Le monomère pur fournit entre 80 °K et 200 °K une raie de largeur constante environ 10 gauss, qui se rétrécit rapidement, augmentant ultérieurement la température ; ce rétrécissement semble se conclure à la température de fusion (234 °K). Le polymère, au contraire, présente une largeur de la raie d'environ 11 gauss entre 80 °K et 120 °K, après quoi, augmentant la température, il montre deux régions de transition, la première à 170 °K environ, la deuxiéme à 390 °K environ.

Les mélanges monomère-polymère donnent lieu à des signaux constitués par une seule raie aux températures inférieures à 170 °K, tandis qu'aux températures supérieures la raie s'est composée de deux parties de largeur différente dont la raie large est attribuable aux protons des chaînes polymériques, l'étroite aux protons des molécules du monomère. Si la température est augmentée, le rétrécissement procède différemment pour chacune des deux parties.

Faisant le rapport entre l'aire limitée par le signal étroit et l'aire limitée par le signal large, on observe qu'il est inférieur au rapport des pourcentages de monomère et de polymère présents dans l'échantillon. On peut expliquer ce désaccord par l'hypothèse que les molécules de monomère les plus voisines aux chaînes polymériques soient moins libres de se mouvoir que celles lointaines et qu'en outre celles-là prennent part aussi aux mouvements des chaînes.

Il est intéressant de noter aussi que la transition que l'on observe à 170 °K environ est plus accentuée dans le polymère pur que dans les divers mélanges. On peut expliquer la plus grande mobilité dans les mélanges à 170 °K environ en considérant que les chaînes polymériques sont separées par des molécules de monomère. Cette situation ne permet pas aux molécules de monomère de solidifier et en même temps elle permet une plus grande liberté de mouvement aux chaînes. Cet effet atteint un maximum pour un pourcentage de polymère d'environ 50%.

La présence de plusieurs types de mouvements rend compliqué le calcul des énergies d'activation pour les concentrations comprises entre 50% et 100%. Pour les concentrations inférieures s'établissent deux mouvements principaux entre 135 °K et 200 °K. Les énergies d'activation respectives augmentent avec la concentration et leurs valeurs sont approximativement 2 kcal/mole et 9 kcal/mole.

Bibliographie

1. N. L. Alpert, *Phys. Rev.* **75**, 398 (1949).
2. L. V. Holroyd, B. A. Mrowca et E. Guth, *Phys. Rev.* **79**, 1026 (1950).
3. R. Newman, *J. Chem. Phys.* **18**, 1303 (1950).

4. F. A. Bovey, G. V. D. Tiers et G. Filipovich, *J. Polymer Sci.* **38**, 73 (1959).
5. A. Odajma, *Rept. Progr. Polymer Phys. Japan* **11**, 119 (1959).
6. A. Odajma, *J. Phys. Soc. Japan* **14**, 777 (1959).
7. G. Chiarotti, G. Cristiani, L. Giulotto et G. Lanzi, *Nuovo Cimento* **12**, 519 (1954).
8. G. Bonera, P. De Stefano et A. Rigamonti, *J. Chem. Phys.* **37**, 1226 (1962).
9. G. Bonera, A. Chierico et A. Rigamonti, *J. Polymer Sci.* **A2**, 2693 (1964).
10. H. S. Gutowsky et G. E. Pake, *J. Chem. Phys.* **18**, 163 (1950).
11. F. Borsa, et G. Lanzi, *J. Polymer Sci.* **A2**, 2623 (1964).

The Recording of Broad-Line NMR Spectra from Solid Polyamides

D. W. JONES and J. E. PEARSON

Department of Chemical Technology, Institute of Technology, Bradford, England

The Robinson NMR Spectrometer

Since NMR spectra from solid samples typically have a frequency spread of several kilocycles per second rather than cycles per second as in a liquid, it is important that a circuit designed for the observation of such broad lines has a high signal-to-noise ratio and is free from microphonics. It is further desirable that the circuit should be able to operate at a low but controllable rf level and, since second moments of the absorption line shapes are required, the line must not be distorted during passage through resonance. Additionally, it may be an asset if the operational frequency can be varied.

The fact that many detection circuits have been suggested is an indication that the above criteria are not satisfied as readily as might be imagined. We have used a variant of the circuit proposed by Robinson (*1*). In essence, this may be regarded as a basic NMR circuit (Fig. 1a) in which the separate rf generator is replaced by feeding back the output from the rf amplifier. A limiter (Fig. 1b) serves to remove microphonic noise, but the circuit is not, of course, a marginal oscillator. It should

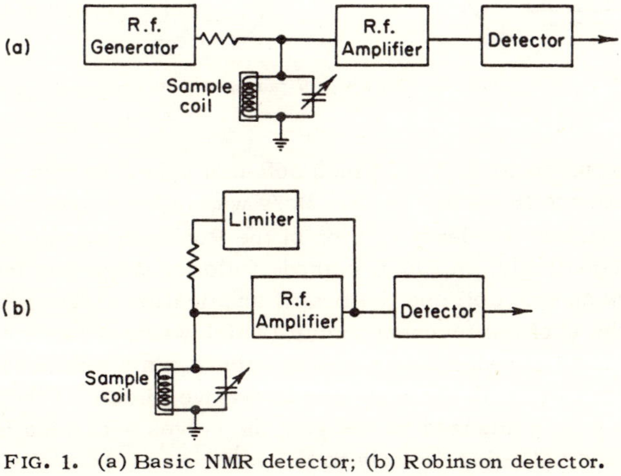

FIG. 1. (a) Basic NMR detector; (b) Robinson detector.

be more sensitive than the Pound–Watkins detector and, like that circuit, can have its operating frequency readily changed.

The large permanent magnet (Fig. 2) of field strength 6800 gauss was built by Swift-Levick from ten $7\frac{1}{2}$ inch Columax segments housed in brass

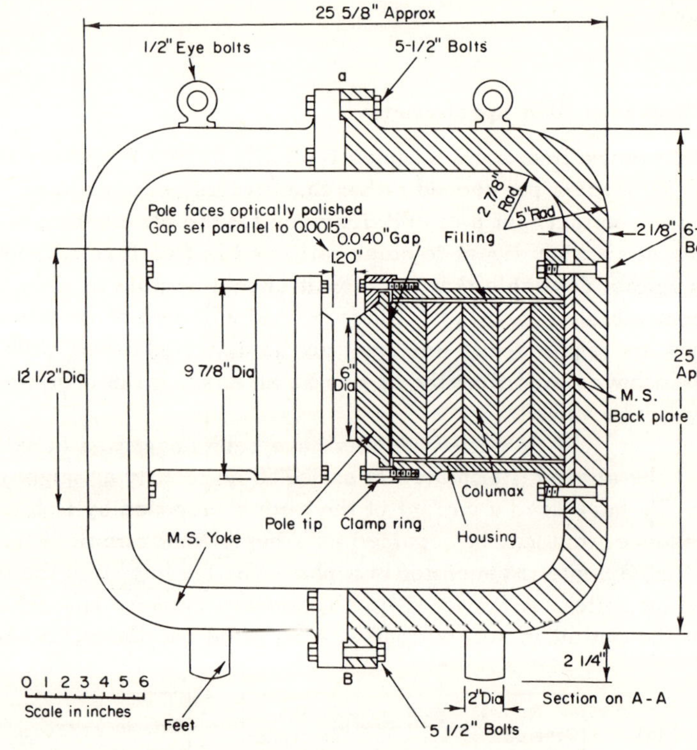

FIG. 2. Drawing of permanent magnet.

cylinders mounted on an $8 \times 2\frac{1}{8}$ inch soft-iron yoke. At this field, the proton resonance frequency of 28.9 Mc/s was rather higher than that for which the original low-level version of the Robinson circuit was intended. Accordingly, an EF 95 cathode follower stage was introduced between the sample coil and the first rf amplication stage in order to maximize the Q of the resonant circuit (which would otherwise be damped by the low input impedance of the first valve). Care was necessary to prevent instability with the resulting five-valve circuit. Vibration and earth loops were minimized by wedging the valves with mica in an aluminum block and soldering directly to the valve pins (2).

Sample Temperature Control

Sample temperature is controlled by the balance of heat supplied by a 10-ohm electrical heater against heat leaking along a 10-inch brass tube which is hard-soldered to a copper refrigerant container (Fig. 3). Immediately below the heater coil, a platinum resistance coil of 4 ft of 46 SWG wire is also wound astatically on the heat leak. Both coils are insulated from the brass rod and are held in place with Autostick fire-resistant cement.

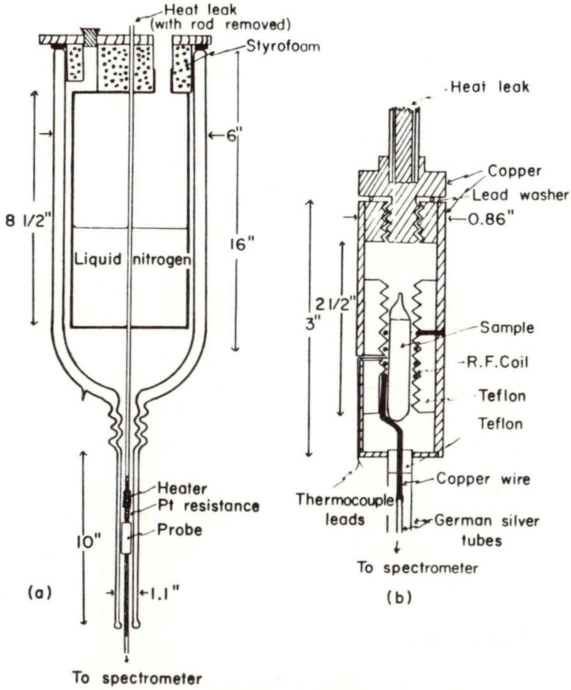

FIG. 3. Heat leak cryostat and probe.

The platinum resistance coil forms one arm of a 50-c/s ac bridge, the output from which is transistor-amplified (Q_1 and Q_2) and rectified at a phase-sensitive detector (Q_3, Fig. 4). After dc amplification (Q_4 and Q_5), a power stage (Q_6) switches on the heater if the platinum resistance becomes too cold. Although the amplifier stage Q_3 has been found to depend slightly on the ambient temperature, the feedback nature of the circuit prevents this from having much effect on the accuracy of temperature control, which is otherwise $\pm 0.2\,^{\circ}$C. In Fig. 4, resistance P can

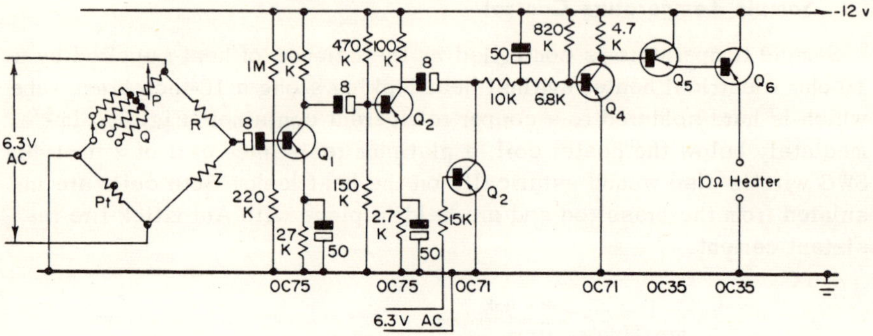

FIG. 4. Circuit of platinum resistance bridge thermostat.

be about 1000 ohms, and $Z = 27.6$ ohms is the value of the platinum resistance at the center of the total temperature range covered. Bridge elements Q (the range selector) and R also depend on the size of the platinum resistance: $R = ZP/(b - a)$, and $Q = aP/(b - a)$, where a and b are respectively the minimum and maximum values of the platinum resistance over the temperature range selected by Q. We have used $R =$ 2.5 kil-ohms, and the four values for Q of 0.536, 1.52, 2.50, and 3.48 kilohms give overlapping coverage of the temperature range $+200$ to $-200\,^{\circ}$C.

The sample coil is unsleeved since it is screwed directly ($\frac{1}{4}$ inch B.S.F.) into a Teflon plug surrounded by a copper cylinder, 3 inches long by 0.86 inches OD. One end of this cylinder forms the termination of the heat leak and the other end carries a pair of concentric German silver leads to the spectrometer. Sample temperature is measured by a copper-constantan thermocouple. Dewar tubing surrounds the whole assembly.

Recording Procedures

Rapid Survey

For a rapid survey of the NMR linewidth, $\Delta H_{m.\,sl.}$, of a polymer specimen over a wide temperature range, the sample is first cooled by allowing liquid oxygen to condense on the probe (*3*). As the sample warms up under manual control (i.e., without utilizing the platinum thermostat just described) at a rate of about 10 degrees per hour, recordings of spectra are made continuously with a sweep time of about 3 minutes. The field sweep is driven by the recorder chart motor.

Second Moment Measurements

When several spectra are required at specific temperatures, as for second moment calculations, the platinum thermostat must be used. Second moments may be computed from ordinates of the differential curve taken at equal intervals of field. A potentiometer (Fig. 5) is mechanically attached to the Kent Mk. III recorder pen and a digital voltmeter (DVM) across this potentiometer yields ordinate values directly every time it is triggered by a cam-operated microswitch on the recorder. The total reference voltage can be supplied by mercury cells or a transistor-stabilized power supply.

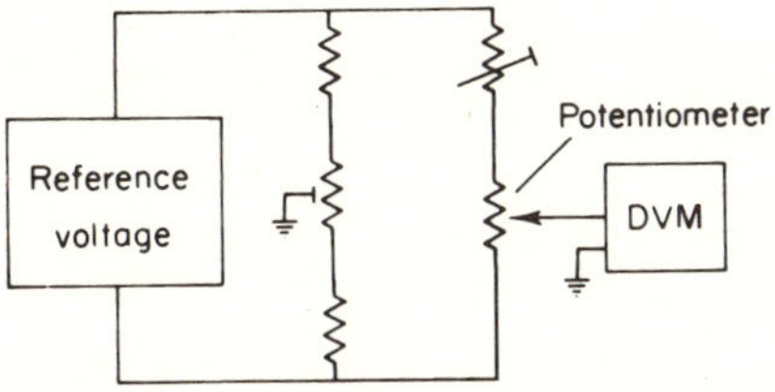

FIG. 5. Use of digital voltmeter for registering ordinates.

Since spectrometer frequency and laboratory field are not locked, successive spectra will not necessarily have coincident abscissas, and so the second moment must be computed separately for each spectrum. If the spectrometer is stabilized, it should be possible to obtain coincident abscissas so that the means of the ordinates can be obtained for several runs at a given temperature and only one second moment computation need then be made for that temperature.

NMR Spectra of Polyamides

Diamine-dioic acid polyamides, $[NH(CH_2)_x—NHCO(CH_2)_{y-2}CO—]_n$, designated Nylon *xy*, have received much more study by NMR than have the polyamides made by condensation of ω-aminocarboxylic acids, $[NH—(CH_2)_{x-1}CO—]_n$, designated Nylon *x*. We have started a study of chip samples of the ω-amino acid Nylons, which have a D.P.[1] of about 50; for Nylon 11, the density measured by flotation was 1.034 gm × cm^{-3}. In view of the previously observed effect of moisture on some NMR parameters in polyamides, all samples were dried carefully for 3 days, over calcium chloride at 100 °C and 10^{-2} mm Hg pressure. Spectra were recorded at an rf field of 3 milligauss, a modulation depth of 1 gauss at 290 c/s, and a sweep rate of 10 gauss min^{-1}. The measured time constants on the lock-in amplifier were 5 sec for spectra below −20 °C, 1 sec above 50 °C, and 3 sec for intermediate temperatures.

In Fig. 6, the full lines represent mean values of linewidth, $\Delta H_{m.sl.}$,

[1] Degree of polymerisation.

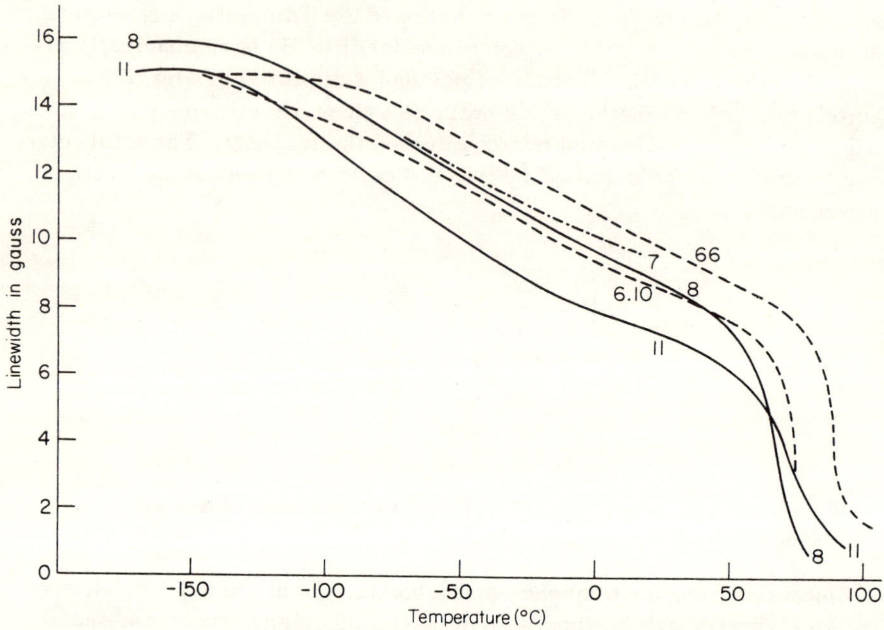

FIG. 6. Variation of linewidths of polyamides with temperature.

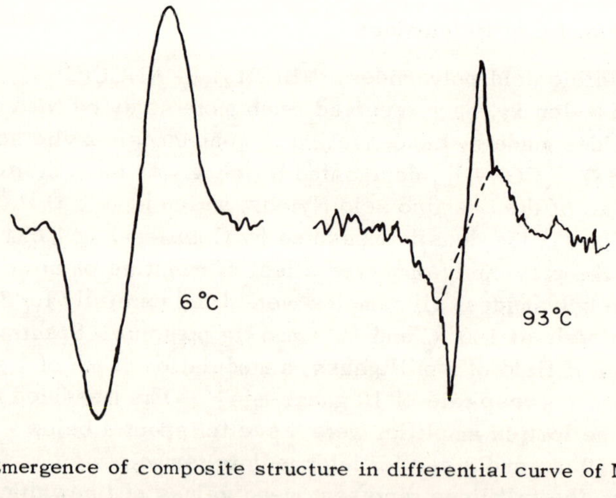

FIG. 7. Emergence of composite structure in differential curve of Nylon 11.

for Nylons 8 and 11 derived with the apparatus described above; partial results from Nylon 7 are also included. For comparison, the dotted curves give representative values for dry Nylons 66 and 610 (*4, 5*).

Both series of polyamides show a long continuous $\Delta H_{m.sl.}$ change from -150°C upwards, and both show a sharp drop in $\Delta H_{m.sl.}$ at 150° or so below the melting point. This sharp narrowing tends to occur at a lower temperature for polyamides containing longer chains of methylene groups. While saturation measurements will be necessary to confirm the observations, there is some evidence (Fig. 7) that composite structure of the differential curve can be detected in Nylons 8 and 11 at temperatures as low as 70°C. Such two-component structure is known to occur in diamine-dioic polyamides with comparable lengths of CH_2 groups at 110°C and above (*6*).

We should like to thank Dr. F. N. H. Robinson for advice on the spectrometer.

References

1. F. N. H. Robinson, *J. Sci. Instr.* **36**, 481 (1959).
2. J. E. Pearson, M. Sc. Thesis, Univ. of Leeds, England, 1964.
3. L. N. Mulay, *Rev. Sci. Instr.* **28**, 279 (1957).
4. A. E. Woodward, J. M. Crissman, and J. A. Sauer, *J. Polymer Sci.* **44**, 23 (1960).
5. D. W. Jones, *J. Polymer Sci.* **59**, 271 (1962).
6. D. W. McCall and E. W. Anderson, *Polymer* **4**, 93 (1963).

Proton Magnetic Resonance in $CuSO_4 \cdot 5H_2O$ Single Crystals

N. J. POULIS and S. WITTEKOEK

Kamerlingh Onnes Laboratorium, Rijks-Universiteit, Leiden, The Netherlands

At low temperatures the resonance frequency of each proton resonance line is given by the formula

$$\nu_t = \frac{\gamma}{2\pi}(\mathbf{H}_t) = \frac{\gamma}{2\pi}(\mathbf{H}_0 + \mathbf{h}_i), \tag{1}$$

where $\mathbf{H}_0$ is the external applied magnetic field and $\mathbf{h}_i$ the internal magnetic field at the position of the protons due to the surrounding Cu ions. From the measurement of the splitting of several resonance lines as a function of temperature it can be concluded that two independent magnetic systems exist. The ions at the (0, 0, 0) positions show the behavior of a system of antiferromagnetic linear chains, the exchange interaction between two neighboring Cu ions in one chain being characterized by $J/k = 1.3\,^\circ$K. The temperature dependence of the inverse splitting $\Delta\nu^{-1}$ for protons near this system is given by line 3 in Fig. 1. The ions at the $(\frac{1}{2}, \frac{1}{2}, 0)$ positions have a much weaker mutual interaction and behave paramagnetically down to temperatures below $0.3\,^\circ$K, with a Curie Weiss constant of about $+0.05\,^\circ$K. The $\Delta\nu^{-1} = f(T)$ for protons near the paramagnetic ions are shown by curves 1 and 2 of Fig. 1. From these two types of proton resonance lines also the linewidth and line shapes were measured as a function of temperature. The results are shown in Fig. 2. The linewidth $\delta\nu$ of the lines belonging to the paramagnetic system remains nearly constant down to $0.3\,^\circ$K. The width $\delta\nu$ of the line belonging to the system of antiferromagnetic linear chains broadens strongly at the lower temperatures. It can be concluded that the dipolar interaction of a proton with a cupric ion of the linear chain system causes a remarkable broadening of the proton resonance line. This Cu–P interaction alone would cause a Lorenzian type of resonance curve with a half-intensity width given by

$$\delta\nu_{\text{Cu-P}} \approx \left(\frac{\gamma_\text{P}}{2\pi}\right)^2 \left(\frac{H_f^2}{\nu_\text{c}}\right), \tag{2}$$

where ν_p is the proton gyromagnetic ratio, H_f = fluctuating part of the local field at the position of the protons, and ν_c is the average frequency

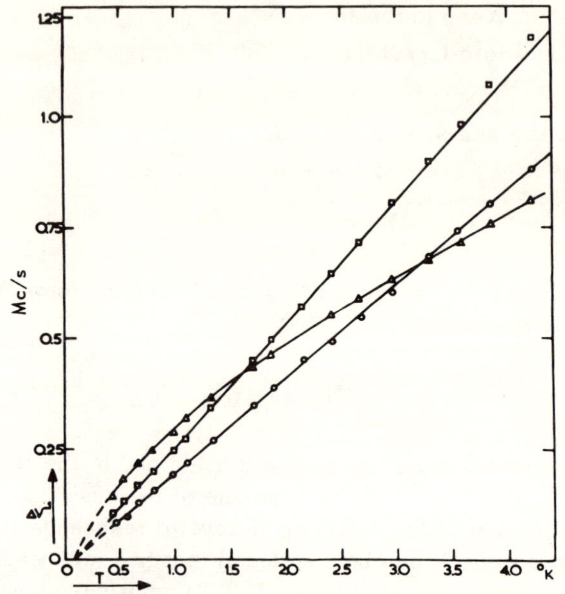

FIG. 1. □ and ○, lines 1 and 2; Δ, line 3.

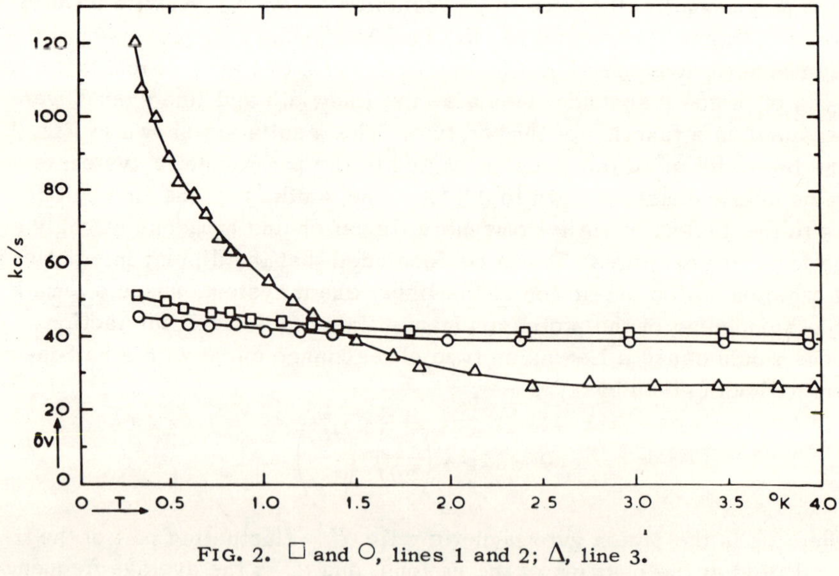

FIG. 2. □ and ○, lines 1 and 2; Δ, line 3.

with which the Cu spins of the linear chain change their orientation. At high temperatures (4 °K), where the magnetic moments in the linear chains are uncorrelated, ν_c equals the exchange frequency which is about 10^{11} c/s as follows from $J/k = 1.3$°K. As H_f is of the order of 500 0e $\delta\nu_{Cu-P}$ will be of the order of a fraction of a kc/s and is negligible compared with the proton–proton linewidth, which is about 25 kc/s. For lower temperatures the magnetic moments in the linear chains become correlated, while they tend to align antiparallel. This increasing correlation causes a decreasing ν_c and thus, as follows from form 2, an increasing δ_{Cu-P}. At very low temperatures ($T < 0.7$ °K) the resonance lines show apart from the line broadening also a line doubling which increases strongly for lower temperatures and becomes of the order of 150 kc/s at $T = 0.3$ °K. This doubling is not due to proton–proton interaction between the protons in one water of hydration as it is strongly temperature dependent. This doubling of the resonance lines appears for the lines belonging to the linear chain system as well as for those belonging to the paramagnetic system. The field as well as the temperature dependences of this doubling leads us to the conclusion that it originates from the different orientations of the time-averaged magnetic moments of the Cu spins in the antiferromagnetic linear chain.

NMR in Solid Alicyclic Hydrocarbons

M. J. R. HOCH[1] and F. A. RUSHWORTH

Department of Physics, University of St. Andrews, St. Andrews, Fife, Scotland

One of the earliest applications of NMR in the field of chemical physics was to the study of the motion of atoms and molecules in the solid state. Measurements of the variation with temperature of the NMR absorption line shape in solids can give information about the various molecular motion and diffusion processes which may be occurring, while the rates of these processes and the barriers hindering them may be found from spin-lattice relaxation time measurements. A review by Andrew (*1*) surveys the field for small molecules.

Experiments of this nature have been made in this laboratory over a number of years, and in this paper we review the results for members of the alicyclic series $(CH_2)_n$ for $n = 3, 4, 5$, and 6. The nature of the processes which can occur is strikingly illustrated by the well-known case of cyclohexane (*2*). The relevant theory requires measurement of the second moment of the absorption line, or the mean square width. This is plotted for C_6H_{12} in Fig. 1, and the narrowing which occurs can be calculated to agree quantitatively with the following processes:

(a) At the lowest temperatures, any motion which may be occurring is too slow to narrow the line.

(b) As the temperature increases, the motions become faster and in general will affect the line when they are occurring at a rate comparable with the linewidth. This is usually in the region 10^4 to 10^5 c/s, and in this case is interpreted as a reorientation process of the molecules about their triad axes.

(c) Above the transition point, much more molecular freedom is allowed, and the molecules rotate freely on their lattice sites.

(d) The further narrowing above 220°K is interpreted as due to self-diffusion of the molecules.

All these stages are not always shown. For instance, in cyclopentane (*7*) the transition points occurred before any molecular reorientation process was fast enough to narrow the broad line, and for cyclobutane (*5*) the diffusion process commenced at the transition point, and there was no region where the molecules were observed to be simply freely rotating on

[1] *Present address*: Department of Physics, University of Natal, Pietermaritzburg, South Africa.

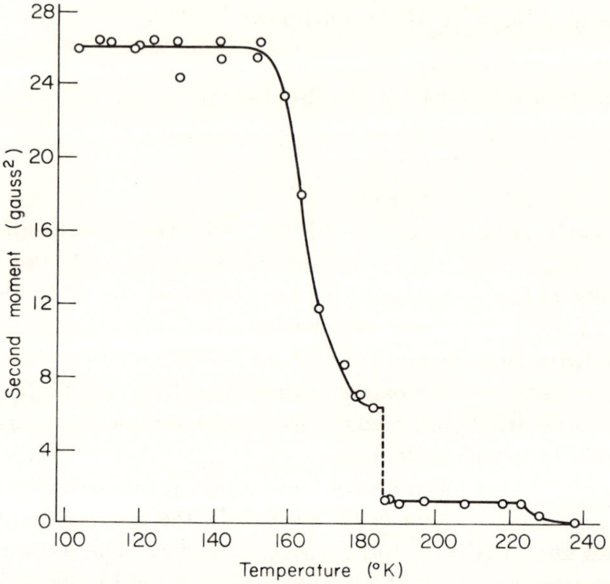

FIG. 1. Variation with temperature of the second moment of the proton magnetic resonance absorption line for polycrystalline cyclohexane (from Andrew and Eades, *2*).

their lattice sites. Rotational solid-state transitions have been reported for these three substances, which is naturally an indication of a considerable change in molecular freedom. Cyclopropane has no such phase transition, the linewidth and second moment variations being shown in Fig. 2. The calculated second moment for the rigid lattice is 19.6 ± 1.3 gauss^2, which compares well with the low-temperature region of the second moment curve. For a reorientation process about the triad axes, the calculated reduced second moment is about 11.3 gauss^2. Although some estimations have to be made in these calculations, mainly owing to the lack of knowledge of the crystal structure, nevertheless they are in good agreement with the experimental values. The only other kind of restricted motion which might occur is molecular reorientation about a preferred diad axis, but this does not produce sufficient reduction in second moment. Above the melting point, of course, the gross increase in molecular freedom in the liquid state decreases the line to a very small value, but we are not concerned with liquid linewidths in broad-line NMR studies.

The variations of the spin-lattice relaxation times T_1 for the alicyclic series where $n = 4$, 5, and 6 are described in the references quoted. The

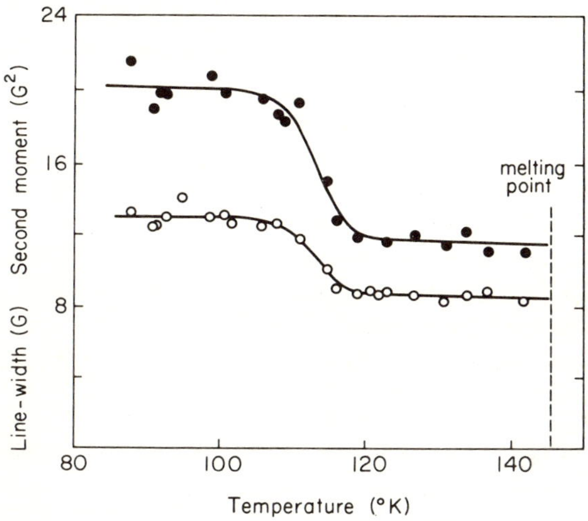

FIG. 2. Variation with temperature of the linewidth ○ and second moment ● for polycrystalline cyclopropane.

general procedure in obtaining values for the barriers hindering motion is to assign a correlation time τ_c to the motion which is assumed to vary with temperature T in the manner $\tau_c = \tau_0 \exp(E_a/RT)$, where E_a is the barrier height. The correlation times can be linked with the measured T_1 values through an expression which follows from assuming a Debye spectral density function for the Fourier intensities of the nuclear position functions,

$$\frac{1}{T_1} = C_1 \left[\frac{\tau_c}{1 + \omega^2 \tau_c^2} + \frac{4\tau_c}{1 + 4\omega^2 \tau_c^2} \right] \tag{1}$$

C_1 being a constant and ω the angular Larmor frequency. This expression must be modified should molecular tumbling be present. If $\tau_c > 1/\omega$ then the above expressions simplify to

$$T_1 = \frac{\omega^2}{2C_1} \tau_0 \exp(E_a/RT),$$

and a plot of log T_1 against $1/T$ gives E_a directly.

The spin-lattice relaxation time for cyclopropane does not show the discontinuities which are manifested for the higher members of the series, and decreases gradually from 300 sec at 90°K to about 0.12 sec at the melting point. This is consistent with a reorientation process about the triad axis. Using the theory outlined above, the plot of T_1 against $1/T$

is indeed a reasonable straight line leading to a value of 4.1 k cal/mole for the barrier hindering this process. The substance melted before the minimum value predicted by Eq. (1) occurred.

As a check on this value of E_a the linewidth transition was synthesized in the manner first described by Gutowsky and Pake (*4*) and very good agreement was obtained between the theoretical curve and the experimental linewidth temperature variation.

It is interesting to compare the results obtained for the activation energies for the reorientation and diffusion processes with the entropies of transition and melting obtained from thermodynamic considerations, and these values are given in Table I.

TABLE I

ENTROPIES AND ACTIVATION ENERGIES GOVERNING MOLECULAR PROCESSES

	Entropies (eu)		Activation energies (kcal/mole)	
	Transition	Melting	Reorientation	Diffusion
C_3H_6	—	8.9	4.1	—
C_4H_8	9.4	1.4	10.5	4.0
C_5H_{10}	9.5 + 0.6	0.8	—[a]	5–6
C_6H_{12}	8.6	2.2	11.0	8.5

[a]Reorientation not fast enough below transition point.

Several points emerge from this table:

(a) In those cases where rotational solid-state transitions are observed, the entropies of transition are much greater than the entropies of melting.

(b) The total entropies of transition and melting for $n > 3$ are almost identical, and appreciably greater than the entropy of melting for $n = 3$. This suggests that the packing of the C_3H_6 molecules in the lattice may be different from the packing of the other three members of the series. It is unfortunate that the crystal structures are not known below the transition points.

(c) The activation energy governing the reorientation process in C_3H_6 is much smaller than for the other members of the series.

A recent theory of fusion by Pople and Karasz (*6*) takes an ideal model of two interpenetrating lattices and allows for order-disorder phenomena in both the orientations and positions of the molecules. Although this model cannot describe accurately the situation considered here, nevertheless it is interesting to note that one of the key parameters in the theory is a factor ν, defined as the ratio of the barrier hindering rotation of a molecule to the barrier for diffusion. A critical value of ν (0.325) is found below which a rotational solid-state transition will occur before melting sets in, and above which the rotation and melting transitions coalesce. Values for the barrier to free rotation cannot be calculated for cyclopropane and cyclobutane, and in the other two members of the series the molecular tumbling and diffusion processes are too interlinked to separate. For similar "globular" molecules, Stejskal and co-workers (*8*) have found barriers to tumbling of between 1 and 3 kcal/mole. From an inspection of the last column of Table I it would seem reasonable to expect the barrier to diffusion for cyclopropane to be less than 4 kcal/mole. The ratio corresponding to ν may therefore well exceed the critical value in the case of cyclopropane and be less than this value for the higher members of the series so far studied. This is, of course, only a very rough calculation, but does give some insight into the lack of a solid-state rotational transition in cyclopropane.

It has been pointed out by Dunning (*3*) that the van der Waals diameters of cyclobutane, cyclopentane, and cyclohexane are appreciably greater than the distances between molecular centers as measured by x-ray diffraction methods. This indicates that any reorientation process which occurs should be a cooperative effect since steric considerations would appear to preclude completely free rotation of individual molecules. The tight packing also suggests that the diffusion process at the higher temperatures is most likely a vacancy-type diffusion rather than of the exchange or interstitial processes.

References

1. E. R. Andrew, *J. Phys. Chem. Solids* **18**, 9 (1961).
2. E. R. Andrew and R. G. Eades, *Proc. Roy. Soc.* (*London*) **A216**, 398 (1953).
3. W. J. Dunning, *J. Phys. Chem. Solids* **18**, 21 (1961).
4. H. S. Gutowsky and G. E. Pake, *J. Chem. Phys.* **18**, 162 (1950).
5. M. J. R. Hoch and F. A. Rushworth, *Proc. Phys. Soc.* **83**, 949 (1964).
6. J. A. Pople and F. E. Karasz, *J. Phys. Chem. Solids* **18**, 28 (1961).
7. F. A. Rushworth, *Proc. Roy. Soc.* (*London*) **A222**, 526 (1954).
8. E. O. Stejskal, D. E. Woessner, T. C. Farrar, and H. S. Gutowsky, *J. Chem. Phys.* **31**, 55 (1959).

Nuclear Magnetic Resonance of Ionic Solutions

C. FRANCONI and F. CONTI
Istituto di Chimica Fisica, Università di Roma, Roma, Italy

Introduction

Over the past few years, various papers have appeared reporting investigations of the structure of ionic solutions by means of NMR using solutions of diamagnetic salts.

A number of these papers have dealt in particular with the shifts of the hydroxylic protons in aqueous solutions (*1–10*). The aim has been to separate specific anion and cation effects on the shift of the solvent proton, and hence to determine shifts characteristic of single ionic species. However, even when put together, these measurements are not as systematic as is required by this type of investigation, and further, the choice of the reference ionic shift by the various authors has always been arbitrary. What is more, the large number of factors determining the final value of the shift renders the task of separating the various superimposed effects very difficult.

For this reason we have undertaken a systematic examination of aqueous solutions of alkali and alkaline earth halides and nitrates, so as to realize the full potentialities of NMR for studying the structure of electrolyte solutions, and in order to differentiate the effect of the solute on the structure of the solvent from the solvent-solute specific interactions.

Experimental

The shifts of the water protons were measured with a Varian V-4300 high-resolution spectrometer at 60 Mc/s by the side-band technique, using a Hewlett & Packard 200 CD audio-oscillator to modulate the radio frequency. The audio frequency was controlled after each measurement by a Venner TS336 frequency counter to an accuracy of 0.1 c/s. At least ten measurements were made, for each experimental point, of the audio frequency needed to superimpose the side-band of the water line on the reference line as observed on the spectrometer's oscilloscope. The maximum error in the reading under the best conditions of homogeneity of the magnetic field was about 0.2 c/s. At least two series of measurements were made for each experimental point in order to check the reproducibility.

Since the position of the resonance of water protons is a function of the temperature [varying practically linearly with temperature over the range 30–100°C, with a temperature coefficient of about 0.5 c/s per deg C (*11*)], it was necessary to take special precautions in order to obtain consistent results. First of all, the temperature of the probe was checked before each series of measurements and the solutions were thermostated at that temperature before being introduced into the probe. Further, the shift of pure water was measured immediately after each measurement of the shift of a solution in order to eliminate errors due to short-term temperature variation. Finally the following measures were taken to stabilize the temperature of the magnet assembly: a cabin of plastic material was built around the magnet, and the interior thermostated to 24 ± 0.5°C by a system of forced air ventilation; the magnet yoke was insulated with polystyrene foam 4 cm thick; and the water cooling the magnet was thermostated at 24 ± 0.1°C. With these precautions the temperature of the probe, as measured regularly with a thermocouple inside the insert, remained at 27.5 ± 0.2°C for the whole time it was employed.

The salts LiI, KI, NaI, NaBr, KBr, RbCl, KCl, NaF, $LiNO_3$, $NaNO_3$, $MgCl_2$, $CaCl_2$, $SrCl_2$, $BaCl_2$, $SrBr_2$, CaI_2, BaI_2, $Ca(NO_3)_2$, and $Sr(NO_3)_2$ were Merck guaranteed products. The salts CaBr, CsI, CsCl, and $CsNO_3$ were BDH "Laboratory Grade" products; NaCl, LiCl, KF, LiBr, and $MgBr_2$, C. Erba RP products; RbI, RbF, CsF, and $RbNO_3$, Light products. The salt BeF_2 was a Fluka "pure" product, $Be(NO_3)_2$ was an Alfa Inorganics product, and $BeCl_2$ a K & K product.

All the salts were dried at 110°C before use. Solutions were made up with distilled water. Solutions of various concentration were obtained by dilution of a master solution whose concentration was controlled by direct weighing beforehand and analytically afterward.

As far as the problem of choice of the reference line is concerned, previous workers have made use of both internal (*3*) and external (*1, 5, 7–10*) reference standards.

There are many advantages in using an internal standard, chief of which is the direct measurement of the shift, but interactions between the reference substance and the solvent may arise that make the measurements liable to systematic errors not easily detected and allowed for (*12*).

One must also bear in mind that there is a shift of the signal of the water solvent on addition of any substance (e.g., a reference standard), which one cannot with certainty sum algebraically with the shifts caused by the various salts. Consider for example $(CH_3)_4NBr$, used by Hertz and Spalthoff (*3*) as an internal standard at a by no means negligible concentration (0.2*M*): this by itself causes a shift of 0.6 c/s (*11*), which ought to be subtracted from the shifts reported by these authors, assuming additivity for such shifts. Further, in their choice of *tert*-butanol as an internal reference (0.3*M*) for other measurements, they did not take into account that its hydroxylic protons exchange rapidly with water, thereby giving rise to a single resonance whose position depends on the concentration and on the characteristic shifts of the two species in solution, which are difficult to measure. For such high concentrations of reference standard, one has in effect a binary mixture instead of a pure solvent, from a chemical-physical point of view. Further, from a spectroscopic point of view, the parameter measured (shift) comes to depend to an unknown extent on the strongly polar reference substance in solution.

In an investigation of the behavior of various substances that might lend themselves as internal standards in aqueous electrolytic solutions, we have measured the shift of water from the peak of the protons of the methyl group of tetramethylammonium chloride, of *tert*-butanol, and of acetone at various concentrations and for different electrolytes at 1*M* concentration. The results are set out in Table I.

TABLE I
INTERNAL REFERENCE TEST FOR AQUEOUS ELECTROLYTIC SOLUTIONS AT 27.5°C

Solute (1*M*)	Internal reference[a]						External ref.[b]
	$(CH_3)_4NCl$		*tert*-Butanol		Acetone		
	C	δ	*C*	δ	*C*	δ	δ_{corr}
NaCl	0.5	0.090	0.5	0.107	0.5	0.082	0.058
	0.3	0.087	0.3	0.112	0.3	0.088	
	0.2	0.083	—	—	—	—	
CsCl	0.5	0.065	0.5	0.087	0.5	0.073	0.047
	0.3	0.082	0.3	0.093	0.3	0.062	
	0.2	0.088	—	—	—	—	
RbI	—	—	0.3	0.165	—	—	0.118
CsI	—	—	0.3	0.142	—	—	0.087

[a]Shift from methyl resonance to water resonance, with respect to pure water, in parts per million. Concentrations in moles/liter.

[b]Shift from cyclohexane resonance to water resonance, with respect to pure water, in parts per million.

It will be seen first of all that the molar shift depends on the concentration of the reference substance to an extent not easily foreseeable. Further, for a given salt, the value of the molar shift depends not only on the concentration but also on the nature of the reference used. The discrepancy between the values of the molar shift measured with different internal standards and those corrected using an external reference are also considerable. In particular, the discrepancy reported by other authors (*8*) for the salts RbI and CsI was confirmed. Thus at first analysis these results confirm the presence of some interaction between these standards and the solutions. This method must hence be further studied before it can be used for measurements of electrolytic solutions.

With regard to the measurement of shifts using an external reference in a coaxial cell, we first checked Zimmerman and Foster's affirmation (*13*) that in such a coaxial system the external fields are the same at the inner sample and in the annulus, since we failed to find any report of a test of such a system (*14*) in the literature.

The test consisted of measuring the shift between the signal of traces of tetramethylsilane in mixtures of apolar solvents of different susceptibility and that of cyclohexane as an external standard in a coaxial cell (*15*). It is implicitly assumed that the interaction of tetramethylsilane with the solvent affects the shift of the latter's protons in practically the same way in the neat solvents as a mixture of them. This assumption would be difficult to prove, but is reasonable in our view.

The results of this check are shown in Table II. It does indeed seem that the formula applied is correct, for one can surely exclude a fortui-

TABLE II

TEST OF COAXIAL CELL[a] FOR EXTERNAL REFERENCE MEASUREMENTS AT 27.5°C

Internal cell solvent	δ_{exp} (ppm)[b]	$\chi_v \cdot 10^6$	Correction (ppm)[c]	δ_{corr} (ppm)
CCl_4	1.190 ± 0.008	-0.687	0.157 ± 0.015	1.345 ± 0.020
DCE + CCl_4 (1 : 1 by volume)	1.150 ± 0.008	-0.704	0.200 ± 0.015	1.352 ± 0.020
DCE	1.080 ± 0.008	-0.750	0.288 ± 0.015	1.365 ± 0.020

[a]Coaxial cell manufactured by Wilmad Glass, Landsville, New Jersey.

[b]Measured shift between the resonance due to the protons of a trace of tetramethylsilane dissolved in the internal cell solvent and the proton resonance of cyclohexane in the annulus.

[c]Correction evaluated for the mixture assuming the additivity of specific susceptibilities of the two solvents. Susceptibilities have been taken from Ref. (*16c*); the density of the mixture has been measured with an accuracy of ±0.001.

tous coincidence of the values of the corrections with the internal shifts of tetramethylsilane dissolved in the various solvents.

Consequently the shifts were measured with an external standard (cyclohexane) by means of a precision coaxial cell (*15*) using the formula

$$\sigma = \frac{\nu_{rif} - \nu}{\nu_{rif}} \cdot 10^6 \quad \text{(ppm)}$$

and correcting the results, expressed as the difference of shift of the protons of the neat solvent and of the solution, by means of the formula

$$\delta = \delta_{wat} - \delta_{sol} = \sigma_{wat} - \sigma_{sol} + \frac{2\pi}{3}\left[\chi_v(\text{wat}) - \chi_v(\text{sol})\right],$$

and trying to obtain a consistent set of values for the susceptibility of the solutions.

Of the authors who have made corrections to their measurements on electrolytic solutions, Fabricand and Goldberg (*5*) have used Pauling's ionic susceptibilities (*16a*); Hindman (*7*), Selwood's ionic susceptibilities (*16b*); and Bergqvist and Forslind (*8*), susceptibilities of salts reported by Foex (*16c*). Notable differences are encountered between these susceptibilities however, so we tried first of all to measure the susceptibility of the solutions by an indirect NMR method (*17*). The volume susceptibilities measured in this way are shown in Table III, column 2, but the error in the reproducibility of these measurements is in general of an order of magnitude greater than that reported (*17*). It will be seen that the value of the susceptibility depends on the homogeneity of the magnetic field at the sample. Column 2 of Table III gives the susceptibility of 1 *M* solutions of salts as measured in this way according to the position of the knob controlling the *Y* gradient of the field, given in arbitrary units in column 3. The susceptibility values are shown to be of poor accuracy and systematically higher than the mean values, given in column 4, calculated for the same solutions from data in the literature for crystalline salts (*16c*), for solutions (*16c*), and of two sets of ionic susceptibilities (*16b*, *16c*).

Thus, in order to correct the experimental data it was necessary to calculate the volume susceptibilities for each concentration and each salt starting from the specific susceptibilities obtained from the four sources mentioned above for each salt as given in the literature (*16b*, *16c*), assuming that specific susceptibilities are additive. The density values needed for the different salts at the different concentrations were interpolated from data given in the International Critical Tables, or else measured with a picnometer to an accuracy about of 0.1%.

The values of the corrections for molar solutions of the various alkali halides and nitrates are given in column 6 of Table III by way of example.

Results

Measurements have been made at five concentrations up to about 1.5 mole/kg solvent for all the alkali halides and nitrates, with the exception of $CsNO_3$. The experimental results are represented in Fig. 1, but

TABLE III
NMR OF PROTONS OF AQUEOUS SOLUTIONS OF ALKALI HALIDES AND NITRATES AT 27.5°C

	NMR							
	$\chi_v \cdot 10^6$ [a]	Y grad.	Exp. average $\chi_v \cdot 10^6$ [a]	$d_{27.5°}$ [a]	Molar correct. [b]	δ_{exp} [b]	δ_{corr} [b]	δ [c]
H_2O	—	—	−0.720	0.998	0.000	0.000	0.000	0.000
LiCl	—	—	−0.729 ± 0.002	1.020	0.01_8	-0.02_3	-0.00_5	-0.00_5
LiBr	−0.749	30	−0.735 ± 0.001	1.058	0.03_0	0.00_2	0.03_2	0.02_8
	−0.749	35						
	−0.761	60						
LiI	—	—	−0.742 ± 0.002	1.094	0.04_7	0.01_0	0.05_7	0.05_3
$LiNO_3$	—	—	−0.715 ± 0.001	1.036	-0.01_0	0.02_7	0.01_7	0.01_7
NaF	—	—	−0.736 ± 0.001	1.043	0.03_2	-0.07_5	-0.04_3	-0.04_3
NaCl	—	—	−0.734 ± 0.001	1.036	0.03_0	0.02_8	0.05_8	0.05_5
NaBr	−0.761	30	−0.740 ± 0.001	1.074	0.04_2	0.05_0	0.09_2	0.08_8
	−0.765	35						
NaI	—	—	−0.748 ± 0.003	1.110	0.05_8	0.06_3	0.12_1	0.11_8
$NaNO_3$	—	—	−0.721 ± 0.001	1.051	0.01_2	0.07_7	0.08_9	0.07_8
KF	—	—	−0.738 ± 0.002	1.051	0.04_0	-0.07_3	-0.03_3	-0.03_3
KCl	—	—	−0.735 ± 0.001	1.042	0.03_2	0.03_7	0.06_9	0.06_5
KBr	−0.757	30	−0.742 ± 0.001	1.081	0.04_5	0.05_5	0.10_0	0.09_0
	−0.765	35						
KI	—	—	−0.748 ± 0.004	1.115	0.06_0	0.06_8	0.12_8	0.12_3
KNO_3	—	—	−0.721 ± 0.001	1.057	0.00_3	0.08_2	0.08_5	0.08_0
RbF	—	—	−0.746 ± 0.002	1.096	0.05_5	-0.08_8	-0.03_3	-0.03_3
RbCl	—	—	−0.740 ± 0.001	1.083	0.04_3	0.02_5	0.06_8	0.06_3
RbBr	−0.757	30	−0.745 ± 0.001	1.121	0.05_2	0.04_0	0.09_2	0.08_6
	−0.765	35						
RbI	—	—	−0.752 ± 0.001	1.157	0.06_7	0.05_2	0.11_9	0.11_0
$RbNO_3$	—	—	−0.725 ± 0.001	1.098	0.01_0	0.06_5	0.07_5	0.07_0
CsF	—	—	—	—	—	—	—	-0.02_7 [d]
CsCl	—	—	−0.747 ± 0.003	1.123	0.05_5	-0.00_8	0.04_7	0.04_5
CsBr	−0.765	35	−0.751 ± 0.003	1.161	0.06_5	-0.00_2	0.06_3	0.06_0
	−0.778	40						
CsI	—	—	−0.760 ± 0.003	1.197	0.08_3	0.00_3	0.08_6	0.08_0
$CsNO_3$	—	—	—	—	—	—	—	0.05_7 [d]

[a] At 1 mole/liter concentration.

[b] Shift expressed in parts per million per mole per liter.

[c] Shift expressed in parts per million per mole per kilogram of solvent. Average values obtained by the least squares method; mean square deviations of the experimental values from the interpolating line less than ±0.006.

[d] Value obtained from a 0.5-mole/kg solvent solution.

only for the three higher concentrations; the lines were drawn according to the least squares method. Their slopes are given in the last column of Table III in parts per million per mole per kilogram of solvent.

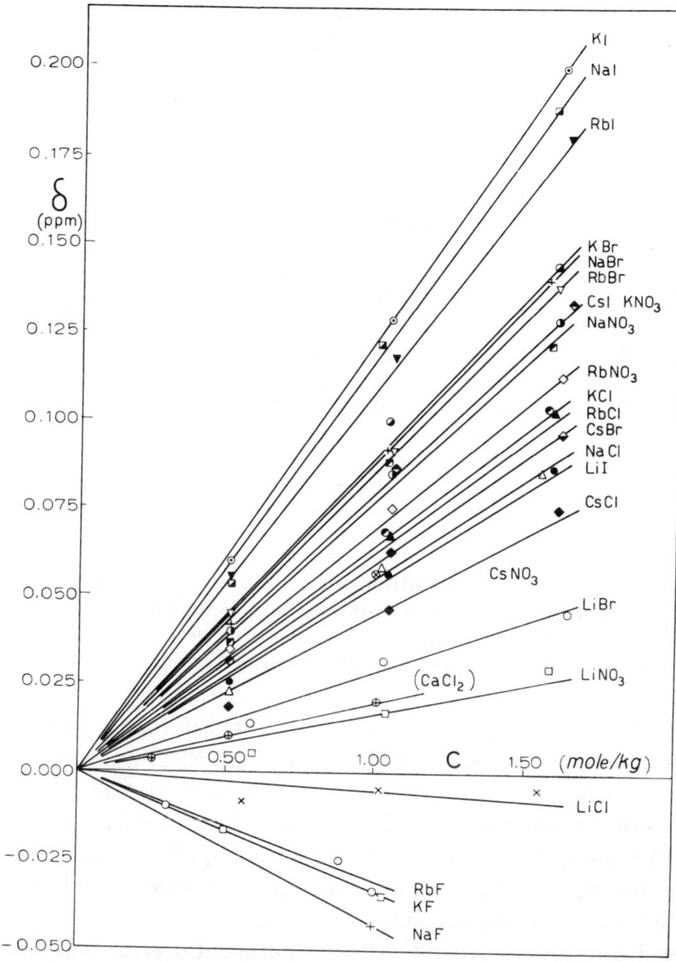

FIG. 1. Alkali halides and nitrates in water; NMR shift of water resonance line (27.5°C, 60 Mc/s).

Measurements were made on molar solutions for the alkaline earth halides and nitrates, and in the case of $CaCl_2$ the linearity of the variations of shift with concentration was verified. The molar shifts, and the molal shifts calculated from these, are given in the last two columns of Table IV.

The linearity of molal shifts with concentration for solutions of alkali halides has been verified within experimental error up to concentrations of about 1.5 mole/kg solvent, as can be seen from Fig. 1.

All the same, it was thought desirable to control the additivity of the molal shifts of the different salts, and hence measurements were carried

TABLE IV

NMR SHIFT OF PROTONS OF AQUEOUS SOLUTIONS OF ALKALINE EARTH HALIDES AND NITRATES AT 27.5°C

	δ_{exp} [a]	$d_{27.5°}$	$\chi_v \cdot 10^6$ [a]	Correct. [b]	δ_{corr} [b]	δ [c]
BeF_2	—	—	—	—	—	-0.27_5 [d]
$BeCl_2$	−0.405	1.050	−0.725 ± 0.001	0.01_0	-0.39_5	-0.38_5
$Be(NO_3)_2$	−0.350	1.078	−0.722 ± 0.005	0.00_3	-0.34_7	-0.32_7
$MgCl_2$	−0.193	1.070	−0.751 ± 0.006	0.06_5	-0.12_8	-0.12_7
$MgBr_2^-$	−0.185	1.145	−0.764 ± 0.003	0.09_2	-0.09_3	-0.09_0
$CaCl_2^-$	−0.085	1.103	−0.770 ± 0.004	0.10_5	0.02_0	0.02_0
$CaBr_2^-$	−0.062	1.158	−0.767 ± 0.003	0.09_8	0.03_7	0.03_5
$Ca(NO_3)_2$	0.015	1.113	−0.729 ± 0.002	0.01_8	0.03_3	0.03_2
CaI_2^-	−0.037	1.235	−0.789 ± 0.007	0.14_5	0.10_8	0.10_2
$SrCl_2^-$	−0.042	1.131	−0.765 ± 0.002	0.09_3	0.05_2	0.05_0
$SrBr_2^-$	−0.008	1.203	−0.773 ± 0.002	0.11_0	0.10_2	0.09_7
$Sr(NO_3)_2$	0.025	1.159	−0.740 ± 0.005	0.04_2	0.06_7	0.06_3
$BaCl_2$	−0.035	1.173	−0.771 ± 0.007	0.10_7	0.07_2	0.07_0
BaI_2^-	0.017	1.320	−0.804 ± 0.011	0.17_5	0.19_2	0.17_5

[a] At 1-mole/liter concentration.

[b] Shift expressed in parts per million per mole per liter.

[c] Shift expressed in parts per million per mole per kilogram of solvent. Average values obtained by the least squares method; mean square deviations of the experimental values from the interpolating line less than ±0.006.

[d] Value obtained from a 0.798-mole/kg solvent solution.

out on equimolal solutions of pairs of salts at different concentrations. Hence the shifts of solutions of NaCl, NaI, and RbCl mixed with other salts were measured to give a total of 15 different mixtures. The concentrations of the salts were 1, 2, and 2.5 moles/kg solvent, respectively, and the experimental results were compared with those calculated by adding the appropriate experimental shifts of solutions of the single salts at the same concentrations.

The results amply confirmed within experimental error that molal shifts of the different salts are additive up to a total concentration of salt of about 5 moles/kg solvent.

Discussion

Before attempting to correlate the shift of the protons of the aqueous solvent with effects exerted on the solvated molecules by the ions and with change of the solvent structure by the solute, it will be useful to make some general observations on shifts produced by these specific effects.

Changes in the structure of water are revealed by a shift of the protons' resonance, since they bring about a change in the number and presumably also the strength of the hydrogen bonds present (*18*). When hydrogen bonds are broken, the resonance shifts to higher fields (diamagnetic shift, which we take as positive), because the additional polarization of the O–H bond by the O atom disappears, and consequently the screening constant of the proton increases.

The observed diamagnetic shift of 4.58 ppm from water just above its freezing point to water at 180°C in the gaseous phase (*18*) must be attributed to this cause. It is well established that the effect of dissolving electrolytes in water on certain physical-chemical properties of pure water (*19*) is similar to that produced by a rise of temperature, and can be correlated qualitatively with hydrogen bond breaking. A structure-breaking diamagnetic shift must hence appear for every electrolytic solution.

Reasonably precise estimates of the distribution of hydrogen bonds among water molecules as a function of temperature have been made from dielectric measurements (*20*). For electrolyte solutions, however, the results of measurements based on this and other methods (e.g., thermodynamic measurements) do not permit a clear separation of effects due to breaking of the ordered structure of water and to the creation of a new order by solvation.

Such a separation may be possible in our case, however, where observed shift is a mean of the values for all the hydroxylic protons, because the polarization induced by the ions depends on the proton-to-ion-center distance and hence on the ionic radius, while the polarization produced by formation of hydrogen bonds is constant and, to a first approximation, independent of the ionic radii of the solute ions.

When a water molecule solvates an ion, either cation or anion, its O–H bonds are polarized, and the diamagnetic screening of the terminal protons is thereby diminished, resulting in a negative (i.e., paramagnetic) shift analogous to that produced on formation of a hydrogen bond. This paramagnetic shift is thus electrostatic in nature, and should hence be expressible in general in a series of terms proportional to reciprocal powers of the average distance from the center of the ion to the electron cloud screening the proton.

A first attempt at analysis of the data should therefore aim at differentiation of these two effects, exploiting the dependence of the latter on the ionic radii.

Provided that the solutions are sufficiently dilute for the shift to vary linearly with concentration, as is true in our case, then the shifts at a convenient concentration, say 1 molal, can be correlated with the ionic radii in the way indicated above. Clearly, however, a treatment of this type must be accompanied by a number of assumptions, based on current understanding of the phenomenon, as to the solvation number of the various ions, the ion-to-molecule distance, etc., all values that are difficult to check.

We proceeded on these lines to try and discover some dependence of the molal shift of aqueous solutions of alkali and alkaline earth halides and nitrates on the radii of the ions in solution.

Values for the radii of ions in solution were obtained by multiplying Pauling's crystallographic radii by 1.21, this being the factor indicated by measurements of ionic partial molar volumes in electrolytic solutions by which ions in solution have been found to be larger than ions in crystals (*21*). The values of these radii are given in Figs. 2 and 3.

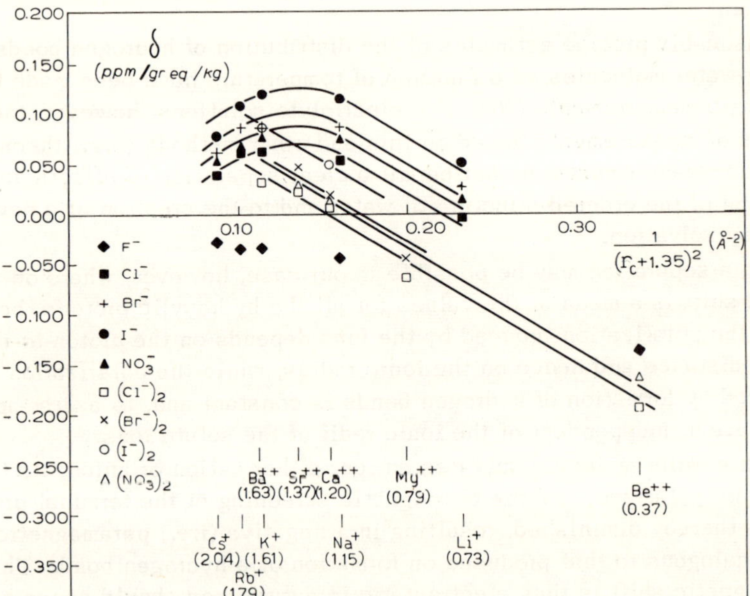

FIG. 2. Alkali and alkaline earth halides and nitrates in water; shift of the O—H resonance (27.5°C, 60 Mc/s).

When the logarithms of molal shifts are plotted against the anionic radii, the fluoride ion included, it is found that the curves joining shifts for salts with a common cation can in reality be considered as a series of parallel lines, for the deviations of the experimental points from the interpolated lines are within experimental error. The slope common to these lines is about −3.

However, when the shifts of the alkaline earth salts are plotted against the reciprocal of the cube of the cation radius, it is not possible to obtain an analogous series of anionic curves, in that the shifts for

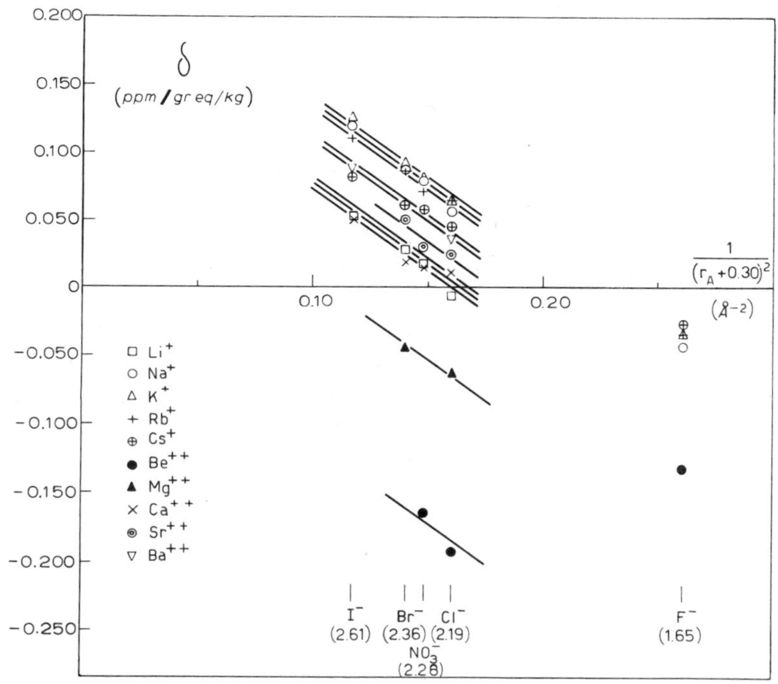

FIG. 3. Alkali and alkaline earth halides and nitrates in water; shift of the O—H resonance (27.5°C, 60 Mc/s).

beryllium salts are not in line with shifts for the other alkaline earth salts with the same anion. Yet an experimental search for the dependence of the shift on the ionic radius must be made precisely by studying the behavior of the shift as a function of the cation radius for the alkaline earth salts, because of the wider range available from Be^{2+} (r_c = 0.37) to Ba^{2+} (r_c = 1.63), and because in the case of anions the range available is too small to be reliable, being limited to I^-, Br^-, and Cl^- since one cannot hold that the behavior of the fluoride ion should be readily fitted into an electrostatic effect of this type.

An attempt can be made in another direction, however, by imagining a solvation model for anions and cations in which the water molecules are in contact with the ion, and by assuming a linear dependence of the shift on a Coulombic ionic field. Let us consider an anion solvated by water molecules having one or two hydrogen atoms in contact with the ion: the proton-to-ion-center distance will be equal to about $(r_a + 0.30)$ A, the factor of 0.30 being added for the covalent radius of the hydrogen atom (*22*). If we now plot the shifts of salts with a common cation against $(r_a + 0.30)^{-2}$ for the various anions, we obtain the curves shown in

Fig. 2, from which it can be seen that there is a linear dependence with the inverse square of the proton-to-ion-center distance within experimental error except for the fluorides, BeF_2 in particular.

Since an alkaline earth salt has double the number of anions at the same concentration, it is necessary to plot the shifts for these salts in parts per million per gram equivalent per kilogram of solvent in order to be able to compare the slopes of lines for these salts with those for the alkali halides. The shifts for the nitrates were inserted after drawing the interpolated line for the bromides, chlorides, and iodides: it will be seen that the shifts for the nitrates fit into this scheme well if the nitrate ion is assumed to have an ionic radius in solution of 2.28 Å.

Similar arguments follow for the cations if we imagine that the water molecules are solvated dipolarly (*19*), when the proton-to-anion-center distance may be taken in first approximation as equal to $(r_c + 1.35)$ Å for all the cations examined.

The relative plots are shown in Fig. 3. Assuming an electrostatic effect is at work, the alkali ions could be compared with the alkaline earth ions by dividing the slopes of the lines for the latter by 2 in order to allow for the double charge; that is, by expressing the shifts in parts per million per gram equivalent per kilogram of solvent again.

The curves of the alkaline earth chlorides and nitrates (Fig. 3) are indeed linear and parallel within experimental error over an ample range of cation radius (0.37–1.63 Å). This effect in itself clearly brings out a cation effect on the screening constant of hydroxylic protons that depends on the inverse square of the proton-to-cation-center distance, just as expected for an electrostatic effect by the cation field on the electron distribution immediately around the proton.

The experimental validity of the hypothesis of a Coulombic interaction on the screening constant of the hydroxylic protons of molecules solvating an alkaline earth halide permits us to extend these conclusions also to anion solvation with a high degree of confidence. This too shows the same behavior as regards the proton-to-anion-center distance, though over a more restricted range of ionic radius (Fig. 2).

It should be noted that curves for alkali halides and nitrates (Fig. 2) do not show a linear dependence on $(r_c + 1.35)^{-2}$. In fact, one can observe an increase of the values of the molal shifts of salts with common anion going from lithium to sodium to potassium, and a decrease from the potassium to rubidium to cesium salts.

It can be noted that the common slope of the curves in the interval between lithium and sodium (Fig. 2) coincides with the common slope of the two curves for the chlorides and nitrates of the alkaline earths, from beryllium to barium. One might therefore reasonably assume that the lithium and sodium halides and nitrates behave the same as the alkaline earths salts with respect to the dependence on the reciprocal of the square of the proton-to-cation-center distance. The experimentally demonstrated validity of our treatment of the molal shifts of the alkaline earth halides and nitrates, down to the radii of Mg^{2+} and Be^{2+}—the latter having half the radius and double the charge of the Li^{+} ion—gives strong support to this hypothesis.

Thus only potassium, rubidium, and cesium would deviate from linearity, which is not too surprising.

The semiempirical preliminary treatment of the molal shifts of ionic solutions outlined here shows the lines by which one may differentiate the effect of the polarization of the O–H bond by the ionic field from all other nonelectrostatic effects on the structure of the solvent, even if the treatment itself is in some way approximate and needs further refinement.

References

1. J. N. Shoolery and B. J. Alder, *J. Chem. Phys.* **23**, 805 (1955).
2. R. G. Axtmann, *J. Chem. Phys.* **30**, 340 (1959).
3. H. G. Hertz and W. Spalthoff, *Z. Elektrokem.* **63**, 1096 (1959).
4. R. G. Axtmann, W. E. Shuler, and B. B. Murray, *J. Phys. Chem.* **64**, 57 (1960).
5. B. P. Fabricand and S. Goldberg, *J. Chem. Phys.* **34**, 1624 (1961).
6. B. Pesce, C. Franconi, and M. Battistini, *Sci. & Tecn.* **5**, 129 (1961).
7. J. C. Hindman, *J. Chem. Phys.* **36**, 1000 (1962).
8. M. S. Bergqvist and E. Forslind, *Acta Chem. Scand.* **16**, 2069 (1962).
9. J. H. Swinehart and H. Taube, *J. Chem. Phys.* **37**, 1579 (1962).
10. A. Fratiello and D. C. Douglass, *J. Chem. Phys.* **39**, 2017 (1963).
11. C. Franconi, unpublished results.
12. A. A. Bothner-By and R. E. Glick, *J. Chem. Phys.* **26**, 1647, 1651 (1957).
13. J. R. Zimmerman and M. R. Foster, *J. Phys. Chem.* **61**, 282 (1957).
14. J. A. Pople, W. G. Schneider, and J. Bernstein, "High Resolution Nuclear Magnetic Resonance." McGraw-Hill, New York 1959, p. 80.
15. Wilmad Glass, Landsville, New Jersey.

16a. L. Pauling, *Proc. Roy. Soc. (London)* **A114**, 181 (1927).

16b. P. W. Selwood, "Magnetochemistry," Wiley (Interscience), 1956, p. 78.

16c. G. Foex, "Constants Sélectionnées Diamagnetisme et Paramagnetisme," Masson, Paris, 1957.

17. K. Frei and H. J. Bernstein, *J. Chem. Phys.* **37**, 1891 (1962).
18. W. G. Schneider, H. J. Bernstein, and J. A. Pople, *J. Chem. Phys.* **28**, 601 (1958).
19. J. D. Bernal and R. H. Fowler, *J. Chem. Phys.* **1**, 515 (1933).
20. G. H. Haggis, J. B. Hasted, and T. J. Buchanan, *J. Chem. Phys.* **20**, 1952 (1952).
21. P. Mukerjee, *J. Phys. Chem.* **65**, 740 (1961).
22. L. Pauling, "The Nature of the Chemical Bond," Cornell Univ. Press, Ithaca, New York, 1960, p. 514.

Electrostatic and Nonelectrostatic Effects in Ionic Solutions as Studied by Nuclear Magnetic Resonance

CAFIERO FRANCONI

Istituto di Chimica Fisica, Università di Roma, Roma, Italy

CAMILLO DEJAK

Istituto Chimico Policattedra, Università di Cagliari, Cagliari, Italy

FILIPPO CONTI

Istituto di Chimica Fisica, Università di Roma, Roma, Italy

Introduction

In a previous paper (*1*) we have reported a consistent set of resonance shifts of water protons in aqueous electrolytic solutions of alkali and alkaline earth halides and nitrates. In a first interpretation of the data it was seen that the shifts for molal concentrations of the salt could be represented as a linear function of the inverse square of the proton-to-anion-center and proton-to-cation-center distances. This dependence led us to think of an electrostatic type of effect on the screening constant of the hydroxylic protons of the aqueous solvent.

The aim of this work is to further develop the previous semiempirical treatment to the general case, to complete the experimental data with measurements on methanol and ethanol solutions, and to discuss the various models proposed for the solvation.

The parameters obtained from the treatment itself will be interpreted on the basis of the proposed models, making use of an empirical separation of the electrostatic effect from those of a different nature.

Results in Aqueous Solution

As was pointed out in the previous work (*1*), when the shifts of molal aqueous solutions of alkaline and alkaline earth halides, except fluorides and the salts of K, Rb, and Cs, are plotted as a function of the inverse square of the proton-to-ion-center distance (assuming it is possible to represent this variable as continuous in treating solvation models), a linear relation is observed over the whole range of the radii of the ions considered [see Fig. 2 of Ref. (*1*; also *2*)].

Such a dependence on the ionic radius allows us to extrapolate the molal shifts of alkali and alkaline earth salts with a common anion to infinite radius of the cation, and thereby to obtain a value for the limiting shift relative to that anion, with all effects dependent on the inverse square of the proton-cation distance removed.

This extrapolation is possible for all the anions with the exception of fluoride. Assuming it is possible to fit the nitrates into this scheme, one can obtain the reasonable value of 2.28 ± 0.04 Å for the radius of the nitrate ion in solution (*1*).

In the same way, extrapolating the shifts for salts having a common cation to infinite anion radius, one obtains a limiting value for the ionic shift relative to that metal cation, free of all effects dependent on the inverse square of the proton-to-anion-center distance.

Of the alkali cations, only lithium and sodium were taken into consideration in the extrapolation leading to the limiting anionic shift, for the shifts of the halides and nitrates of potassium, rubidium, and cesium are not consistent with the extrapolations based on any other cations. However, different behavior by the heavier alkali ions is not completely unexpected, for it is a well-attested experimental fact.

Thus the lines extrapolated to infinite cation radius pass through only the two points for sodium and lithium. Knowing that the behavior of the latter ion is not always analogous to that of the other alkali ions, it is clear that the results for the alkali halides by themselves would not permit a trustworthy extrapolation. However, extension of the measurements to alkaline earth salts has allowed us to confirm the validity of such extrapolations made within the stated assumptions.

It may be seen in fact in Fig. 2 of the previous paper (*1*) that plots of the molal shifts of the alkaline earth salts have double the slope of plots for the alkali halides.

Since this effect is linear with concentration over the range examined (*1*), we can ascribe it to the twice as great anionic concentration. In fact, when the values for the molal shifts of the alkaline earth salts are divided by 2, the slopes of all extrapolations to infinite anion radius become identical, within experimental error. The same happens for the slopes of extrapolations to infinite cation radius, which thereby become parallel to the analogous lines drawn for the shifts of the halides of lithium and sodium.

Bearing in mind that the alkaline earth cations carry a double charge, and that the contribution to the shift dependent on the radius appears to depend on the first power of the electric field exerted by the ion on the proton, the similarity of these slopes may be considered with a good degree of certainty to be not a coincidence, but rather a confirmation of the hypothesis put forward here.

This parallelism of the extrapolations for alkali halides to infinite cationic radius with those for the alkaline earth halides going from strontium to beryllium provides further assurance that the behavior of lithium as regards these effects is not abnormal.

These extrapolations were accordingly carried out using data given in the previous work (*1*). The values for the limiting molal anionic and cationic shifts obtained in this way are given in Table I.

TABLE I

LIMITING IONIC SHIFTS OF THE O-H RESONANCE OF SOLUTIONS OF ALKALI AND ALKALINE EARTH HALIDES IN WATER, METHANOL AND ETHANOL

	Water (27.5 °C)		Methanol (29.5 °C)		Ethanol (29.5 °C)	
	δ_i^a	δ_i^b	δ_i^a	δ_i^b	δ_i^a	δ_i^b
Alkali halides						
I^-	0.27_8	0.15_6	0.68	0.22	0.60	0.14
Br^-	0.25_2	0.14_1	0.60	0.20	0.49	0.11
NO_3^-	0.24_4	0.13_7	—	—	—	—
Cl^-	0.22_6	0.12_7	—	—	—	—
Li^+	0.19_3	0.10_8	0.41	0.13	0.63	0.14
Na^+	0.25_4	0.14_2	0.61	0.20	0.79	0.18
K^+	0.25_9	0.14_6	0.66	0.21	—	—
Rb^+	0.25_5	0.14_3	—	—	—	—
Cs^+	0.23_0	0.12_9	—	—	—	—
Alkaline earth halides						
$(I^-)_2$	0.40_4	0.11_4	—	—	—	—
$(Br^-)_2$	0.35_3	0.09_9	—	—	—	—
$(NO_3^-)_2$	0.33_9	0.09_5	—	—	—	—
$(Cl^-)_2$	0.30_8	0.08_7	—	—	—	—
Be^{2+}	0.01_5^c	0.00_4	—	—	—	—
Mg^{2+}	0.25_0	0.07_0	—	—	—	—
Ca^{2+}	0.38_7	0.10_7	—	—	—	—
Sr^{2+}	0.42_9	0.12_1	—	—	—	—
Ba^{2+}	0.45_4	0.12_7	—	—	—	—

[a]Limiting ionic shift in parts per million per mole per kilogram.

[b]Limiting ionic shift in parts per million per gram equivalent percentage.

[c]Obtained assuming a spherical radius in solution of 2.28 ± 0.04 Å for the NO_3^- ion.

The extrapolation lines were corrected by a least squares treatment, starting from the assumption of a common slope for all the lines and taking account of the statistical weights relative to the errors of each experimental value. The deviations of the experimental points from the corrected lines are shown to be insignificant when analyzed with the Fisher *F* test. When the results of this extrapolation are compared with those for individual extrapolations of the lines (i.e., not assuming a common slope), the differences are not significant with respect to the extrapolation errors according to Student's *t* test.

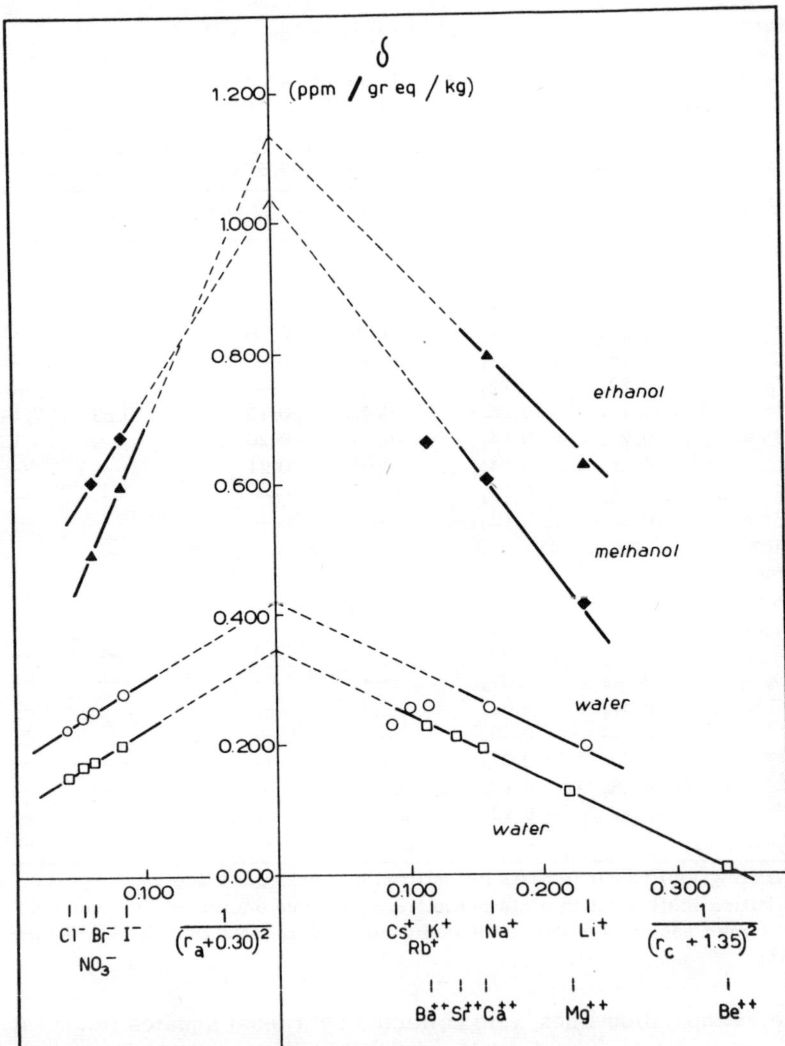

FIG. 1. Electrostatic effects in the limiting shifts in water, methanol, and ethanol.

However, a given limiting anionic shift from which the paramagnetic contribution linearly dependent on the electrostatic effect of the cations has been subtracted still contains a paramagnetic contribution linearly dependent on the electrostatic effect of the anion itself. It is hence necessary to extrapolate the limiting anionic shifts again, this time to infinite anionic radius, in order to obtain a shift completely free of electrostatic effects. The same could be done, vice versa, for cation limiting shifts. It is easily shown, however, that the two limiting shifts thus obtained would be identical if all the experimental errors could be eliminated. Perfect coincidence of limiting shifts has been assumed in the statistical treatment, in order to have a consistent set of results.

These extrapolations are shown in Fig. 1. In this way, one common value for the limiting molal shift is determined for the alkali halides, and a second limiting value for the alkaline earth halides.

A limiting salt shift of this type, extrapolated from the ionic shifts, will amount to that diamagnetic contribution to the shift produced by the salt which does not vary with the inverse square of the proton-to-ion-center distance for cations and anions. This contribution turns out to be considerable (0.42 ± 0.01 ppm/mole/kg solvent for the alkali halides and 0.68 ± 0.01 ppm/mole/kg solvent for the alkaline earth halides).

The deviations from linearity of the limiting ionic shifts for potassium, rubidium, and cesium are paramagnetic, and may be due to a smaller limiting diamagnetic contribution in the case of these salts, or else to a different electrostatic paramagnetic contribution, or to some other cause. The deviations from linearity in parts per million per mole per kilogram of solvent are, respectively, -0.04_7 for potassium, -0.06_4 for rubidium, and -0.12_7 for cesium.

In contrast, the deviations of the fluorides from the scheme proposed are diamagnetic. Thus F^-, together with K^+, Rb^+, and Cs^+—only four species out of the fifteen examined—must be considered anomalous.

However, before discussing these data for aqueous solvent, it will be useful to consider the behavior of electrolytic solutions of solvents having hydroxyl groups of a similar electronic structure, but different chemico-physical and structural properties. We have therefore undertaken preliminary measurements of solutions of electrolytes in methanol and ethanol; the results are reported below.

Results for the Solvents Methanol and Ethanol

Measurements similar to those conducted in water in the previous work (*1*) have been carried out in methanol and ethanol. Owing to the low solubility of many of the alkali salts examined, measurements have been confined to those salts whose concentrations at saturation were greater than 0.1 mole/kg solvent.

The solvents used were Merck products employed without further purification. Also used was A.S.D. methanol that had been distilled twice and then further distilled over $AgNO_3$. The amount of any remaining water was checked, using Fisher's method, and was in all cases less than 0.1%. The same methods as previously described (*1*) were used for the preparation of the solutions.

The linearity of the shift with concentration was checked within the limits of experimental error for LiCl in methanol and for NaI in ethanol, with four experimental points, at concentrations less than 1 mole/kg solvent.

The salts and the concentrations at which the measurements were carried out are given in Table II. In the same table are given the experi-

TABLE II

NMR SHIFT OF O-H PROTONS OF SOLUTIONS OF ALKALI HALIDES IN METHANOL AND ETHANOL AT 29.5 °C

	C (mole/kg)	δ_{exp} (ppm)	Corr. (ppm)	δ_{corr} (ppm)	δ (ppm/mole/kg)
Methanol					
LiCl	1.001	−0.120	0.043	−0.077	-0.07_6
LiBr	0.208	−0.010	0.012	0.002	0.00_8
LiI	0.730	−0.040	0.070	0.030	0.04_2
NaBr	1.000	0.110	0.057	0.167	0.16_6
NaI	1.000	0.160	0.065	0.225	0.22_5
KBr	0.134	0.022	0.012	0.033	0.24_8
KI	0.806	0.170	0.058	0.228	0.28_1
Ethanol					
LiCl	1.000	−0.140	0.038	−0.102	-0.10_2
LiBr	1.000	−0.047	0.047	0.000	0.00_0
LiI	1.000	0.007	0.081	0.088	0.08_8
NaBr	0.116	0.013	0.008	0.021	0.18_0
NaI	0.949	0.157	0.056	0.213	0.22_5

mental shifts at the concentrations there indicated, as measured at 29.5 °C with a Varian A-60 spectrometer according to the method used for the aqueous solutions (*1*); the table also contains the corrections and the corrected shifts relative to the individual solutions, and the molal shifts calculated from these. These shifts can be interpreted graphically in the same way as discussed before in more detail for the aqueous solutions.

In fact it is possible to construct similar curves against the inverse square of the proton-to-anion-center and proton-to-cation-center distances (Figs. 2 and 3) for the two solvents, using the same hypotheses as for anionic and cationic solvation.

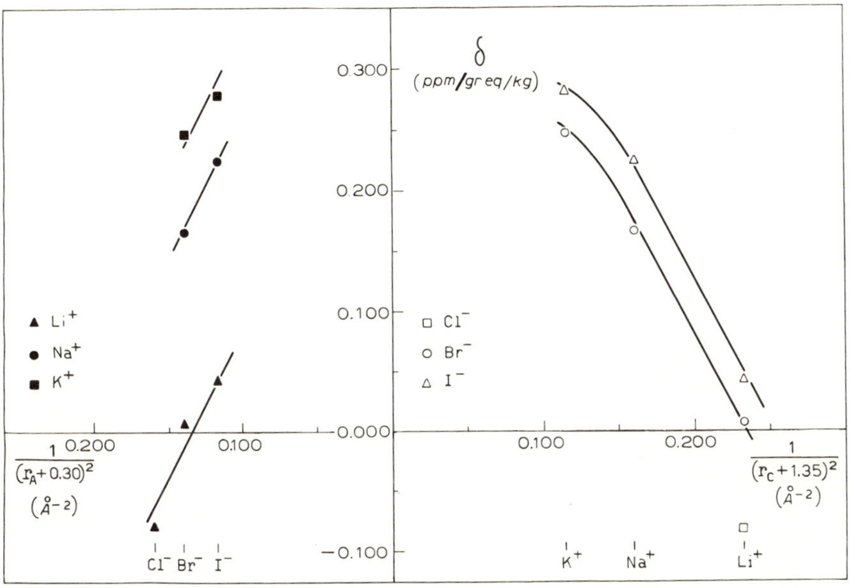

FIG. 2. Alkali halides in methanol; shift of the O—H resonance (29.5°C, 60 Mc/s).

The discussion of these graphs follows the same lines used in evaluating the results for aqueous solutions. Despite the scarcity of the experimental data, one observes here also that potassium salts deviate from the lines through the values obtained for lithium and sodium salts. Further the linear relation of the shifts to the inverse square of the proton-to-anion-center distance is brought out well by the shifts of lithium salts in both solvents.

Here also there arises the problem of the meaningfulness of the curves relating to lithium and sodium. For this reason it would be useful to have data also for alkaline earth halides in these solvents, in order to check the generality of the slope of the lithium-sodium section of the curve. Measurements to this end are in progress.

Assuming that the hypotheses made for water are valid also in the case of these two solvents, one can deduce the limiting shifts for individual ionic species free of all the paramagnetic electrostatic contributions due to the ions of opposite sign by extrapolating the four series of

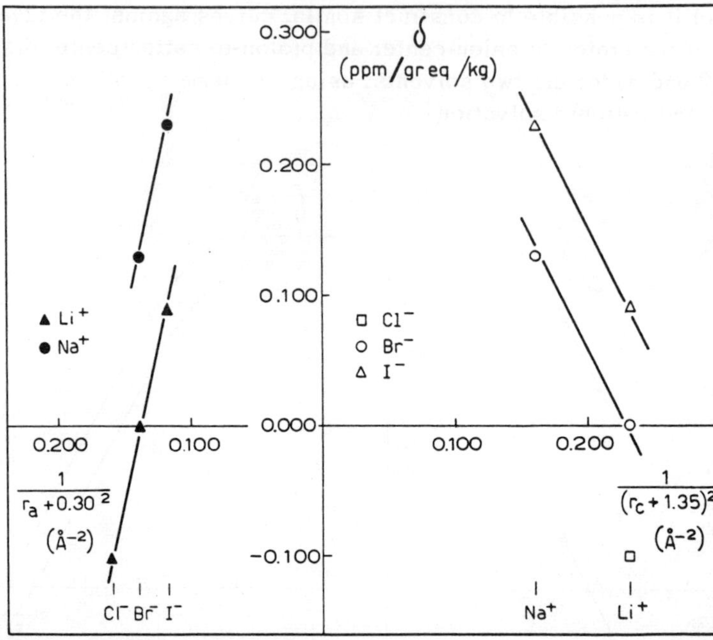

FIG. 3. Alkali halides in ethanol; shift of the O—H resonance (29.5°C, 60 Mc/s).

curves represented in Figs. 2 and 3 to infinitely large ionic radius. The results thus obtained are given in Table I. The paramagnetic electrostatic contributions due to the anion or cation itself must be subtracted from the anion or cation limiting shift in order to obtain the limiting shift for the alkali halides in each of the two solvents, in the same way as carried out for aqueous solvent.

These extrapolations are shown in Fig. 1, together with those relating to aqueous solvent. The deviation from linearity of the limiting ionic shift for potassium is paramagnetic (0.05 ppm/mole/kg).

Discussion of the Results

Analysis of the results for the solutions of alkali and alkaline earth halides in water and in alcohols has shown for almost all the solutions examined that the shifts can be divided into a diamagnetic contribution, not dependent on the ionic radius, and two paramagnetic terms proportional to the inverse square of the proton-to-ion-center distance, within experimental error. This type of separation, although based on a semiempirical type of extrapolation, has the advantage of not requiring any arbitrary hypothesis: Other authors have made unjustifiable assumptions for the molal shifts, for example,

$\delta_{K^+} = \delta_{Cl^-}$ [Hertz and Spalthoff (*3*)],

$\delta_{ClO_4^-} = 0.85$ [Shoolery and Adler (*4*)],

$\delta_{NH_4^+} = 0$ [Hindman (*5*)], and

$\delta_{Cl^-} = 0$ [Fabricand and Goldberg (*6*)],

in their analysis of their experimental data.

However, before beginning a semiquantitative treatment of the results for water and the other solvents, it will be well to consider a factor that could complicate the interpretation of the experimental results to no purpose. From Fig. 1 it appears that at the same concentrations (1 mole/kg solvent) the effects for the alcohols are very much greater, both for the diamagnetic effect (limiting shift) and for the paramagnetic effect (slope of the line), as compared with those for water. It is clear, however, that such behavior might depend in part also on the choice of the unit of measurement of the concentration. In fact, if the shifts for the different

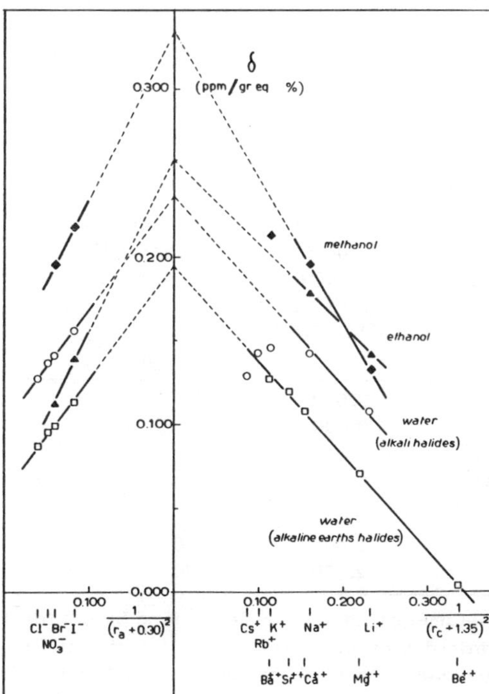

FIG. 4. Electrostatic effects in the limiting ionic shifts in water, methanol, and ethanol.

solvents were to be referred not to the molality but to the molar fraction, using, for example, the units of parts per million per molar fraction percentage, one would obtain nondimensional relation that would permit comparison of the data under identical molecular concentration conditions in all three solvents. Further, expressing the shift in gram equivalent fraction percentage would also take advantage of the effect on the molal shift of the double anionic concentration and of the double cation charge of the alkaline earth salts. The values of the limiting ionic shifts in these units for each solvent are given in Table I. These values have been used in Fig. 4 to obtain again the limiting salt shifts for each type of salt and for the different solvents. As can be seen, the diamagnetic and paramagnetic effects for the different solvents are now more easily compared.

The semiempirical treatment of the experimental data leads in turn to Fig. 4. From this figure one sees first all that the experimental data reported in Tables III and IV of Ref. (*1*) can be fitted to a linear formula:

$$\delta = \delta_0 - \delta' - \frac{S_a}{(r_a + 0.30)^2} - \frac{S_c}{(r_c + 1.35)^2}. \tag{1}$$

The δ values thereby obtained (ppm/mole/kg solvent) and the slopes S (ppm/mole/kg Å^2) are given in Table III.

TABLE III[a]

	Water (27.5°C)		Methanol (29.5°C)	Ethanol (29.5°C)
	Alkali halides[b]	$BeCl_2$, $Be(NO_3)_2$ $MgCl_2$, $MgBr_2$ $CaCl_2$, $Ca(NO_3)_2$, $CaBr_2$, $SrCl_2$, $Ca(NO_3)_2$, $CaBr_2$ $BaCl_2$	LiBr, LiI NaBr, NaI KBr, KI	LiBr, LiI NaBr, NaI
δ_0	0.420	0.690	1.04	1.14
δ'	0.04_7 (K salts) 0.06_4 (Rb salts) 0.12_7 (Cs salts) 0.00 (Other salts)	0.000	0.05 (K salts)	
S_a	1.200	2.400	3.10	4.60
S_c	1.000	2.000	2.70	2.20

[a]Values of the parameters have been determined by the least squares method. The average quadratic deviations of the single experimental points from the interpolating line are ±0.006 for alkali halides and ±0.013 for alkaline earth halides. The determination of the parameters is obviously made with more precision, since they are average values. Shifts in parts per million per mole per kilogram; slopes in parts per million per mole per kilogram-angstrom^2.

[b]Fluorides not included.

As a first step toward understanding the results it is convenient to discuss separately the nature of the diamagnetic and the paramagnetic terms.

Discussion of the Diamagnetic Term

It is well established that a shift to higher fields, i.e., a diamagnetic shift, of the resonance of the hydroxylic protons of a solvent is due to an increase in the screening of the protons due to the elimination of strong intermolecular actions through breakdown of the liquid structure. Ions have been known for some time to have just such an effect, which reveals itself even in the simplest thermodynamic experiments. Indeed, Bernal and Fowler (*7*) introduced the term "structural temperature" of an ionic solution after comparing the effects of ions and of a rise of temperature. The destruction of the liquid structure of the solvent directly affects the protons whose shift is measured, and is hence interpreted as a breaking of hydrogen bonds, similar to what takes place as the temperature rises and with the various changes of state.

Measurements of the shift in the liquid state (at the lowest temperature practicable) and in the gaseous state have been made for both water and ethanol by other workers. We have made similar measurements for methanol. Table IV gives the hydroxylic proton shift in the gaseous state (δ_g) and at a temperature near the melting point (δ_m) for the three solvents mentioned, and the difference between them. The shifts for each solvent at room temperature (δ_t) are also given in the table, as well as the difference between this shift and that for the gaseous state.

The table also contains the ratios $(\delta_t - \delta_g)/(\delta_m - \delta_g) \cdot 100$ for each solvent, which may be taken as a first approximation to the percentage of hydrogen bonds not yet destroyed by the thermal agitation at room temperature. It may be noted that the percentage obtained for water (93%) is in fair agreement with that given by Haggis *et al.* (*10*), obtained from dielectric constant measurements. It should be noted that methanol and ethanol have considerably lower values for the percentage of unbroken hydrogen bonds, as expected.

If we interpret the diamagnetic contribution of the shift of the various electrolytes as due entirely to hydrogen bond breaking, just as for the analogous diamagnetic shift with rise of temperature, we can calculate approximately the additional number of bonds broken for each anion-cation pair in solution. This estimate must evidently be approximate, as we cannot be sure that the diamagnetic shift measured for the various solvents in the gaseous state at the comparatively low temperature (about 160°C) is genuinely the shift for molecules completely dissociated; neither can we be sure that a change of molecular parameters does not

TABLE IV

Solvent	Melt. point (°C)	Boil. point (°C)	Liquid state δ_m (ppm)	δ_t (ppm)	Gas, δ_g (ppm)	$\delta_m - \delta_g$	$\delta_t - \delta_g$	$\frac{\delta_t - \delta_g}{\delta_m - \delta_g} \cdot 100$
H_2O	0	100	5.18 (0 °C)[a]	4.87 (27 °C)[c]	0.6 (180 °C)[a]	4.58	4.27	93%
CH_3OH[c]	−94.9	64.7	6.09 (−95 °C)	4.93 (27 °C)	1.28 (160 °C)	4.71	3.65	79%
CH_3CH_2OH[b]	−116.9	78.3	5.5 (−114 °C)	4.16 (27 °C)	−0.65 (195 °C)	6.15	4.81	78%

[a]From Ref. (8).
[b]From Ref. (9).
[c]This work.

take place during change of state such as to prohibit correlation between the shifts measured in the two states for a given solvent (*11*). We have assumed that uncertainties of this type are small enough to be ignored.

The diamagnetic contribution to the shift produced by each salt for the various solvents (δ_0) has been expressed in Table III in parts per million per mole per kilogram.

Assuming that the shift can be attributed entirely to hydrogen bond breaking, it will be better to use instead the apparent mean number of solvent molecules freed of hydrogen bonds per molecule of salt in solution. To obtain this number, the shift δ_0 expressed in parts per million per gram equivalent percentage should be divided by the shift Δ—also diamagnetic—produced by breaking the hydrogen bonds of one molecule of the solvent. The Δ value is obtained immediately for each solvent as $(\delta_m - \delta_g)/99$ from the values of $(\delta_m - \delta_g)$ given in Table IV, assuming as before that $(\delta_m - \delta_g)$ represents the diamagnetic contribution resulting on breakage of all hydrogen bonds of the solvent. The mean values $\bar{n}$ obtained are: 5.1 molecules of water freed of hydrogen bonds per molecule of alkaline earth halide; 7 molecules of methanol and 4.2 molecules of ethanol, per molecule of alkali halide. Thus the diamagnetic shifts produced by the alkali halides in the various solvents correspond roughly to the liberation from hydrogen bonding of 4–6 molecules of solvent per molecule of salt. Only for alkaline earth does this number reach the value of 8.3 molecules of water per molecule of salt. Here we must remember, however, that there are two anions per molecule and that contribution to the breakage will hence be double; further, the cations have a double charge. At this point we may note that within experimental error the number of hydrogen bonds broken depends neither solely on the cationic charge (when it would be 10), nor on the number of cations (when it would be 5). Excluding the possibility that the breakdown is due exclusively to the anions, for their dimensions are greater than the interstitial holes in the tetrahedral lattice of hydrogen-bonded water molecule aggregates, contributions from both ionic species must be considered, owing not only to their number, but in part also to their ionic charge.

So far, we have talked in terms of molecules of solvent freed of hydrogen bonds. But while this picture is adequate for the alcohols, which can be assumed to have one hydrogen bond per molecule, it must be remembered that for water two hydrogen bonds will be formed per molecule. Further, we can deduce from Table IV that about 93% of the hydrogen bonds in water persist at room temperature; and that the broken bonds are variously distributed between molecules, so that at equilibrium there will be molecules of water having 4, 3, 2, 1, and no hydrogen bonds. The breakage of the hydrogen bonds for a further 5.1 molecules each of which has two such bonds, i.e., of another 10.2 bonds, for each molecule of salt introduced will completely upset this equilibrium, especially in the region around the ions, which will then re-establish itself in a completely different manner, in that part of the broken bonds will be substituted by solvation bonds between ions and solvent molecules.

The solvated molecules directly attached to the ions will be subjected to a polarization effect described shortly, of a size proportional to the intensity of the electric field exerted on the electron cloud screening the proton. As for the molecules forming hydrogen bonds with the first solvation shell, they will not directly influence the experimental shifts.

Finally we must consider the deviations from rectilinearity of the results obtained for the ions K^+, Rb^+, and Cs^+. Considering the direction of the deviation, it might be possible to attribute it to a stronger solvation. But this would be in contradiction to the electrostatic nature of solvation and with conclusions drawn from other experiments. A more plausible explanation is a reduction in the hydrogen bond breaking capacity of these ions because their size does not permit them to enter the meshes of the network of tetrahedrally coordinated water molecule aggregates. It can be deduced from Fig. 4 that for the water solutions of potassium, rubidium, and cesium salts, about 1/2, 1, and 3/2 fewer molecules of water are freed, i.e. 1, 2, and 3 fewer hydrogen bonds are broken, respectively, per pair of ions. For solutions of potassium salts in methanol, about one less molecule of solvent is freed per pair of solute ions.

Discussion of the Paramagnetic Term

The paramagnetic term, which shifts the resonance signal to lower fields, can be separated into a cationic and an anionic contribution, proportional to the inverse square of the proton-to-ion-center distance. This proportionality is consistent with an electrostatic effect, on which basis the various possible models of ionic solvation must be considered in a study of the polarization of the electronic charge around the hydroxylic bond by the ionic field.

First let us consider the phenomenon of solvation, and in particular of hydration, of anions and cations. Solvation can occur in two different ways for each ion, as shown in Fig. 5. The difference between the two cation models lies not so much in the distance of the proton from the ion center, the small difference in which would have little effect on the shift, but rather in the number of hydrogen bonds that the solvated water molecule can form. In the dipolar case (a), only two hydrogen bonds can be formed, through the two protons joining themselves to two other oxygen

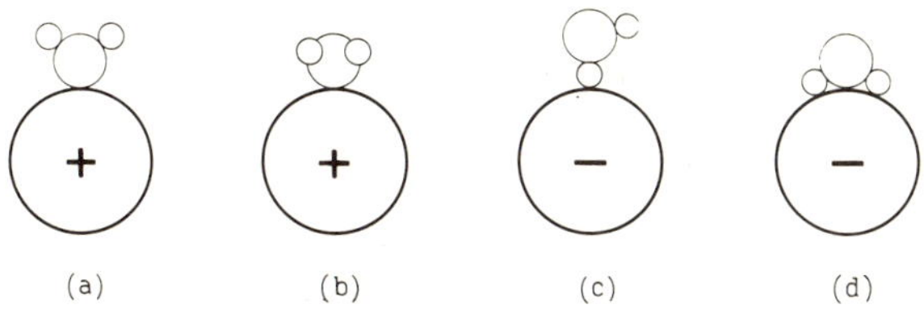

FIG. 5. Hydration models of ions: (a) cationic dipolar, Bernal and Fowler (*7*); (b) cationic tetrahedral, Verwey (*14*); (c) anionic tetrahedral (*7, 14*); (d) anionic dipolar, Van Elteren (*15*).

atoms of other molecules, while in the tetrahedral case (b), a third hydrogen bond can be formed between the free electron pair of the solvated molecule and a proton of another water molecule. Energetic considerations would favor the latter model, therefore. Similar arguments apply to the two anion models, with the difference that here, in the tetrahedral model (c), only one proton per molecule of water is in close contact with the ion and the other proton lies at a much greater distance, whereas in the dipolar model (d) both the protons are in contact with the anion.

This factor might be thought to rule out model (c), which does not meet the experimental requirements of a paramagnetic effect clearly proportional to the electric field at a distance of $(r_a + 0.30)$ Å, i.e., to protons in contact with the anion. One must remember, however, that in this case the external proton of the solvated molecule will not only have its O–H bond out of line with the electric field but also will be subjected to a screening ionic field which will lead to a diamagnetic contribution roughly proportional to $(r_a + 1.35)^{-2}$, and to a paramagnetic contribution proportional, according to Hindman (*5*), to $(r_a + 1.35)^{-4}$. One must also take into account the constant paramagnetic contribution due to possible hydrogen bonds which this external proton might form with molecules of the second hydration shell. Given the present impossibility of evaluating all the effects on this proton, model (c) cannot be considered in obvious disagreement with experiment.

The problem becomes simpler when we turn to the alcohols, where there is only one hydroxyl, and so only one possible orientation for solvation of the anion. In fact, the dipole in this case is directed along the O–H bond, thereby excluding some model of a different type from (c).

The steric hindrance to solvation must also be considered, bearing in mind the possibility of rotation around bonds in the solvating molecules.

As for the shift due to the effect of the ionic field on the screening constant of the hydroxylic protons of molecules in the first solvation shell, Hindman (*5*) provided a formula for an approximate evaluation of this effect, which is based on theoretical derivations of other authors (*12, 13*).

This formula contains a term proportional to the first power of the component of the ionic field parallel to the O–H bond, and a second term proportional to the square of the perpendicular component. For the moment we shall avail ourselves of this formula for a first evaluation of the effect, leaving for a future paper a full theoretical examination of the problem of the perturbation of the screening constant of hydroxylic protons by an ionic field. The formula obtained by Hindman (5) for the shift of the resonance of a hydroxylic proton in an ionic field is

$$\delta = -1.5 \cdot 10^{-18}\left(\frac{\lambda}{R} E_{\parallel} + \frac{1}{2} E_{\perp}^{2}\right) [\text{ppm}], \tag{2}$$

where λ/R^2 is the field at the proton arising from the partial ionic character of the bond, estimated by Hindman at 2.10^6 esu (i.e., charge equal to 39% of the electronic charge, $d_{O-H} = 0.97$ Å). $E_{\parallel}$ and $E_{\perp}$ are the components of the ionic field at the proton along and perpendicular to the direction of the O–H bond.

This equation may now be expressed in terms of parameters derived from the solvation models, namely, d, the proton-to-ion-center distance in angstroms; θ, the angle made by the O–H bond with the direction of the ionic field, $1/K$, the proportion of hydroxylic protons directly involved ($K = 1$ or 2); and h, the number of solvating molecules per solvated ion.

Since we assume there is one pair of singly charged ions in solution per 99 solvent molecules, Eq. (2) then becomes

$$\delta = \left[\frac{0.144}{d^2} \cos\theta + \frac{0.173}{d^4} \sin^2\theta\right]\frac{h}{K}\left(\frac{10^{-4}}{99}\right), \tag{3}$$

where δ is now the individual shift in parts per million per gram equivalent percentage characteristic of a singly charged ionic species.

The maximum and minimum values possible for $\cos\theta$ and $\sin^2\theta$ can be obtained by simple geometrical considerations from the extreme values for the anion and cation radii for the four solvation models represented in Fig. 5.

In model (c) for anion solvation, the two terms relative to the external proton of the solvated molecule could be ignored, for they are of opposite sign; the linear term is in this case diamagnetic and of the order of 20% of the total shift, while the quadratic term is smaller but paramagnetic. The difference between these two terms is very likely balanced out by a paramagnetic term, not easily estimated, due to possible formation of hydrogen bonds by this proton with molecules of the second hydration shell.

Also for anion solvation model (d), where the quadratic term is now diamagnetic, it is possible to make a purely linear approximation, at which there is a variation of the slopes of the lines producing an error of $0.0004 \cdot h$ ppm at maximum. Only for the fluoride ion would such an approximation not be valid, excluding for the moment factors other than simple electrostatic and structure-breaking ones.

For both cation models it is more difficult to ignore the $E_{\perp}^2$ term. More especially for the bivalent ions, from Ba^{2+} to Be^{2+}, there is a variation of d^2 that goes from 5 to over 15% of the total value, i.e., a deviation of 0.003 ppm for barium, which may be ignored, and a more appreciable deviation of 0.03 ppm for beryllium. Since the experimental relationships are linear, it is probable that also this quadratic term, like the corresponding anionic one, is in fact balanced out by the presence of hydrogen bonds formed by these protons with oxygen atoms of other solvent molecules. In fact, these protons would have a greater possibility of hydrogen bonding owing to the presence of the coordinating cation field.

If we consider that even in neat water at room temperature there are very few molecules with less than three hydrogen bonds (*10*), in electrolytic solutions it is still less probable that first-shell solvating molecules, subjected to the polarizing influence of the cation field—that increases the polarity of their O–H bonds—fail to employ both their hydrogen atoms in hydrogen bonding to oxygen atoms of second-layer solvating molecules, hydrogen bonding already being facilitated by the direct electrostatic effect of the cations.

However, for the electron pair on the oxygen atom of a molecule solvating a cation according to model (b) the effect of the field is to polarize the charges in a sense unfavorable to the formation of hydrogen bonds, thereby diminishing the probability of their formation compared with a normal molecule. On the other hand, the number of molecules with all bonds intact is already reduced at normal temperature compared with that just above the melting point, and will diminish still further as a result of structure breaking by the ions. It will thus become rather difficult for this lone electron pair to form such a bond—making the fourth hydrogen bond for the molecule—which would justify the preference for model (b) over (a). Such a model becomes even more improbable for the alcohols, as a result of steric hindrance.

The slopes of the single lines in Fig. 4, both anionic and cationic, are given in Table III (ppm $mole^{-1}$ kg $Å^2$). One can see immediately that the distinction between monovalent and bivalent cations disappears when the double charge of the latter is taken into consideration, which confirms the electrostatic nature of the interaction. Approximate values for the solvation numbers h for the two cation models (a) and (b) can be calculated from Eq. (3), ignoring the quadratic term, and turn out to be quite reasonable (see Table V) apart from somewhat high values for methanol,

TABLE V[a]

	Cations		Anions	
Solvent	h, model (a)	h, model (b)	h, model (c)	h, model (d)
Water	5	6	9[b]	6
Methanol	7	—	7	—
Ethanol	4	—	7	—

[a] $K = 1$ in all cases examined unless otherwise stated.
[b] $K = 2$.

for which a larger steric hindrance should rather lead to a reduction in this number compared with water. A reduction does seem to be obtained for ethanol. It should be noted, however, that the increase in solvation number going from water to methanol could be eliminated by using model (b) for water and model (a) for the alcohols.

The anion solvation numbers can also be obtained from Eq. (3) according to model (c), taking $K = 2$ for water, only one of whose protons intervenes; these numbers turn out to be notably high (see Table V). If anion solvation model (d) is considered instead, which has two protons and hence $K = 1$, the value of h for water improves, but still remains higher than the values obtained with other methods.

The choice of model (d) for the anions can be justified by arguments similar to those that led us to hold model (a) more probable than model (b) for the cations, but here it seems to us still less possible to tackle the problem of the choice between the two models, considering that all the values of h for the anions turn out to be too large.

The reason for this disagreement can probably be sought in the approximate nature of Eq. (3); it will be necessary to see if a more detailed quantum mechanical treatment can account for a higher effect for each solvating molecule, and hence lower solvation number.

Considering the difficulties in the interpretation of anionic solvation, it seems superfluous to attempt similar treatment for the fluoride ion results.

Making a balance now of the hydrogen bonds broken and of those employed in solvation, for each molecule of solute added, although with a proper caution in discussing the anionic solvation, it appears that the total number of broken bonds diminishes with concentration. This should not seem strange when we recall that formation of solvation shells follows destruction of the structure of normal water.

It will, however, be interesting to see what the effect is on the parameters appearing in this paper of varying the degrees of freedom of the solvent molecules by changing the temperature. Experiments to this end are in progress.

References

1. C. Franconi and F. Conti, in this volume, p. 349.
2. P. Mukerjee, *J. Phys. Chem.* **65**, 740 (1961).
3. H. G. Hertz and W. Spalthoff, *Z. Elektrochem.* **63**, 1096 (1959).
4. J. N. Shoolery and B. J. Alder, *J. Chem. Phys.* **23**, 805 (1955).
5. J. C. Hindman, *J. Chem. Phys.* **36**, 1000 (1962).
6. B. P. Fabricand and S. Goldberg, *J. Chem. Phys.* **34**, 1624 (1961).
7. J. D. Bernal and R. H. Fowler, *J. Chem. Phys.* **1**, 515 (1933).
8. W. G. Schneider, H. J. Bernstein, and J. A. Pople, *J. Chem. Phys.* **28**, 601 (1958).
9. W. G. Schneider, "Symposium on Hydrogen Bonding, Ljubljana, 1957" (D. Hadzi, ed.), Pergamon Press, Oxford, England, p. 55.
10. G. H. Haggis, J. B. Hasted, and T. J. Buchanan, *J. Chem. Phys.* **20**, 1952 (1952).
11. J. I. Musher, *J. Chem. Phys.* **35**, 1989 (1961).
12. T. W. Marshall and J. A. Pople, *Mol. Phys.* **1**, 199 (1958).
13. A. D. Buckingham, *Can. J. Chem.* **38**, 300 (1960).
14. E. J. W. Verwey, *Rec. Trav. Chem.* **61**, 127 (1942).
15. J. F. Van Elteren, *Proc. Akad. Wetensch, Amsterdam* **44**, 930 (1941).

Subject Index

(References to papers in French are given in English)

A

B

C

T

V

W

Z